OXFORD STUDIES
IN
NUCLEAR PHYSICS

GENERAL EDITOR
P. E. HODGSON

OXFORD STUDIES IN NUCLEAR PHYSICS
General Editor: P. E. Hodgson

INTERACTING BOSON MODELS OF NUCLEAR STRUCTURE

D. BONATSOS

*Institute of Theoretical Physics,
University of Tübingen, West Germany*

CLARENDON PRESS · OXFORD
1988

To STAVROULA

PREFACE

This book was started as an extension of the lecture notes given at the University of Oxford during the Trinity term 1987.

The aim of the book is to introduce the reader to the algebraic description of nuclear collective motion in terms of the Interacting Boson Model (IBM) and its various extensions. This approach has enjoyed much popularity over the last decade since it combines relative mathematical simplicity with considerable predictive power, and still is the subject of very active research. It must therefore be stated that the literature has been covered up to August 1987.

Every effort has been made to keep the book self-contained. The reader who is not familiar with elementary group theory, which is essential for the subject discussed in this book, should first familiarize himself with the appendices, where a brief account of the group theoretical concepts needed for understanding the present book is given. It is then suggested that he gives emphasis on chapter 1, where the various group theoretical concepts are used for the first time in this book and many details are given. Once one has a "working knowledge" of the appendices and chapter 1, he will find it quite easy to proceed to the rest of the chapters of this book. By the time one finishes this book, he should have a rather complete picture of the IBM approach to nuclear collectivity and he should be able to do his own research in this area.

A large number of references has been given at the end of each chapter. Although these lists are far from complete, the author believes that the research-oriented reader will find them particularly helpful, since they contain numerous interesting theoretical developments as well as experimental tests of the model predictions, which had to be omitted from this book because of obvious lack of space. References appearing in conference proceedings are labelled by the place and year of the conference in the listings at the end of each chapter. A list of conferences, including publication details of the proceedings, is given at the end of the book.

The author is most indebted to Abe Klein, his Ph.D. thesis supervisor at the University of Pennsylvania, for the essential contributions he has made to his knowledge and understanding not only of the subject of the present book, but of physics in general. His boundless enthusiasm for physics still provides a great source of inspiration for the author.

Over the years the author has learnt a lot on the subject from many colleagues, so that any effort to present here a list of names would have been fruitless. An exception is made, however, for Franco Iachello, since from his lecture notes, research articles, and private discussions the author has gotten much profit, being at the same time impressed by their obvious pedagogical virtues.

The author is greatly indebted to Dr. P. E. Hodgson, the General Editor of this series, for his constant encouragement and numerous useful suggestions and comments. Without his enthusiasm this project would have been never undertaken. He is also very grateful to Dr. D. M. Brink for numerous inspired questions and enlightening comments.

Finally, the author is very grateful to his companion in life, Stavroula Begni, for her valuable encouragement and selfless support, without which this work would have not existed.

Oxford   D. B.
1987

CONTENTS

0

INTRODUCTION

0.1 The Nuclear Shell Model

The fundamental model of nuclear structure is the **shell model**, proposed by Maria Goeppert Mayer and independently by J. H. D. Jensen around 1950 (Mayer and Jensen 1955). In its simplest form, the so-called single-particle model, it is assumed that each nucleon (proton or neutron) moves independently in a spherically symmetric potential that represents the average interaction with the other nucleons of the nucleus. This oversimplified model can explain the existence of magic numbers of nuclear shell structure and describe spins and parities of low-lying states of closed major shell nuclei or nuclei consisting of closed major shells plus or minus one particle. But in order to obtain a reasonable picture of the spectrum of nuclei with several nucleons outside closed major shells, one has to introduce two-body residual interactions, which cause configuration mixing. In this extended form of the model it is assumed that closed major shells are inert and nuclear properties come from the behavior of the valence nucleons. Very good results have been obtained using this model for sd shell nuclei. However, the dimension of the configuration space increases very rapidly in higher shells, so that the calculation becomes forbiddingly large even for today's computing machines. Furthermore, even if these large scale calculations become possible some (happy?) day, the question remains of extracting the useful features out of the bulk of computer output. It is thus apparent that in order to achieve a practical description of nuclei with many valence nucleons, one has to consider models employing a few – hopefully physically meaningful – **collective parameters**, able to include the most important physical features of these nuclei and give predictions in reasonable agreement with experiment.

0.2 Quasi-bands

Fortunately, low-lying spectra of nuclei away from closed major shells indeed show a relatively simple structure, which can be related to the occurence of collective phenomena, which can then be described in terms of a few collective parameters. It turns out that their spectra can be arranged into **bands** of collective states, the members of such

1

a band being connected by strong electric quadrupole transitions. In analogy to molecular spectra, these spectra are called **vibrational** if the interlevel spacing is almost constant, or **rotational** if the spacing obeys a J(J+1) rule, where J is the nuclear spin of the state. (Notice that it is customary in nuclear physics to use the term nuclear spin for the angular momentum). The existing experimental data for such bands in even–even nuclei have been tabulated by M. Sakai and can be found in Sakai (1984). A rich variety of nuclear data can be found in Lederer and Shirley (1978).

0.3 Rotational spectra

In an ideal rotational spectrum, a level sequence

$$J^\pi = 0^+, 2^+, 4^+, 6^+, \ldots \tag{0.1}$$

based on the ground state is observed, called the **ground state band**. In addition, one observes bands with the same sequence based on excited 0^+ states, called β-bands, as well as bands with the sequence

$$J^\pi = 2^+, 3^+, 4^+, 5^+, 6^+, \ldots \tag{0.2}$$

based on excited 2^+ states, called γ-bands. The levels of the ground state band are described by the formula

$$E(J) = \frac{J(J+1)}{2\Theta}, \tag{0.3}$$

where Θ is the moment of inertia of the nucleus, which is assumed to be constant in first approximation. However, in order to get closer agreement to experimental data, one finds that he has to include higher order terms in this formula, namely

$$E(J) = \frac{J(J+1)}{2\Theta} - B(J(J+1))^2 + \cdots. \tag{0.4}$$

In nuclei close to the rotational limit the quantity $2\Theta B$ is of order 10^{-3}. Rotational spectra occur in even–even nuclei with permanent quadrupole deformation. Such nuclei are encountered away from closed major shells, usually near the middle of the shells.

Example 1 The lowest part of the experimental spectrum of the nucleus ^{172}Yb is given below (Sakai 1984). In addition to the ground state band (denoted as gsb), three β-bands (denoted as β_1, β_2, β_3), as well as two γ-bands (denoted as γ_1, γ_2) are known. All energies are in MeV. The symbol (-) means that this state is not allowed in the corresponding band, while lack of any number or symbol means that the state has not yet been observed experimentally.

L	gsb	β_1	β_2	β_3	γ_1	γ_2
0	0	1.04293	1.40487	1.79405	-	-
2	0.078750	1.11785	1.47676	1.84980	1.46586	1.60842
3	-	-	-	-	1.54906	1.70057
4	0.260283	1.2865	1.63250	1.975	1.65791	1.80304
5	-	-	-	-		
6	0.53984	1.5375		2.156		

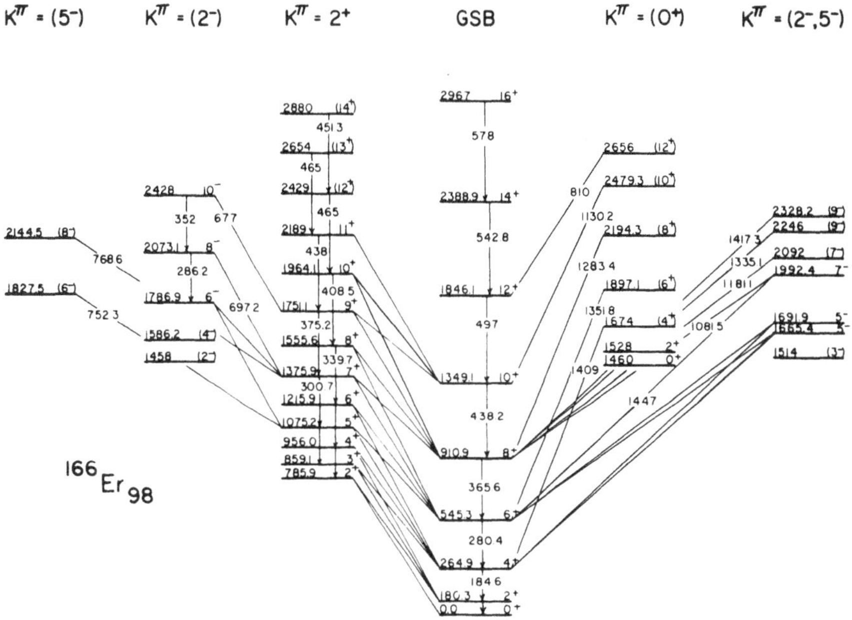

Fig. 0.1 Experimental level scheme of the deformed nucleus ^{166}Er. (Taken from Fields *et al.* (1985)).

Example 2 The experimental level scheme of the rotational nucleus ^{166}Er is shown in Fig. 0.1. All energies are given in keV. In addition to the ground state, β_1 and γ_1 bands, bands of negative parity are observed as well. Negative parity bands will be discussed later.

0.4 Vibrational spectra

In an ideal vibrational spectrum, one expects to see a one-phonon 2^+ excitation at energy $h\omega$, a two-phonon triplet 0^+, 2^+, 4^+ at twice this energy, a three-phonon quintet 0^+, 2^+, 3^+, 4^+, 6^+ at an energy nearly $3h\omega$, and so on. In reality, however, rather strong anharmonicities occur, having as a result that the expected degeneracies are broken and only one or two members of each multiplet can be experimentally identified. What is usually seen is the highest angular momentum state of each multiplet (which lies lowest in energy).

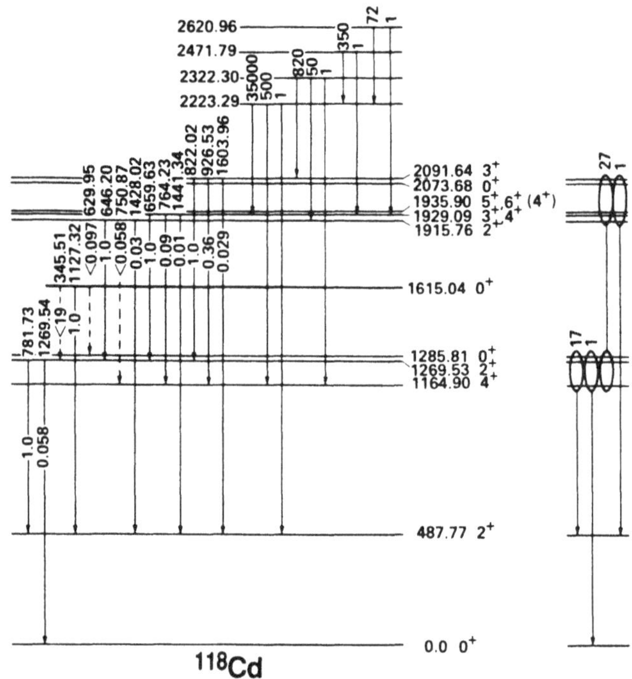

Fig. 0.2 Experimental level scheme of the vibrational nucleus ^{118}Cd. (Taken from Aprahamian *et al.* (1987)).

Vibrational spectra are grouped into bands, in the same way as rotational spectra. They occur in spherical even–even nuclei, lying near closed shells. In lowest approximation, the levels of the ground state band are described by the formula

$$E(J) = aJ, \qquad (0.5)$$

where $a = \hbar\omega$. However, in order to get closer agreement with experiment, one has to include higher order terms. One possible extension is provided by the Anharmonic Vibrator Model (AVM) (Das *et al.* 1970)

$$E(J) = aJ + bJ(J - 2). \qquad (0.6)$$

Example 1 The lowest part of the experimental spectrum of ^{150}Gd is given below (Sakai 1984). The same notation as above is used.

L	gsb	β_1	γ_1
0	0	1.2072	-
2	0.63805	1.5185	1.4305
3	-	-	1.9880
4	1.28843	1.7001	2.0800
5	-	-	
6	2.116		

Example 2 The experimental level scheme of the nucleus ^{118}Cd is shown in Fig. 0.2. All energies are given in keV. The two-phonon triplet (1164.90, 1269.53, 1285.81 keV) is clearly seen. This is the first nucleus in which the three-phonon quintet (the 0^+, 2^+, 3^+, 4^+, 6^+ states near 2000 keV) has been observed.

0.5 The ratio R_4

A useful measure of collectivity is the ratio

$$R_4 = \frac{E(4)}{E(2)}, \qquad (0.7)$$

where by $E(J)$ we denote the excitation energy of the member of the ground state band having angular momentum J. In the vibrational limit, where states with energy proportional to J are expected, this ratio has the value 2. For rotational nuclei, where the J(J+1) rule is expected to be valid, this ratio becomes equal to $\frac{10}{3} = 3.333$. Usually the region $2 \leq R_4 \leq 2.4$ is called the **vibrational** or **near-vibrational** region, the region $3 \leq R_4 \leq 3.333$ is called the **rotational** or **near-rotational** region, while the region $2.4 \leq R_4 \leq 3$ is called the **transition region**. The transition region contains nuclei with structure intermediate between vibrational and rotational.

Examples The nucleus ^{172}Yb, considered above, has $R_4 = 3.305$, thus it is a good example of a rotational nucleus. The nucleus ^{150}Gd has $R_4 = 2.019$, thus it is a good example of a vibrational nucleus. In Fig. 9.1 parts of the experimental level schemes of ^{224}Th, ^{226}Th, and ^{228}Th are shown. The values of the R_4 ratio are 2.898, 3.153, and 3.224, respectively. Thus the last two nuclei are rotational, while the first one is nearly rotational. In Fig. 9.2 parts of the experimental level schemes of ^{142}Ba, ^{144}Ba, and ^{146}Ba are shown. The corresponding R_4 values are 2.321, 2.659, and 2.838. Thus the first of these nuclei is nearly vibrational, the second is clearly transitional, and the third is nearly rotational.

0.6 The transition from vibrational to rotational region

A transition from vibrational to rotational character is observed as one moves away from closed major shells. As an example, consider the $_{64}$Gd isotopes, listed below along with their R_4 ratios (Sakai 1984)

nucleus	R_4
^{146}Gd$_{82}$	1.326
^{148}Gd$_{84}$	1.806
^{150}Gd$_{86}$	2.019
^{152}Gd$_{88}$	2.194
^{154}Gd$_{90}$	3.015
^{156}Gd$_{92}$	3.239
^{158}Gd$_{94}$	3.288
^{160}Gd$_{96}$	3.298

The first of these nuclei is a semi-magic nucleus, since 82 is a magic number. Thus this nucleus has no valence neutrons. The next nucleus, ^{148}Gd, has only two valence neutrons. Collectivity is not yet well-developed in these nuclei, thus single-particle effects are important. This is shown by their R_4 ratios as well, which are less than 2. ^{150}Gd and ^{152}Gd are clearly vibrational, while ^{154}Gd–^{160}Gd are clearly rotational, with the rotational character becoming more profound as the number of valence neutrons increases.

0.7 The geometrical collective model of Bohr and Mottelson

The first comprehensive phenomenological model able to give a satisfactory description of these features was the collective model of

A. Bohr and B. R. Mottelson, proposed in 1952 (Bohr and Mottelson 1975). In this geometrical model the nucleus is assumed to have a well-defined surface and able to undergo small surface and shape oscillations. One usually considers the quadrupole deformation as the most important mode of collective motion. Keeping only the corresponding term in the Hamiltonian, one simply obtains the Hamiltonian of a five-dimensional quadrupole harmonic (to lowest order) oscillator. This Hamiltonian can be quantized by introducing quadrupole boson creation and annihilation operators. It becomes then, to lowest order, proportional to the boson number operator, thus predicting degeneracy within each phonon multiplet and equispaced separation among the multiplets, features which, as we have already seen, roughly characterize vibrational spectra.

For the description of nuclear rotations the theory assumes a permanent ellipsoidal shape for the deformed nucleus. In this case the natural frame of reference is the body-fixed system (called the intrinsic frame). Choosing the body-fixed axes to be the principal axes of the ellipsoid, one introduces new variables β and γ, where β is a measure of the total deformation of the nucleus and γ is related to its shape, which can be a prolate (cigar-like) or oblate (pancake-like) axially symmetric ellipsoid or triaxially deformed (with no axis of symmetry). Quantizing the resulting expression for the Hamiltonian, one can express the collective part of the kinetic energy (in the adiabatic approximation) as a sum of β and γ vibrational energy, which for the special case of the ground state band reproduces the $J(J+1)$ rule which, as mentioned above, is in rough agreement with experiment. In order to get closer agreement with the experimental data, one has to include the weak coupling of vibrational modes to the rotational motion in the form of perturbative correction terms.

Some further discussion of the β and γ parameters is appropriate at this point. β vanishes for sperical shapes, while it is different from zero for deformed shapes. The above mentioned β-bands arise from vibrations of the β degree of freedom. Similarly, γ-bands arise from vibrations of the γ degree of freedom. γ vanishes for prolate shapes and has the value $\pi/3$ for oblate shapes, while values $0 < \gamma < \pi/3$ correspond to triaxial shapes. Nuclei in the transition region ($2.4 \leq R_4 \leq 3$) are thought to be triaxial γ-unstable nuclci, continuously changing their shape from prolate to oblate and *vice versa*.

0.8 Algebraic collective models

So far we have seen that collective spectra of nuclei can be described using a geometrical model. Alternatively, algebraic models can be used for the same purpose. We give here a brief account of such models.

The first algebraic model in nuclear structure was the SU(3) model introduced in 1958 by J. P. Elliott (Elliott 1958a, 1958b, Elliott and Harvey 1963, Harvey 1968). This model makes use of the SU(3) symmetry of the harmonic oscillator, thus it is applicable only in the sd-shell region, where this symmetry is still present. In higher shells the spin–orbit interaction completely destroys this symmetry, thus the model is not applicable any more. An extension of the Elliott SU(3) model in higher shells was later achieved by introducing the concepts of pseudo-spin and pseudo-SU(3) symmetry (Ratna Raju *et al.* 1973, Draayer *et al.* 1982).

A new approach was proposed in 1974 by A. Arima and F. Iachello, known as the Interacting Boson Model (IBM). In this model a description of low-lying collective states in medium and heavy even nuclei in terms of bosons is attempted. These bosons are supposed to be correlated fermion pairs outside closed shells. In the original version of the model, by now called IBM-1, only monopole (J=0) and quadrupole (J=2) bosons are used. This model and its various extensions and generalizations will be the subject of the present book.

It must be emphasized here that several other algebraic models of collective nuclear spectra exist. We simply mention here the Symplectic Model of nuclear structure (Rowe 1985, Carvalho *et al.* 1986, Le Blanc *et al.* 1986), algebraic models using the pseudo-SU(3) symmetry (Draayer *et al.* 1983, 1984, 1985, Bonatsos and Klein 1984, 1985, 1986a, 1986b, Bonatsos *et al.* 1986, Castaños *et al.* 1987) the Ginocchio model (Ginocchio 1980, Arima *et al.* 1981), and its recent extension, the Fermion Dynamic Symmetry Model (Wu *et al.* 1986, Halse 1987, Halse and Pan 1987).

References

Aprahamian, A, Brenner, D S, Casten, R F, Gill, R L, and
    Piotrowski, A, 1987. *Phys. Rev. Lett.,* **59,** 535.
Arima, A, Yoshida, N, and Ginocchio, J N, 1981. *Phys. Lett.,*
    101B, 209.
Bohr, A, and Mottelson, B R, 1975. *Nuclear Structure,* Vol II.
    Benjamin, Reading.
Bonatsos, D, 1985. *La Rábida, 1985,* 615.
Bonatsos, D, and Klein, A, 1984. *Drexel, 1984,* 635.
Bonatsos, D, and Klein, A, 1985. *Phys. Rev.,* **C31,** 992.
Bonatsos, D, and Klein, A, 1986a. *Dubrovnik 1986,* **2,** 1001.

Bonatsos, D, and Klein, A, 1986b. *Ann. Phys.*, **169**, 61.

Bonatsos, D, Klein, A, and Zhang, Q Y, 1986. *Phys. Rev.*, **C34**, 686.

Carvalho, J, Le Blanc, R, Vassanji, M, Rowe, D J, and McGrory, J B, 1986. *Nucl. Phys.*, **A452**, 240.

Castaños, O, Draayer, J P, and Leschber, Y, 1987. *Oaxtepec, 1987*, 111.

Das, T K, Dreizler, R M, and Klein, A, 1970. *Phys. Rev.*, **C2**, 632.

Draayer, J P, and Rosensteel, G, 1985. *Nucl. Phys.*, **A439**, 61.

Draayer, J P, and Weeks, K J, 1983. *Phys. Rev. Lett.*, **51**, 1422.

Draayer, J P, and Weeks, K J, 1984. *Ann. Phys.*, **156**, 41.

Draayer, J P, Weeks, K J, and Hecht, K T, 1982. *Nucl. Phys.*, **A381**, 1.

Elliott, J P, 1958a. *Proc. R. Soc. London* Ser A, **245**, 128.

Elliott, J P, 1958b. *Proc. R. Soc. London* Ser A, **245**, 562.

Elliott, J P, and Harvey, M, 1963. *Proc. R. Soc. London* Ser A, **272**, 557.

Fields, C A, Hicks, K H, and Peterson, R J, 1985. *Nucl. Phys.*, **A440**, 301.

Ginocchio, J N, 1980. *Ann. Phys.*, **126**, 234.

Halse, P, 1987. *Phys. Lett.*, **186B**, 119.

Halse, P, and Pan, Z Y, 1987. *Phys. Rev.*, **C35**, 774.

Harvey, M, 1968. In *Advances in Nuclear Physics* (ed. M. Baranger and E. Vogt) Vol 1, p 67. Plenum, New York.

Le Blanc, R, Carvalho, J, Vassanji, M, and Rowe, D J, 1986. *Nucl. Phys.*, **A452**, 263.

Lederer, C M, and Shirley, V S, 1978. *Tables of Isotopes* (7th edition. Wiley, New York.

Mayer, M G, and Jensen, J H D, 1955. *Elementary Theory of Nuclear Structure*. Wiley, New York.

Ratna Raju, R D, Draayer, J P, and Hecht, K T, 1973. *Nucl. Phys.*, **A202**, 1973.

Rowe, D J, 1985. *Rep. Prog. Phys.*, **48**, 1419.

Sakai, M, 1984. *At. Data Nucl. Data Tables*, **31**, 399.

Wu, C L, Feng, D H, Chen, X G, Chen, J Q, and Guidry, M W, 1986. *Phys. Lett.*, **168B**, 313.

THE INTERACTING BOSON MODEL-1 (IBM-1)

1.1 Introduction to BMW (Boson Model World !)

In the simplest version of the Interacting Boson Model (IBM), it is assumed that low-lying collective states in medium and heavy even–even nuclei away from closed shells are dominated by excitations of the valence protons and the valence neutrons (i.e. particles outside the major closed shells at 2, 8, 20, 28, 50, 82, and 126) only, while the closed-shell core is inert. Furthermore, it is assumed that the particle configurations which are most important in shaping the properties of the low-lying states are these in which identical particles are coupled together forming pairs of angular momentum 0 and 2. In addition, these proton (neutron) pairs are treated as bosons. Proton (neutron) bosons with angular momentum L=0 are denoted by s_π (s_ν) and are called s-bosons, while proton (neutron) bosons with angular momentum L=2 are denoted by d_π (d_ν) and are called d-bosons. Finally, taking into account the particle–hole conjugation, the number of valence proton (neutron) pairs, denoted by N_π (N_ν), is counted from the nearest closed shell, i.e. if less than half of the shell is filled, the number of pairs of particles is considered as the number of bosons, while if the shell is more than half-filled, the number of bosons is considered equal to the number of pairs of holes.

Example : The nucleus ^{152}Sm has 62 protons and 90 neutrons. Since the relevant mid-shells for protons and neutrons are located at 66 and 104, respectively, both its proton bosons and its neutron bosons correspond to pairs of particles, taking the values $N_\pi = (62 - 50)/2 = 6$ and $N_\nu = (90 - 82)/2 = 4$. On the other hand, the nucleus ^{176}Os has 76 protons and 100 neutrons, thus its proton bosons will correspond to pairs of holes, while its neutron bosons will correspond to pairs of particles, taking the values $N_\pi = (82 - 76)/2 = 3$ and $N_\nu = (100 - 82)/2 = 9$. A third nucleus (^{118}Xe) is analyzed in Fig. 1.1.

A detailed description of nuclear properties would require to treat the proton bosons and the neutron bosons separately. This is done in the version of the model which is called IBM-2, or proton–neutron IBM (p–n IBM), which we will consider later (in Chapter 5). In the simplest version of the model, called IBM-1, no distinction between protons and neutrons is made. As a result, in IBM-1 a nucleus is treated as a system of $N = N_\pi + N_\nu$ bosons. In this chapter we will deal exclusively with IBM-1.

One may wonder if, after neglecting the difference between protons and neutrons, it is still possible to get any kind of reasonable description of nuclear properties in the IBM-1 framework. It turns out that many features of low-lying spectra can be adequately described in this model, the underlying reason for this adequacy being the existence of a new kind of symmetry, called F-spin symmetry, which we will deal with later (in Chapters 5 and 7).

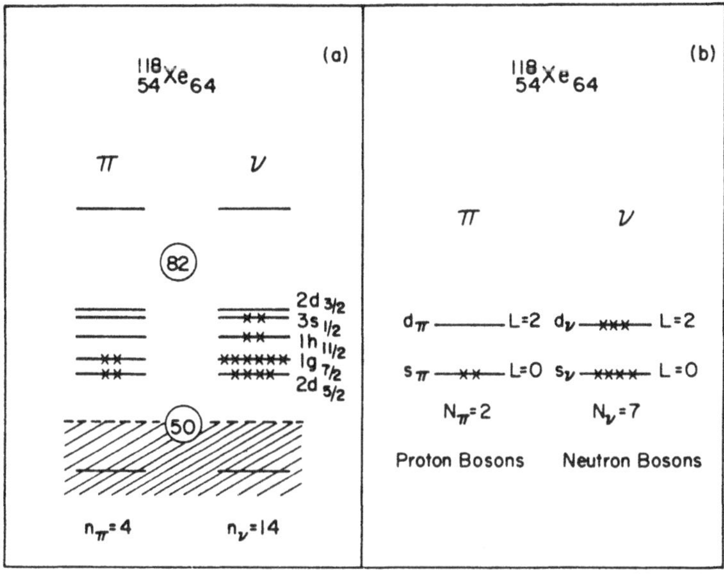

Fig. 1.1 (a) Schematic representation of the shell-model problem for the nucleus ^{118}Xe. n_π and n_ν are the number of valence protons and the number of valence neutrons outside the closed shell. (b) Schematic representation of the IBM-1 problem replacing the shell-model problem for ^{118}Xe. (Taken from Iachello and Talmi (1987)).

One may also wonder why only proton–proton pairs (bosons) and neutron–neutron pairs (bosons) are allowed in the model, while proton–neutron pairs are excluded. The reason is that in medium and heavy nuclei the valence protons and the valence neutrons occupy different major shells, so that the formation of proton–neutron pairs becomes very improbable. In lighter nuclei, though, valence protons and neutrons do occupy the same major shell (e.g. in the so-called sd-shell nuclei, both valence protons and valence neutrons lie in the sd major shell). For the description of light nuclei, proton–neutron pairs do become important. They are included in extended versions of the model, called IBM-3 and IBM-4, which will be discussed later (in Chapter 7).

Another possible question is to what extent bosons of angular momentum 0 and 2 only would suffice. It turns out that although s and d bosons are enough in providing satisfactory description of several nuclear properties, the description of other properties requires the introduction of bosons with angular momentum 4, denoted by g_π (g_ν) and called g-bosons. The extension of IBM-1 including the g-boson is called the sdg-IBM, and will be discussed in Chapter 4.

The models mentioned so far can only describe nuclear levels of positive parity. For the description of levels of negative parity, negative parity bosons must be used. Three different algebraic approaches to relatively low-lying nuclear levels of negative parity will be given in Chapters 3, 8 and 9. An algebraic approach to the higher-lying giant dipole resonances will be discussed in Chapter 10.

The models mentioned so far are suitable for describing even–even nuclei only. For the description of odd nuclei, the Interacting Boson-Fermion Model (IBFM) has been introduced. In IBFM-1 no distinction between protons and neutrons is made, while in IBFM-2 this additional degree of freedom is explicitly taken into account. These models will be discussed in Chapter 11.

Sometimes the term Interacting Boson Approximation (IBA) is used, supposed to have exactly the same content as the term IBM, while avoiding a somewhat disturbing coincidence of acronyms! It must be emphasized that IBM is equivalent to the Truncated Quadrupole Model (TQM) of Paar *et al.* (Janssen *et al.* 1974, Jolos *et al.* 1975, Paar 1978, 1979, 1981, Paar *et al.* 1984, Kyrchev 1980, Kyrchev and Paar 1983, 1986, Lopac and Paar 1984).

Finally, a word of warning: One should **not** think of the IBM bosons as simple fermion pairs. In fact they correspond to very complicated correlated fermion pairs. This point will be further discussed later, in the section about the microscopic justification of the model.

1.2 Operators and definitions

In describing nuclear properties, IBM is using the so-called second quantization language (which is neither second nor quantized, but this is not our subject here !). Following Iachello (1979, 1980, 1981) we introduce boson creation (s^+, d_μ^+) and annihilation (s, d_μ) operators (where $\mu = -2, -1, 0, 1, 2$), satisfying the non-vanishing commutation relations

$$[s, s^+] = 1, \tag{1.1}$$

$$[d_\mu, d_\nu^+] = \delta_{\mu\nu}, \tag{1.2}$$

while all other possible commutators vanish. The six operators s^+, d_μ^+ will be denoted altogether by $b_{l\mu}^+$, $l = 0, 2 \equiv s, d$. Their commutation relations can then be summarized as

$$[b_{l\mu}, b_{l'\mu'}^+] = \delta_{ll'}\delta_{\mu\mu'}. \tag{1.3}$$

In the following, instead of the boson annihilation operators $b_{l\mu}$, it will be convenient to use the operators

$$\tilde{b}_{l\mu} = (-1)^{l+\mu}b_{l,-\mu}. \tag{1.4}$$

In particular, this implies

$$\tilde{d}_\mu = (-1)^\mu d_{-\mu}, \tag{1.5}$$

and

$$\tilde{s} = s, \tag{1.6}$$

thus the only reason to introduce $\tilde{s}$ is to keep the equations symmetric. The reason for introducing these tilded operators is that although the boson creation operators $b_{l\mu}^+$ transform as spherical tensors under rotations, the boson annihilation operators $b_{l\mu}$ do not. The tilded annihilation operators, however, do transform as spherical tensors under rotations.

With tensor operators one can form tensor products. The **tensor product** of two operators $T_{\kappa_1}^{k_1}$, $T_{\kappa_2}^{k_2}$ is defined as

$$[T^{k_1} \otimes T^{k_2}]^{k_3} = \sum_{\kappa_1 \kappa_2} (k_1\kappa_1 k_2\kappa_2 | k_3\kappa_3)T_{\kappa_1}^{k_1}T_{\kappa_2}^{k_2}, \tag{1.7}$$

where $(k_1\kappa_1 k_2\kappa_2 | k_3\kappa_3)$ are the well known Clebsch–Gordan coefficients. The **scalar product** of two operators T^k and U^k is defined as

$$(T^k \odot U^k) = (-1)^k\sqrt{2k+1}[T^k \otimes U^k]_0^0 = \sum_\kappa (-1)^\kappa T_\kappa^k U_{-\kappa}^k. \tag{1.8}$$

Example

$$(d^+ \odot \tilde{d}) = \sqrt{5}[d^+ \otimes \tilde{d}]_0^0 = \sqrt{5} \sum_{\mu_1,\mu_2} (2\mu_1 2\mu_2|00)d_{\mu_1}^+ \tilde{d}_{\mu_2}$$

$$= \sum_{\mu_1} d_{\mu_1}^+ d_{\mu_1} = n_d, \tag{1.9}$$

where n_d is the number of d-bosons.

Note The reader who is not very familiar with angular momentum algebra can find the details about Clebsch-Gordan coefficients and all that in Brink and Satchler (1968) or Edmonds (1960). However, the reader is warned to use only one of the two, because of minor differences in conventions which can cause **big** trouble !

1.3 The Lie algebra U(6)

Let us now introduce the operators

$$G_\kappa^k(ll') = [b_l^+ \otimes \tilde{b}_{l'}]_\kappa^k, \tag{1.10}$$

where $l, l' = 0, 2 \equiv s, d$. It so happens that the commutation relations of these operators among themselves are the same as the commutation relations of the Lie algebra of the group U(6) of unitary transformations in 6 dimensions (discussed in the Appendices). These operators are thus identified as the generators of the algebra U(6). Since they are the building blocks of the most general Hamiltonian one can write in the IBM-1 model, one says that the Hamiltonian has the group structure of U(6). There are in total $36 = 6^2$ of these generators, which written down explicitly read

$$G_0^0(ss) = [s^+ \otimes \tilde{s}]_0^0, \tag{1.11}$$

$$G_0^0(dd) = [d^+ \otimes \tilde{d}]_0^0, \tag{1.12}$$

$$G_\kappa^1(dd) = [d^+ \otimes \tilde{d}]_\kappa^1, \tag{1.13}$$

$$G_\kappa^2(dd) = [d^+ \otimes \tilde{d}]_\kappa^2, \tag{1.14}$$

$$G_\kappa^3(dd) = [d^+ \otimes \tilde{d}]_\kappa^3, \tag{1.15}$$

$$G_\kappa^4(dd) = [d^+ \otimes \tilde{d}]_\kappa^4, \tag{1.16}$$

$$G_\kappa^2(ds) = [d^+ \otimes \tilde{s}]_\kappa^2, \tag{1.17}$$

$$G^2_\kappa(sd) = [s^+ \otimes \tilde{d}]^2_\kappa. \tag{1.18}$$

1.4 The most general IBM-1 Hamiltonian

Once the full algebraic structure of the problem has been identified , one can immediately write down the Hamiltonian which is supposed to describe the nucleus under study. If one makes the assumption that only one-body and two-body terms suffice to describe the low-lying states, it follows that only two-boson and four-boson terms are present in the Hamiltonian, which takes the form

$$H = \epsilon_s(s^+ \odot \tilde{s}) + \epsilon_d(d^+ \odot \tilde{d})$$

$$+ \sum_{L=0,2,4} \frac{1}{2}\sqrt{2L+1}c_L[[d^+ \otimes d^+]^L \otimes [\tilde{d} \otimes \tilde{d}]^L]^0$$

$$+ \frac{1}{\sqrt{2}}\tilde{v}_2[[d^+ \otimes d^+]^2 \otimes [\tilde{d} \otimes \tilde{s}]^2 + [d^+ \otimes s^+]^2 \otimes [\tilde{d} \otimes \tilde{d}]^2]^0$$

$$+ \frac{1}{2}\tilde{v}_0[[d^+ \otimes d^+]^0 \otimes [\tilde{s} \otimes \tilde{s}]^0 + [s^+ \otimes s^+]^0 \otimes [\tilde{d} \otimes \tilde{d}]^0]^0$$

$$+ u_2[[d^+ \otimes s^+]^2 \otimes [\tilde{d} \otimes \tilde{s}]^2]^0 + \frac{1}{2}u_0[[s^+ \otimes s^+]^0 \otimes [\tilde{s} \otimes \tilde{s}]^0]^0. \tag{1.19}$$

This Hamiltonian contains nine parameters, two of them multiplying the one-body terms (ϵ_s, ϵ_d), and 7 multiplying the two-body terms (c_0, c_2, c_4, $\tilde{v}_0$, $\tilde{v}_2$, u_0, u_2). Remember, however, that the total number of bosons N, which equals the total number of valence nucleon pairs, is strictly conserved in the **IBM** framework. Using this constraint, one can rewrite the above Hamiltonian in terms of $9 - 1 = 8$ parameters only, in the form

$$H = \epsilon_s N + \frac{1}{2}u_0 N(N-1) + \epsilon'(d^+ \otimes \tilde{d})$$

$$+ \sum_{L=0,2,4} \frac{1}{2}\sqrt{2L+1}c'_L[[d^+ \otimes d^+]^L \otimes [\tilde{d} \otimes \tilde{d}]^L]^0$$

$$+ \frac{1}{\sqrt{2}}\tilde{v}_2[[d^+ \otimes d^+]^2 \otimes [\tilde{d} \otimes \tilde{s}]^2 + [d^+ \otimes s^+]^2 \otimes [\tilde{d} \otimes \tilde{d}]^2]^0$$

$$+ \frac{1}{2}\tilde{v}_0[[d^+ \otimes d^+]^0 \otimes [\tilde{s} \otimes \tilde{s}]^0 + [s^+ \otimes s^+]^0 \otimes [\tilde{d} \otimes \tilde{d}]^0]^0, \tag{1.20}$$

where

$$N = (s^+ \odot \tilde{s}) + (d^+ \odot \tilde{d}). \tag{1.21}$$

Using the relations

$$(d^+ s^+) \odot (\tilde{d}\tilde{s}) = n_n n_s, \tag{1.22}$$

$$s^+ s^+ \tilde{s}\tilde{s} = n_s(n_s - 1), \tag{1.23}$$

where n_s (n_d) is the number operator of the s (d) bosons, it is easy to see that the new parameters are connected to the old parameters through the equations

$$\epsilon' = \epsilon_d - \epsilon_s + \frac{1}{\sqrt{5}} u_2 (N - 1) - \frac{1}{2} u_0 (2N - 1), \tag{1.24}$$

$$c'_L = c_L + u_0 - 2u_2. \tag{1.25}$$

The first two terms in the new form of the Hamiltonian will have the same contribution to any excited state of the nucleus, since they depend only on the total number of bosons N. Thus they do not affect the excitation energies, but only the binding energies of the various nuclei. Since in IBM-1 we are interested in the various excited levels of a single nucleus and not in the binding energies of different nuclei, we can omit these two terms. Then we are left with a Hamiltonian with 6 free parameters ($\epsilon', c'_0, c'_2, c'_4, \tilde{v}_0, \tilde{v}_2$). With the assumptions made, this is the most general IBM-1 Hamiltonian to be used to fit the low-lying spectra of medium and heavy nuclei.

1.5 Chains of subalgebras of U(6)

Once the Hamiltonian of the physical system has been identified, one wishes to diagonalize it. In order to be able to do so, one must know the states. The technique usually used is to label the states by the quantum numbers characterizing the various irreps of a chain of subalgebras of the algebra associated with the symmetry of the Hamiltonian, U(6) in our case. Thus the next step is to identify all possible subalgebras of the full U(6) algebra. A subalgebra is generated by a subset of the generators of the full algebra, which happens to close under commutation (i.e. the commutator of any two generators belonging to the subalgebra has to be expressible in terms of generators belonging to the same subalgebra only). It turns out that in the present case there are three possible chains of subalgebras.

1.5.1 Chain I

A) From the 36 generators given above, delete $G_0^0(ss)$, $G_\kappa^2(ds)$, and $G_\kappa^2(sd)$. The remaining 25 operators happen to be the generators of the algebra U(5), the group of unitary transformations in 5 dimensions. Notice that only d-bosons are employed by these 25 operators.

B) From the remaining 25 operators, delete $G_0^0(dd)$, $G_\kappa^2(dd)$, $G_\kappa^4(dd)$. The remaining 10 operators close under the algebra O(5), the orthogonal group in 5 dimensions.

C) From the remaining 10 operators delete $G_\kappa^3(dd)$. The remaining three operators close under the algebra of O(3), the well known rotation group, i.e. the operators $G_\kappa^1(dd)$ are proportional to the usual angular momentum operators. In fact it is true that

$$\vec{L} = \sqrt{10}[d^+ \otimes \tilde{d}]^1. \tag{1.26}$$

D) Finally, delete from the 3 remaining operators the operators $G_{+1}^1(dd)$ and $G_{-1}^1(dd)$. The remaining operator, $G_0^1(dd)$, is the generator of the algebra O(2), the group of rotations around the z–axis.

Thus, one possible chain of subalgebras is:

$$U(6) \supset U(5) \supset O(5) \supset O(3) \supset O(2). \tag{1.27}$$

1.5.2 Chain II

A) Consider the 9 operators

$$G_0^0(ss) + \sqrt{5}G_0^0(dd) = [s^+ \otimes \tilde{s}]_0^0 + \sqrt{5}[d^+ \otimes \tilde{d}]_0^0, \tag{1.28}$$

$$G_\kappa^1(dd) = [d^+ \otimes \tilde{d}]_\kappa^1, \tag{1.29}$$

$$G_\kappa^2(ds) + G_\kappa^2(sd) \mp \frac{\sqrt{7}}{2}G_\kappa^2(dd) = [d^+ \otimes \tilde{s} + s^+ \otimes \tilde{d}]_\kappa^2 \mp \frac{\sqrt{7}}{2}[d^+ \otimes \tilde{d}]_\kappa^2. \tag{1.30}$$

These operators close under commutation and form the Lie algebra U(3).

B) From the operators given above delete the first one. The remaining 8 operators are the generators of the group SU(3).

C) As in chain I, the 3 operators $G_\kappa^1(dd)$ form the subalgebra O(3).

D) As in chain I, the operator $G_0^1(dd)$ forms the subalgebra O(2).

Thus , a second possible chain of subalgebras is:

$$U(6) \supset U(3) \supset SU(3) \supset O(3) \supset O(2). \qquad (1.31)$$

The reader will have noticed the double sign appearing in the quadrupole generator. Both choices for this sign allow the algebra to close, satisfying the same SU(3) commutation relations. Although the two choices are mathematically equivalent, each sign has been assigned in applications a different physical meaning. Actually, as we will see later, the SU(3) symmetry describes nuclei which are permanently deformed. The minus sign has been used to indicate prolate (cigar-like) deformation, while the positive sign has been assigned to oblate (pancake-like) deformation.

1.5.3 Chain III

A) Consider the 15 operators

$$G_\kappa^1(dd) = [d^+ \otimes \tilde{d}]_\kappa^1, \qquad (1.32)$$

$$G_\kappa^3(dd) = [d^+ \otimes \tilde{d}]_\kappa^3, \qquad (1.33)$$

$$G_\kappa^2(ds) + G_\kappa^2(sd) = [d^+ \otimes s + s^+ \otimes \tilde{d}]_\kappa^2. \qquad (1.34)$$

These operators close under commutation, yielding the Lie algebra of O(6), the orthogonal group in 6 dimensions.

B) From the operators given above, delete the last 5. The remaining 10 operators form the algebra O(5).

C) As in the previous two chains, the 3 operators $G_\kappa^1(dd)$ form the subalgebra O(3).

D) As in the previous two chains, the operator $G_0^1(dd)$ forms the subalgebra O(2).

Thus , a third possible chain is

$$U(6) \supset O(6) \supset O(5) \supset O(3) \supset O(2). \qquad (1.35)$$

Moral : *Groups were born free . We put them in chains ! (S. Pittel).*

It is possible to show that these are the only possible chains one can obtain for this problem, if one insists that O(3) must be contained in the chain (this requirement is equivalent to demanding that the angular momentum L be a good quantum number).

1.6 Classification of states

Once a group chain has been identified, its first important application is in constructing a basis in which the Hamiltonian can be diagonalized. In order to do this, one needs to know the labels which characterize the irreducible representations (irreps) of the various groups which appear in the chain. Furthermore, one needs to know which values of the quantum number(s) characterizing the irreps of a subgroup G' are allowed for given values of the quantum number(s) characterizing the irreps of the larger group G.

Example Consider the well known group of rotations in the 3-dimensional space, O(3). Its irreps are characterized by the values of angular momentum, L. This group is known to have an O(2) sub group, corresponding to rotations around the z–axis. The irreps of this subgroup are known to be characterized by the z–component of angular momentum, M. Thus, states in the group chain $O(3) \supset O(2)$ are characterized by the classification scheme $|LM>$. It is well known that for a given value of the angular momentum, L, the allowed values of M are $M = -L, -L+1, \ldots, +L$.

The situation is apparently more complex in the case of the three group chains described before. The general procedure by means of which one can find the quantum numbers characterizing the irreps of the various groups, as well as the irreps of a subgroup contained in a given irrep of the larger group (the latter known as **building up process**), can be found in the Appendices. Here only final results will be quoted, which are necessary in carrying on the discussion. The fact that we are dealing with a system of identical bosons causes certain simpifications, as it will be evident below.

1.6.1 Chain I

A) In general the irreps of U(6) are labelled by six quantum numbers (as shown in Appendix 2.2). Then the corresponding Young tableaux are denoted as $[f_1, f_2, \ldots, f_6]$. But in the present case we are dealing with a system of identical bosons, thus only the most symmetric irreps (i.e. the one-row irreps) can occur. Thus the irreps of U(6) are labelled by only one quantum number, the total number of bosons N, and are denoted by [N].

B) In general the irreps of U(5) are labelled by five quantum numbers (Appendix 2.2). In the present case, as a consequence of the Bose statistics, only one quantum number is necessary. Thus the irreps of U(5) are labelled by the number of d-bosons, n_d. They are denoted by (n_d). The values of n_d allowed for a given value of N are

$$n_d = 0, 1, \ldots, N. \tag{1.36}$$

C) In general the irreps of O(5) are labelled by two quantum numbers (Appendix 2.3). Because of the Bose statistics, only one quantum number is required here. Thus the irreps of O(5) are labelled by the quantum number v, which is called the seniority and is defined as the number of boson pairs not coupled to zero angular momentum. The values of v contained in each n_d are

$$v = n_d, n_d - 2, \ldots, 1 \quad or \quad 0 \quad ; \quad n_d = odd \quad or \quad even. \tag{1.37}$$

D) The irreps of O(3) are labelled by the angular momentum quantum number L. But, when going from O(5) to O(3) an additional problem arises, namely that more than one state with a given value of L (characterizing an O(3) irrep) is contained in a given representation v of O(5). When this occurs, one says that the step from O(5) to O(3) is not fully decomposable and an additional quantum number is needed to characterize the states. Identifying this additional quantum number and finding its values is one of the most difficult problems in group theory. Anyway, in this particular case the extra quantum number, denoted by n_Δ, is chosen as the number of boson triplets not coupled to zero angular momentum. The values of L contained in each irrep n_d of $U(5)$ are then given by the following algorithm. First, partition n_d as

$$n_d = 2n_\beta + 3n_\Delta + \lambda, \tag{1.38}$$

where

$$n_\beta = \frac{n_d - v}{2}, \tag{1.39}$$

$$n_\beta = 0, 1, \ldots, \frac{n_d}{2} \ or \ \frac{n_d - 1}{2}. \tag{1.40}$$

Then,

$$L = \lambda, \lambda + 1, \lambda + 2, \ldots, 2\lambda - 2, 2\lambda. \tag{1.41}$$

Notice that $2\lambda - 1$ is missing!

E) The irreps of O(2) are labelled by the z-component of angular momentum, M. The values of M allowed for a given value of L are well known to be

$$-L \leq M \leq L. \tag{1.42}$$

Thus the complete classification scheme for chain I is

$$|[N](n_d)vn_\Delta LM > . \tag{1.43}$$

Example Consider the case $n_d = 3$. Then, the allowed values of seniority are $v = 1, 3$. For $v=1$ it turns out that $n_\Delta = 0$ and $L = 2$. For $v = 3$ one can either have $n_\Delta = 0$ and $L = 6, 4, 3$, or one can have $n_\Delta = 1$ and $L = 0$.

1.6.2 Chain II

A) As in subsec. 1.6.1, the irreps of U(6) are labelled by the total number of bosons N, and are denoted by [N].

B) The irreps of SU(3) are characterized by two quantum numbers, (λ, μ). The values of (λ, μ) contained in each [N] are given by

$$[N] = (2N, 0)$$

$$\oplus(2N - 4, 2) \oplus (2N - 8, 4) \oplus \ldots \oplus \left\{ \begin{matrix} (0, N) \\ (2, N - 1) \end{matrix} \right\} \left\{ \begin{matrix} if & N = even \\ if & N = odd \end{matrix} \right\}$$

$$\oplus(2N-6, 0)\oplus(2N-10, 2)\oplus\ldots\oplus \left\{ \begin{matrix} (0, N - 3) \\ (2, N - 4) \end{matrix} \right\} \left\{ \begin{matrix} if & N - 3 = even \\ if & N - 3 = odd \end{matrix} \right\}$$

$$\oplus(2N-12, 0)\oplus(2N-16, 2)\oplus\ldots\oplus \left\{ \begin{matrix} (0, N - 6) \\ (2, N - 7) \end{matrix} \right\} \left\{ \begin{matrix} if & N - 6 = even \\ if & N - 6 = odd \end{matrix} \right\}$$

$$\oplus \ldots \tag{1.44}$$

C) The irreps of O(3) are labelled again by L. The step from SU(3) to O(3) is not fully decomposable . The simplest choice of the additional quantum number needed to classify the states is due to Elliott (Elliott 1958a, 1958b). The extra quantum number required is called K. The values of L contained in each (λ, μ) in Elliott's basis are given by the following algorithm

$$L = K, K + 1, K + 2, \ldots, K + max\{\lambda, \mu\}, \tag{1.45}$$

where

$$K = integer = min\{\lambda, \mu\}, min\{\lambda, \mu\} - 2, \ldots, 1 \quad or \quad 0, \tag{1.46}$$

with the exception of $K = 0$, for which

$$L = max\{\lambda, \mu\}, max\{\lambda, \mu\} - 2, \ldots, 1 \quad or \quad 0. \tag{1.47}$$

Elliott's basis has the drawback that it is not orthogonal. An orthogonal basis can be constructed from it in a straightforward manner, called Vergados' basis (Vergados 1968). In this orthogonal basis the extra quantum number χ is used instead of K. K and χ are related in the following way. Let K_1, K_2, ..., K_n be the Elliott quantum numbers which occur in a given SU(3) irrep, (λ, μ), where $K_1 < K_2 < \ldots < K_n$. In the Vergados basis, the same SU(3) irrep will contain the sequence of quantum numbers $\chi_1, \chi_2, \ldots, \chi_n$, where $\chi_1 < \chi_2 < \ldots < \chi_n$. The two sequences of quantum numbers, K_n and χ_n, are exactly the same, but the values of L contained in each χ_i are different from those contained in the corresponding K_i. In particular, if a given value L occurs in a given SU(3) irrep, (λ, μ), only once, it belongs to the lowest possible χ. If it occurs twice, it belongs to the two lowest possible χ's, etc. The only exception occurs when $\chi = 0$, for which the allowed L values are restricted to be even or odd, for λ even or odd, respectively. In this case the forbidden L-values are assigned to the next higher value of χ. In the following, the Vergados' basis will be used.

D) The irreps of O(2) are again labelled by the z-component of angular momentum, M.

Thus the complete classification scheme for chain II is

$$| [N](\lambda, \mu)\chi L M > . \tag{1.48}$$

Example Consider $[N] = 3$. The allowed SU(3) irreps are (6,0), (2,2), (0,0). In the (6,0) irrep, one has $\chi = 0$ and $L = 6, 4, 2, 0$. In the (0,0) irrep, one has $\chi = 0$ and $L = 0$. In the (2,2) irrep one has either $\chi = 0$, which contains $L = 4, 2, 0$, or $\chi = 2$, which contains $L = 3, 2$.

1.6.3 Chain III

A) As in subsec. 1.6.1, the irreps of U(6) are again labelled by the total number of bosons N, and they are denoted by [N].

B) In general the irreps of O(6) are labelled by three quantum numbers (Appendix 2.3), the corresponding Young tableaux being labelled as $(\sigma_1, \sigma_2, \sigma_3)$. In the present case, because of the Bose statistics, only one quantum number is necessary. Thus the irreps of O(6) are labelled by the quantum number σ, and they are denoted by (σ). The values of σ contained in each [N] are given by

$$\sigma = N, N - 2, \ldots, 0 \quad or \quad 1, \quad for \quad N = even \quad or \quad odd. \tag{1.49}$$

C) In general the irreps of O(5) are labelled by two quantum numbers (Appendix 2.3), the corresponding Young tableaux being labelled as (τ_1, τ_2). In the present case, because of the Bose statistics, only one quantum number is needed. Thus the irreps of O(5) are labelled by the quantum number τ. The values of τ contained in each (σ) are given by

$$\tau = \sigma, \sigma - 1, \ldots, 0. \tag{1.50}$$

D) The irreps of O(3) are again labelled by L. But, once more, the step from O(5) to O(3) is not fully decomposable. One needs a further quantum number, called ν_Δ. The values of L contained in each τ are found by the following algorithm. First, partition τ as

$$\tau = 3\nu_\Delta + \lambda, \tag{1.51}$$

where

$$\nu_\Delta = 0, 1, \ldots. \tag{1.52}$$

Then take

$$L = 2\lambda, 2\lambda - 2, \ldots, \lambda + 1, \lambda. \tag{1.53}$$

Notice that $2\lambda - 1$ is missing !

E) The irreps of O(2) are again labelled by M.

Thus the complete classification scheme for chain III is

$$|[N](\sigma)\tau\nu_\Delta LM > . \tag{1.54}$$

Example Consider $[N] = 2$. Then one can have either $\sigma = 0$, which implies $\tau = 0$, $\nu_\Delta = 0$, $L = 0$, or one can have $\sigma = 2$, which contains the following cases:

    i) $\tau = 2$, $\nu_\Delta = 0$, L=4,2,

    ii) $\tau = 1$, $\nu_\Delta = 0$, L=2,

    iii) $\tau = 0$, $\nu_\Delta = 0$, L=0.

Having constructed a classification scheme of the states, we can now diagonalize the most general Hamiltonian which can be written using the generators of U(6) as building blocks. Any of these chains can be used for this purpose, since all three bases are complete. However, before considering the most general Hamiltonian and its numerical diagonalization, it is instructive to study those cases where the eigenvalue problem can be solved analytically. This task will be undertaken in the following section.

1.7 Dynamical symmetries

The technique allowing us to find analytic solutions of the eigenvalue problem of the Hamiltonian is again based on group theory.

Two concepts are important at this point: Casimir operators and dynamical symmetries.

1.7.1 Casimir operators

Let G_κ^k be the generators of the Lie algebra G. One can identify some operators C, called **Casimir operators**, which have the property

$$[C, G_\kappa^k] = 0 \quad for \quad any \quad k, \kappa. \tag{1.55}$$

Thus by definition the Casimir operators of the group G commute with all the generators of the group G. There exists at least one Casimir operator (apart from the unit operator) for each of the Lie algebras under study. If one has a Hamiltonian H built up exclusively from generators of G, in all possible combinations, this Hamiltonian will commute with C. Hence the alternative name **Casimir invariants**.

Example 1 The algebra U(6), considered earlier, has a linear Casimir operator

$$C_{1U6} = G_0^0(ss) + \sqrt{5}G_0^0(dd). \tag{1.56}$$

This operator commutes with all the generators of U(6), and it is nothing else but the total number of bosons

$$C_{1U6} = [s^+ \otimes \tilde{s}]^0 + \sqrt{5}[d^+ \otimes \tilde{d}]^0 = n_s + n_d = N. \tag{1.57}$$

The statement that C_{1U6} commutes with all the generators of U(6) guarantees that the total number of bosons will be conserved by any IBM Hamiltonian, written in terms of generators of U(6) only. Conservation of the total number of bosons is a basic requirement in all IBM models. In the above, we used the notation C_{1Un} for a linear Casimir operator of the group U(n). Notice that only the unitary groups U(n) have linear Casimir operators.

Example 2 Besides linear Casimir operators, there are in addition quadratic Casimir operators. For example, consider the algebra O(3), generated by the operators $G_\kappa^1(dd)$. Its quadratic Casimir operator is defined as

$$C_{2O3} = (G^1(dd) \odot G^1(dd)). \tag{1.58}$$

Since $G^1(dd)$ is proportional to the angular momentum operator $\vec{L}$, the Casimir operator C_{2O3} is proportional to $\vec{L} \cdot \vec{L}$. It is well known from simple angular momentum algebra that this last operator commutes with all the components of the angular momentum operator. The notation C_{2On} has been used to denote a quadratic Casimir operator of the orthogonal group O(n).

1.7.2 Casimir operators of sugroups of U(6)

In what follows, we will need all the linear and quadratic Casimir operators of the subgroups of U(6) appearing in the group chains already studied. Among them, only U(5) has a linear Casimir operator, while all of them have quadratic Casimir operators. Finding the expressions of the Casimir operators in terms of boson operators is a straightforward problem , thus only the results will be quoted here :

$$C_{1U5} = (d^+ \odot \tilde{d}) = n_d, \tag{1.59}$$

$$C_{2U5} = (d^+ \odot \tilde{d})(d^+ \odot \tilde{d}) + 4(d^+ \odot \tilde{d}) = n_d(n_d + 4), \tag{1.60}$$

$$C_{2O5} = 4([d^+ \otimes \tilde{d}]^1 \odot [d^+ \otimes \tilde{d}]^1) + 4([d^+ \otimes \tilde{d}]^3 \odot [d^+ \otimes \tilde{d}]^3), \tag{1.61}$$

$$C_{2O3} = 20([d^+ \otimes \tilde{d}]^1 \odot [d^+ \otimes \tilde{d}]^1), \tag{1.62}$$

$$C_{2SU3} = \frac{4}{3}([d^+ \otimes \tilde{s} + s^+ \otimes \tilde{d}]^2 - \frac{\sqrt{7}}{2}[d^+ \otimes \tilde{d}]^2)$$

$$\odot([d^+ \otimes \tilde{s} + s^+ \otimes \tilde{d}]^2 - \frac{\sqrt{7}}{2}[d^+ \otimes \tilde{d}]^2) + 5([d^+ \otimes \tilde{d}]^1 \odot [d^+ \otimes \tilde{d}]^1), \tag{1.63}$$

$$C_{2O6} = 2N(N+4) - 2[(d^+ \odot d^+) - (s^+ \odot s^+)][(\tilde{d} \odot \tilde{d}) - \odot(\tilde{s} \odot \tilde{s})]. \tag{1.64}$$

In phenomenological analyses, it is customary to introduce the physically more transparent operators

$$P = \frac{1}{2}(\tilde{d} \odot \tilde{d}) - \frac{1}{2}(\tilde{s} \odot \tilde{s}), \tag{1.65}$$

$$L = \sqrt{10}[d^+ \otimes \tilde{d}]^1, \tag{1.66}$$

$$Q = [d^+ \otimes \tilde{s}]^2 + [s^+ \otimes \tilde{d}]^2 - \frac{\sqrt{7}}{2}[d^+ \otimes \tilde{d}]^2, \tag{1.67}$$

$$T_3 = [d^+ \otimes d]^3, \tag{1.68}$$

$$T_4 = [d^+ \otimes \tilde{d}]^4. \tag{1.69}$$

P is the pairing operator, L is the angular momentum operator, Q is the quadrupole operator.

Then one gets for the Casimir operators the more compact form

$$C_{2O5} = 4[\frac{1}{10}(L \odot L) + (T_3 \odot T_3)], \qquad (1.70)$$

$$C_{2O3} = 2(L \odot L), \qquad (1.71)$$

$$C_{2SU3} = \frac{2}{3}[2(Q \odot Q) + \frac{3}{4}(L \odot L)], \qquad (1.72)$$

$$C_{2O6} = 2N(N + 4) - 8(P^+ \odot P). \qquad (1.73)$$

Warning Casimir operators are determined up to a constant factor . Different authors use different values for these constant factors . The results given here are consistent with the definitions of Iachello (1979).

1.7.3 The Hamiltonian in terms of Casimir operators

Assume now, as in Sec. 1.4, that we are interested only in the low-lying excitation spectra of medium and heavy nuclei, and not in their binding energies. It is then possible to show that the most general IBM Hamiltonian can be written entirely in terms of the Casimir invariants of the groups U(5), O(5), O(3), SU(3), and O(6) (i.e. the subgroups of U(6) appearing in the chains studied before). Furthermore, assume (as in Sec. 1.4) that the Hamiltonian contains only one-body and two-body terms. Then it is possible to see that the most general IBM Hamiltonian can be entirely written in terms of only linear and quadratic Casimir invariants of the above groups. (Notice that linear Casimir operators are one-body operators, quadratic Casimir operators are two-body operators, etc.). In doing so, one only needs the expressions of the various Casimir invariants involved in terms of the boson operators. These have been given in the previous subsection. Then the Hamiltonian gets the form

$$H = \epsilon''' C_{1U5} + \alpha' C_{2U5} + \beta' C_{2O5} + \gamma' C_{2O3} + \delta' C_{2SU3} + \eta' C_{2O6}, \qquad (1.74)$$

where ϵ''', α', β', γ', δ', η' are six independent free parameters, which are fitted to the data. It is not surprising that the new H contains as many free parameters as the old H did, since no existing constraints were removed or new constraints added at this point. The new form of the Hamiltonian is more useful for practical purposes, because it allows for the easy identification of dynamical symmetries, which we will discuss in the next subsection.

A few comments are appropriate at this point. Remember that this Hamiltonian contains two assumptions: First, terms contributing only to the binding energies have been omitted. Since we are interested in excitation spectra and the relative positions of the levels are not affected by these terms, this assumption is a legitimate one. Second, only one-body and two-body terms have been included in H. Although it has been argued that these terms suffice for the description of low-lying spectra, this assumption is questionable, as we will discuss later on. One may also have noticed that the group O(2) has not been employed in writing down the most general Hamiltonian. This was done because O(2) does not play any role unless the nucleus is placed in an external magnetic field. Thus O(2) will be neglected henceforth in this discussion.

Using the phenomenological operators of the previous subsection, it is easy to see that the most general Hamiltonian can be put in the form

$$H = \epsilon'' n_d + a_0(P \odot P) + a_1(L \odot L) + a_2(Q \odot Q)$$

$$+a_3(T_3 \odot T_3) + a_4(T_4 \odot T_4). \tag{1.75}$$

This form is oftenly used for phenomenological analyses. The reason for this preference is that it has been found that usually one or two terms of this Hamiltonian suffice to get a good description of the spectrum. To obtain the relations between the free parameters of this form of H and the ones of the Hamiltonian written in terms of the Casimir operators, one only needs the expressions of the Casimir operators as functions of the phenomenological operators, given in the previous subsection.

1.7.4 Exact diagonalization of special IBM Hamiltonians

Casimir operators have a very useful property: they are diagonal in the representation provided by the corresponding group. This property allows us to identify all possible special cases for which the eigenvalue problem of the IBM-1 Hamiltonian can be solved analytically. This situation will occur whenever the Hamiltonian can be written in terms only of Casimir operators of a complete chain of subgroups of U(6), since in that case H will be diagonal in the basis provided by the classification scheme for the corresponding chain. When this situation occurs, one says that the Hamiltonian has a dynamical symmetry. In the case of IBM-1, since there are three chains of subgroups, three possible dynamical symmetries occur. As it is clear from the expression of H given above, these dynamical symmetries correspond to the vanishing of some coefficients in this equation.

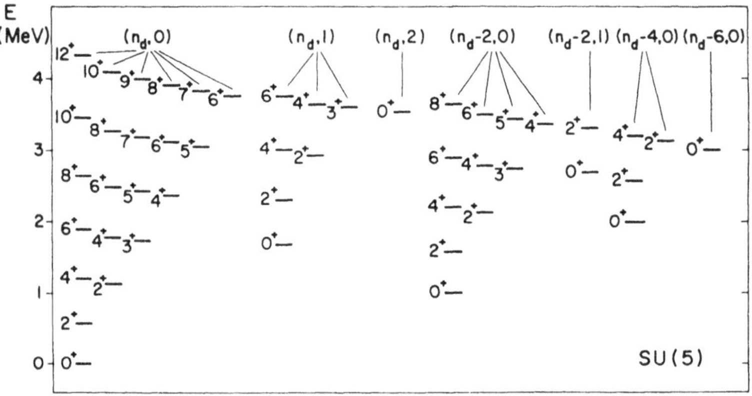

Fig. 1.2 A typical spectrum with U(5) symmetry and N=6. In parentheses the quantum numbers v and n_Δ appear, while L is denoted next to each energy level. (Taken from Iachello (1979)).

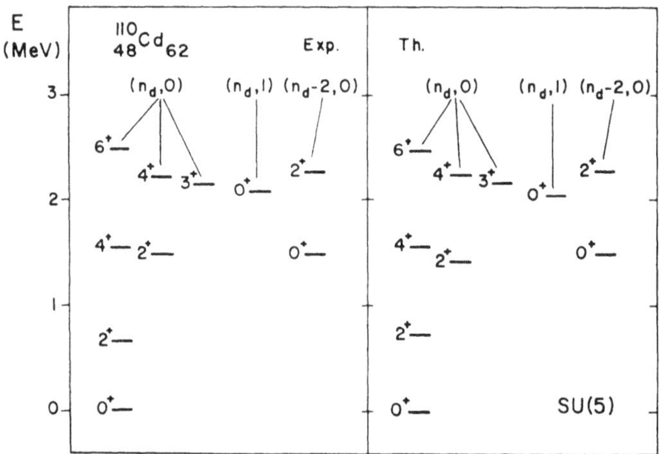

Fig. 1.3 Comparison between the experimental spectrum of the nucleus ^{110}Cd and the predictions of the U(5) IBM-1 Hamiltonian. (Taken from Iachello (1979)).

The three possible dynamical symmetries will now be discussed in detail.

1.7.5 Dynamical symmetry I

This dynamical symmetry occurs when the entire Hamiltonian can be written in terms of the Casimir operators of the groups appearing in the group chain

$$U(6) \supset U(5) \supset O(5) \supset O(3) \supset O(2). \qquad (1.76)$$

Thus this symmetry corresponds to the vanishing of δ' and η' in H. The corresponding Hamiltonian is then

$$H^{(I)} = \epsilon''' C_{1U5} + \alpha' C_{2U5} + \beta' C_{2O5} + \gamma' C_{2O3}. \qquad (1.77)$$

In order to find the eigenvalues of $H^{(I)}$ in the representation $|[N](n_d)vn_\Delta LM >$ one needs to know the expectation values of the various Casimir operators appearing in it. These are given by straightforward techniques. The relevant general formulas are given in the Appendices. The resulting expression is

$$< H^{(I)} >= \epsilon''' n_d + \alpha' n_d(n_d + 4) + 2\beta' v(v + 3) + 2\gamma' L(L + 1). \qquad (1.78)$$

A typical spectrum generated by this Hamiltonian is shown in Fig. 1.2. In Fig. 1.3 the predictions of this Hamiltonian are compared to the experimental energy levels of a vibrational nucleus. Notice that the levels of the ground state band are almost equispaced, as expected for a vibrational nucleus. In addition, notice that the odd-spin levels of the γ_1 band do not occur near the energy midway between their even-spin neighbors, but they are strongly displaced upwards. This effect is called **odd–even staggering** and we will deal again with it later.

1.7.6 Dynamical symmetry II

The relevant group chain here is

$$U(6) \supset SU(3) \supset O(3) \supset O(2). \qquad (1.79)$$

This dynamical symmetry corresponds to the vanishing of ϵ''', α', β' and η' in the most general H. Thus the Hamiltonian in this case becomes

$$H^{(II)} = \delta' C_{2SU3} + \gamma' C_{2O3}. \qquad (1.80)$$

In the representation $|[N](\lambda,\mu)\chi LM >$ the expectation value of $H^{(II)}$ is given by

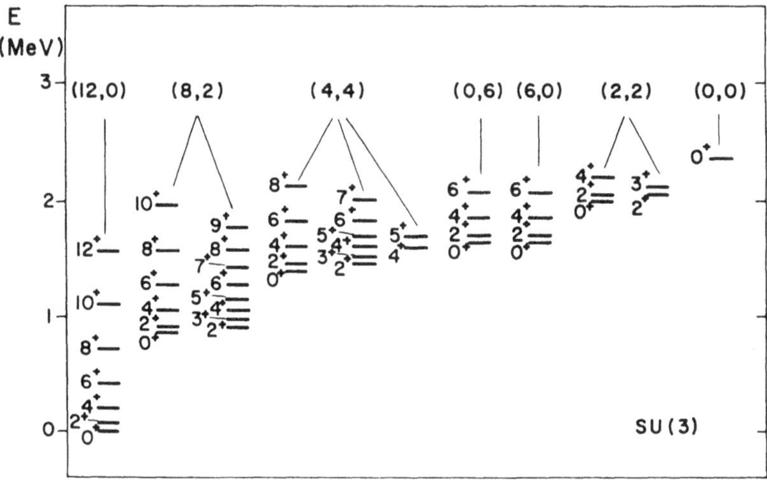

Fig. 1.4 A typical spectrum with SU(3) symmetry and N=6. In parentheses the quantum numbers λ and μ appear. (Taken from Iachello (1979)).

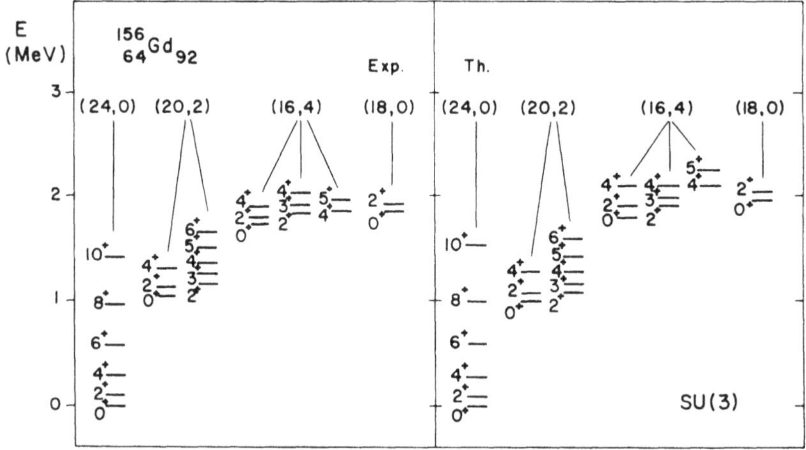

Fig. 1.5 Comparison between the experimental spectrum of the nucleus ^{156}Gd and the predictions of the SU(3) IBM-1 Hamiltonian. (Taken from Iachello (1979)).

$$< H^{(II)} >= \delta' \frac{6}{9}[\lambda^2 + \mu^2 + \lambda\mu + 3(\lambda + \mu)] + \gamma'2L(L+1). \quad (1.81)$$

Of course, any linear combination of C_{2SU3} and C_{2O3} can be used to generate the solution (remember that finally the unknown numerical coefficients will have to be fitted to the experimental data). An equivalent form of the Hamiltonian, which is frequently used, is

$$H^{(II)} = -\kappa 2(Q \odot Q) - \kappa'(L \odot L), \quad (1.82)$$

leading to eigenvalues

$$< H^{(II)} >= (\frac{3}{4}\kappa - \kappa')L(L+1) - \kappa[\lambda^2 + \mu^2 + \lambda\mu + 3(\lambda + \mu)]. \quad (1.83)$$

The relations between the constants δ', γ', and κ, κ' can be easily found, since it can be proved that

$$C_{2SU3} = 2(Q \odot Q) + \frac{3}{4}(L \odot L). \quad (1.84)$$

The second form of the Hamiltonian is physically more transparent: The first term represents the quadrupole–quadrupole interaction, while the second term is the square of the angular momentum operator. Similar rewriting of the Hamiltonian can, of course, be performed in the case of the other two dynamical symmetries as well.

A typical spectrum generated by this Hamiltonian is shown in Fig. 1.4. In Fig. 1.5 the predictions of this Hamiltonian are compared to the experimental spectrum of a rotational nucleus. Notice that within each band the energy of the levels increases following very closely the $L(L+1)$ rule, as expected for rotational nuclei. Also notice that in the SU(3) limit of IBM-1 the ground state band always belongs to an SU(3) irrep with $\mu = 0$ (actually, this is always the (2N,0) irrep), while the γ_1 and the β_1 bands always belong to an SU(3) irrep with $\mu = 2$ (which is always the (2N-4,2) irrep), the γ_2, β_2, and the first K=4 bands always belong to an SU(3) irrep with $\mu = 4$(always the (2N-8,4) irrep), etc. Furthermore, notice that bands belonging to the same SU(3) irrep are predicted to be degenerate (levels of these bands having the same angular momentum should also have the same excitation energy). Thus, for example, the γ_1 and β_1 bands are predicted to be degenerate in the SU(3) limit.

This condition is experimentally known to hold for some deformed nuclei, like ^{156}Gd.

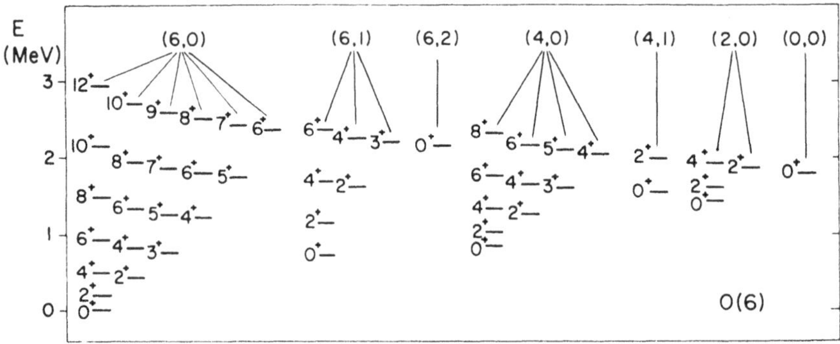

Fig. 1.6 A typical spectrum with O(6) symmetry and N=6. In parentheses the quantum numbers σ and ν_Δ appear. (Taken from Iachello (1979)).

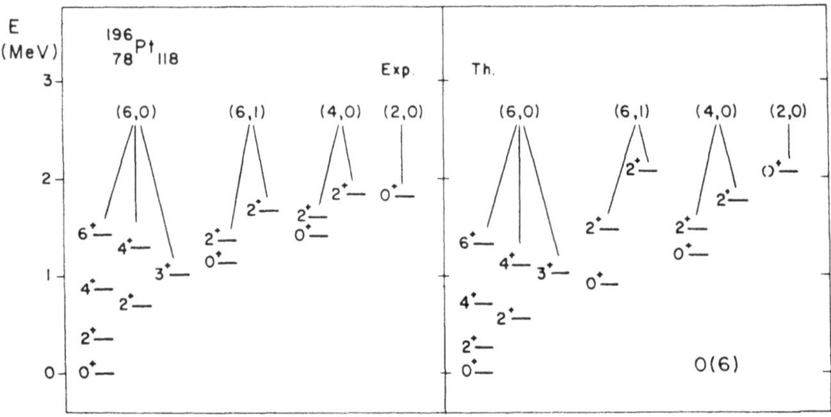

Fig. 1.7 Comparison between the experimental spectrum of the nucleus ^{196}Pt and the predictions of the O(6) IBM-1 Hamiltonian. (Taken from Iachello (1979)).

Why is it true that the ground state band belongs to the (2N,0) irrep? A physical argument can be given. It turns out that the SU(3) symmetry is appropriate for the description of deformed nuclei. For these nuclei, the quadrupole–quadrupole interaction is expected to be strong. Thus, one should expect that the irrep which maximizes the quadrupole–quadrupole interaction should lie lowest in energy. From the relation between C_{2SU3} and $(Q \odot Q)$ given above, it is clear that the $(Q \odot Q)$ interaction is maximized when the C_{2SU3} is maximized. It has been customary to call the irrep of SU(3) maximizing C_{2SU3} the most leading irrep, the irrep giving the second largest value of C_{2SU3} the second most leading irrep, etc. It is easy to check that the (2N,0) irrep gives the maximum value to C_{2SU3}. Thus it is the most leading irrep and the states lowest in energy (i.e. the ground state band) should belong to it. One can then easily check that (2N-4,2) is the second most leading irrep, etc.

1.7.7 Dynamical symmetry III

The group chain here is

$$U(6) \supset O(6) \supset O(5) \supset O(3) \supset O(2). \qquad (1.85)$$

This symmetry corresponds to the vanishing of the coefficients ϵ''', α' and δ'. The corresponding Hamiltonian is

$$H^{(III)} = \beta' C_{2O5} + \gamma' C_{2O3} + \eta' C_{2O6}. \qquad (1.86)$$

In the representation $|[N](\sigma)\tau\nu_\Delta LM >$ the expectation value of $H^{(III)}$ is

$$< H^{(III)} >= \beta' 2\tau(\tau + 3) + \gamma' 2L(L + 1) + \eta' 2\sigma(\sigma + 4). \qquad (1.87)$$

A typical spectrum generated by this Hamiltonian is shown in Fig. 1.6. In Fig. 1.7 the predictions of this Hamiltonian are compared to the experimental spectrum of a soft (γ-unstable) nucleus. As in the U(5) case, strong odd–even staggering is observed in the γ_1 band. We will further discuss this effect in Chapter 2.

1.8 Geometric analysis of IBM-1

So far the algebraic properties of the group U(6) and its subgroups have been discussed. This study led to a classification of nuclear spectra in terms of the various limiting symmetries contained in the group U(6). However, in addition to some algebraic structure, Lie groups also have geometric structures. Although the algebraic structure is relatively simple and easy to use in applications, the geometric properties of Lie groups are also of interest, because they are easier to visualize. In fact, the geometric analogs provide a convenient way to get some physical insight into abstract algebraic concepts.

In this Section some results of the geometrical analysis of IBM-1 will be quoted. Geometric variables are usually introduced through the use of **intrinsic** or **coherent states**, although other methods also exist. Detailed accounts of such analyses will be given in Chapter 13.

In the geometric limit, the shapes of nuclei are expressed in terms of the variables β and γ (details can be found in Preston and Bhaduri (1975)). β is zero for a spherical nucleus, while it takes non-zero values for deformed nuclei. γ is zero for axially symmetric prolate nuclei and is equal to $\pi/3$ for axially symmetric oblate nuclei, while it takes values between zero and $\pi/3$ for nuclei of triaxial shape. When the expression of the energy as a function of β and γ is known, one is interested in the minimum of this function in the (β, γ) plane, since the values of β and γ at the minimum will correspond to the equilibrium shape (the preferred shape) of the nucleus.

In the case of the U(5) limit of IBM-1 (Dynamical Symmetry I), we will see in Chapter 13 that the Hamiltonian in terms of geometric variables reads

$$E^I = \epsilon_d \frac{N\beta^2}{1 + \beta^2} + f_1 N(N-1) \frac{\beta^4}{(1 + \beta^2)^2}, \qquad (1.88)$$

where N is the total number of bosons. This energy functional is independent of γ and has a minimum at $\beta = 0$. Thus it corresponds to a spherical shape. The U(5) limit of IBM-1 is thus appropriate for describing spherical or near-spherical nuclei, where vibrations are the major mechanism of excitations (this is why these nuclei are also called vibrational nuclei).

In the case of the SU(3) limit of IBM-1 (Dynamical Symmetry II), we will see in Chapter 13 that the Hamiltonian in terms of geometrical variables reads

$$E^{II} = -\kappa \left[\frac{N}{1 + \beta^2} \left(5 + \frac{11}{4}\beta^2\right) \right.$$

$$\left. + \frac{N(N-1)}{(1 + \beta^2)^2} \left(\frac{\beta^4}{2} + 2\sqrt{2}\beta^3 \cos 3\gamma + 4\beta^2\right) \right] - \kappa' \frac{6N\beta^2}{1 + \beta^2}. \qquad (1.89)$$

This energy functional has a minimum at $\gamma = 0$ and $\beta \neq 0$. Thus it corresponds to an axially symmetric prolate shape.

In the case of the O(6) limit of IBM-1 (Dynamical Symmetry III), we will see in Chapter 13 that the Hamiltonian in terms of geometrical variables reads

$$E^{III} = C_1 \frac{N\beta^2}{1 + \beta^2} + C_2 N(N-1) \left(\frac{1 - \beta^2}{1 + \beta^2}\right)^2. \qquad (1.90)$$

This energy functional is γ-independent and has a minimum for a value of $\beta \neq 0$. It corresponds to a deformed body with γ-instability. One can think of this as being a very soft nucleus continuously changing its shape from prolate to oblate and vice versa.

We remark that no shapes with permanent triaxial deformation (i.e. shapes with $\beta \neq 0$ and $\gamma \neq 0$, $\pi/3$) occur in the framework of IBM-1. We will see how such shapes can be incorporated in the IBM-1 description in Chapter 2.

1.9 Electromagnetic transitions

1.9.1 The transition operators

Besides excitation energy spectra, IBM is able to describe electromagnetic transition rates as well. In order to do so, one has to specify the transition operators in terms of the boson operators. In lowest order it is assumed that the transition operators will contain only one-body terms. Clearly, the most general form of such an operator in IBM-1 will be

$$T_m^l = \alpha_2 \delta_{l2}[d^+ \otimes \tilde{s} + s^+ \otimes \tilde{d}]_m^2 + \beta_l[d^+ \otimes \tilde{d}]_m^l + \gamma_0 \delta_{l0}\delta_{m0}[s^+ \otimes \tilde{s}]_0^0. \quad (1.91)$$

It is clear that the first term can be present only in the case of $l = 2$ transitions, while the last term can be present only in the case of $l = 0$ transitions. This is assured by the Kronecker deltas accompanying them. In the special cases of electric monopole, quadrupole, and hexadecapole transitions, the specific form of the transition operator is respectively

$$T_0^{E0} = \beta_0[d^+ \otimes \tilde{d}]_0^0 + \gamma_0[s^+ \otimes \tilde{s}]_0^0, \quad (1.92)$$

$$T_m^{E2} = \alpha_2[d^+ \otimes \tilde{s} + s^+ \otimes \tilde{d}]_m^2 + \beta_2[d^+ \otimes \tilde{d}]_m^2, \quad (1.93)$$

$$T_m^{E4} = \beta_4[d^+ \otimes \tilde{d}]_m^4. \quad (1.94)$$

In the case of magnetic dipole and octupole transitions the relevant transition operators are respectively

$$T_m^{M1} = \beta_1[d^+ \otimes \tilde{d}]_m^1, \quad (1.95)$$

$$T_m^{M3} = \beta_3[d^+ \otimes \tilde{d}]_m^3. \quad (1.96)$$

With the restrictions that only s and d bosons are present and that only one-body terms are included in the transition operators, these are the only possible transitions in IBM-1.

1.9.2 Electromagnetic transition rates

Once the transition operators have been identified, electromagnetic transition rates can be calculated in the usual way, by taking the reduced matrix element of the corresponding transition operator between the initial and the final state. The symbol for this reduced matrix element is

$$< J_f ||T^l|| J_i > . \tag{1.97}$$

Then, by definition, the $B(El)$ and $B(Ml)$ values are

$$B(l; J_i \rightarrow J_f) = \frac{1}{2J_i + 1} | < J_f ||T^l|| J_i > |^2. \tag{1.98}$$

In the general case, the calculation has to be done numerically. However, as in the case of the excitation energies, analytical expressions can be found in the cases of the three dynamical symmetries. The results for the electric quadrupole transition rates will be given in the following subsections. Before discussing them, a few remarks are appropriate:

i) In the case of electric monopole transitions, one can exploit the conservation of the total number of bosons $N = n_s + n_d$ to rewrite the transition operator as

$$T^{E0} = \gamma_0 N + (\frac{\beta_0}{\sqrt{5}} - \gamma_0)n_d. \tag{1.99}$$

The first term will not induce transitions, because it is diagonal in any representation. Thus E0 transitions are induced only by the second term. Their transition rates will depend on the matrix elements of the number of d-bosons, n_d.

ii) In the case of magnetic dipole transitions, the transition operator is proportional to the angular momentum operator. Since L is a good quantum number in all cases under study (i.e. the angular momentum operator is always diagonal), it turns out that no M1 transitions are allowed at this level of approximation. One can obtain non-vanishing magnetic dipole transitions in the IBM-1 framework by allowing higher order terms in the M1 transition operator (like $[(d^+ s)^2 (d^+ \tilde{d})^1]^1$ and $[(d^+ \tilde{d})^2 (d^+ \tilde{d})^1]^1)$. In the IBM-2 framework, however, where the distinction between protons and neutrons is not ignored, M1 transitions occur even with the lowest-order M1 transition operator. Thus one may think that the need for higher order terms in the IBM-1 description arises from ignoring the proton–neutron degree of freedom.

1.9.3 E2 transitions in Dynamical Symmetry I

When taken between the states $|[N](n_d)vn_\Delta LM>$ of this limit, the T^{E2} operator (1.93) has selection rules

$$\Delta n_d = 0, \pm 1. \tag{1.100}$$

From Fig. 1.2 it can be seen that the ground state band is characterized by the quantum numbers n_d, $v = n_d$, $n_\Delta = 0$, $L = 2n_d$. It can be seen that the B(E2) values along the ground state band are

$$B(E2; n_d + 1, v = n_d + 1, n_\Delta = 0, L' = 2n_d + 2$$

$$\rightarrow n_d, v = n_d, n_\Delta = 0, L = 2n_d)$$

$$= \alpha_2^2 \frac{L+2}{2} \frac{2N-L}{2}. \tag{1.101}$$

Notice that only the first term of T^{E2} contributes. This is due to the fact that the states in this limiting symmetry are characterized by a fixed number of d-bosons. In particular

$$B(E2; 2_1^+ \rightarrow 0_1^+) = \alpha_2^2 N. \tag{1.102}$$

1.9.4 E2 transitions in Dynamical Symmetry II

For calculations in this limit, it is convenient to rewrite T^{E2} as

$$T_m^{E2} = \alpha_2 Q_m^2 + \alpha_2'(Q')_m^2, \tag{1.103}$$

where Q_m^2 is the already known generator of SU(3)

$$Q_m^2 = (d^+ \otimes \tilde{s} + s^+ \otimes \tilde{d})_m^2 - \frac{\sqrt{7}}{2}(d^+ \otimes \tilde{d})_m^2, \tag{1.104}$$

and

$$(Q')_m^2 = (d^+ \otimes \tilde{d})_m^2, \tag{1.105}$$

$$\alpha_2' = \beta_2 + \frac{\sqrt{7}}{2}\alpha_2. \tag{1.106}$$

It turns out that Q_m^2 is much larger than $(Q')_m^2$ in regions where this symmetry applies. Thus the second term can be omitted. The selection rules for the first term between states $|[N](\lambda,\mu)\chi LM>$ are

$$\Delta\lambda = 0, \Delta\mu = 0. \tag{1.107}$$

This is so because Q_m^2 is a generator of SU(3), so it cannot connect different SU(3) irreps. From Fig. 1.4 it can be seen that the ground state band is characterized by the quantum numbers $\lambda = 2N$, $\mu = 0$, $\chi = 0$, L. The B(E2) values along this band turn out to be

$$B(E2; (\lambda = 2N, \mu = 0), \chi = 0, L' = L + 2$$

$$\rightarrow (\lambda = 2N, \mu = 0), \chi = 0, L)$$

$$= \alpha_2^2 \frac{3(L+2)(L+1)}{4(2L+3)(2L+5)}(2N - L)(2N + L + 3). \qquad (1.108)$$

In particular

$$B(E2; 2_1^+ \rightarrow 0_1^+) = \frac{\alpha_2^2}{5} N(2N + 3). \qquad (1.109)$$

Comparing this result with the corresponding result in Dynamical Symmetry I one sees a change of the N dependence of the quantity $B(E2; 2_1^+ \rightarrow 0_1^+)$ from N in Dynamical Symmetry I to N^2 in Dynamical Symmetry II. This difference is in fact observed experimentally. Deformed nuclei (where the SU(3) symmetry applies) show much higher $B(E2; 2_1^+ \rightarrow 0_1^+)$'s than spherical nuclei (where the U(5) symmetry applies).

1.9.5 E2 transitions in Dynamical Symmetry III

In the regions where this symmetry applies, it turns out that the first term of the transition operator (1.93) is the dominant one. Thus the second term will be omitted (i.e. we consider $\beta_2 = 0$). The remaining operator, taken between states $|[N](\sigma)\tau\nu_\Delta LM >$ of this limit, has selection rules

$$\Delta\sigma = 0, \Delta\tau = \pm 1. \qquad (1.110)$$

The first of these selection rules arises from the fact that the remaining operator is a generator of the group O(6), thus it cannot connect different O(6) irreps (characterized by σ). From Fig. 1.6 it can be seen that the ground state band in this limiting symmetry is characterized by the quantum numbers $\sigma = N$, τ, $\nu_\Delta = 0$, $L = 2\tau$. The B(E2) values for transitions between members of this band are found to be

$$B(E2; \sigma = N, \tau + 1, \nu_\Delta = 0, L' = 2\tau + 2 \rightarrow \sigma = N, \tau, \nu_\Delta = 0, L = 2\tau)$$

$$= \alpha_2^2 \frac{L+2}{8(L+5)} (2N - L)(2N + L + 8). \qquad (1.111)$$

In particular

$$B(E2; 2_1^+ \rightarrow 0_1^+) = \frac{\alpha_2^2}{5} N(N + 4). \qquad (1.112)$$

1.10 Broken symmetries

1.10.1 Numerical calculations

The three limiting cases discussed so far are useful because they provide analytical expressions for excitation energies and transition rates, which can be easily tested against the experimental results. More details can be found in Arima and Iachello (1976, 1978b, 1979). However, very few nuclei show behavior near to one of these three limits. Most nuclei have spectra showing behavior intermediate between these three limits. In order to describe them, one has to include in the Hamiltonian terms belonging to more than one limiting symmetries. Dynamical symmetries are then broken, and no analytical solutions can be found. Excitation energies and transition rates have to be calculated numerically. A computer program, named PHINT, has been written for that purpose (O. Scholten, 1976, University of Groningen, The Netherlands). By now this program has been distributed to many universities and research centers. Here we will not attempt to describe this program. We will only give an example of disagreement between the predictions of IBM-1 and the experimental findings, demonstrating the need of breaking a symmetry.

1.10.2 Broken SU(3) symmetry

Consider, as an example, the SU(3) limit of IBM-1. The following discrepancies between theory and experiment exist

i) We have seen that in this limit the γ_1 band and the β_1 band belong to the same SU(3) irrep, $(2N - 4, 2)$. As a result, the levels of the two bands are predicted to be degenerate. This condition is fulfilled only for very few of the many deformed nuclei in the rare earth and the actinide region. For the majority of them, the two bands are not degenerate.

ii) In this limit the ground state band belongs to the $(2N, 0)$ irrep, while the β_1 and γ_1 bands belong to the $(2N - 4, 2)$ irrep. Remember that in this limit the T^{E2} transition operator can connect only states within the same irrep. As a result, the model predicts strong B(E2) values for $\gamma_1 \rightarrow \beta_1$ transitions, while the $\gamma_1 \rightarrow ground$

and the $\beta_1 \rightarrow ground$ transitions should be forbidden. Experimentally (Casten, von Brentano and Haque 1985) the situation is as follows: $\beta_1 \rightarrow ground$ transitions are indeed very weak (good !), $\gamma_1 \rightarrow ground$ transitions are rather strong (bad !), and $\gamma_1 \rightarrow \beta_1$ transitions are difficult to measure (in the few cases known, their values are intermediate between the two previous cases). It is clear that the prediction of vanishing $\gamma_1 \rightarrow ground$ transitions is not fulfilled.

By breaking the SU(3) symmetry (i.e. by allowing the Hamiltonian and the T^{E2} operator to include terms other than the ones allowed in the SU(3) limit) both of these difficulties can be overcome. We will now explain briefly how this is achieved.

i) The Hamiltonian usually used in the SU(3) limit is

$$H = -\kappa(Q \cdot Q) + \kappa'(L \cdot L), \qquad (1.113)$$

where

$$Q_m = (d^+\tilde{s} + s^+\tilde{d})_m^2 + \chi(d^+ \otimes \tilde{d})_m^2. \qquad (1.114)$$

When $\chi = -\sqrt{7}/2$, Q is a generator of SU(3). Thus the whole Hamiltonian has the SU(3) symmetry. When $\chi = 0$, Q is a generator of O(6), thus the Hamiltonian is not an SU(3) Hamiltonian any more. When χ takes values between zero and $-\sqrt{7}/2$, the spectrum will have a form intermediate between the SU(3) and O(6) symmetries, and the degeneracy of the γ_1 and β_1 bands will be broken.The calculations have to be done numerically. An example of results for various actinides obtained through use of the computer program ("code", if we wish to be really up-to-date !) PHINT are shown in Fig. 1.8. Clearly, the degeneracy between the γ_1 and β_1 bands is broken.

ii) The electric quadrupole transition operator can be written as

$$T_m^{E2} = e[(d^+\tilde{s} + s^+\tilde{d})_m^2 + \chi_{E2}(d^+ \otimes \tilde{d})_m^2]. \qquad (1.115)$$

Although this operator looks very similar to Q, there is no physical reason for the two operators to be the same. Thus the relative coefficient χ_{E2} can in general be different from the coefficient χ appearing in Q. If $\chi_{E2} = -\sqrt{7}/2$, the transition operator is a generator of SU(3) and it obeys the SU(3) selection rules. If χ takes values different from $-\sqrt{7}/2$, the SU(3) symmetry is broken and its selection rules are not valid any more. Thus one can fit the free parameter χ_{E2} to the $\gamma_1 \rightarrow ground$ transitions, allowing them to be different from zero.

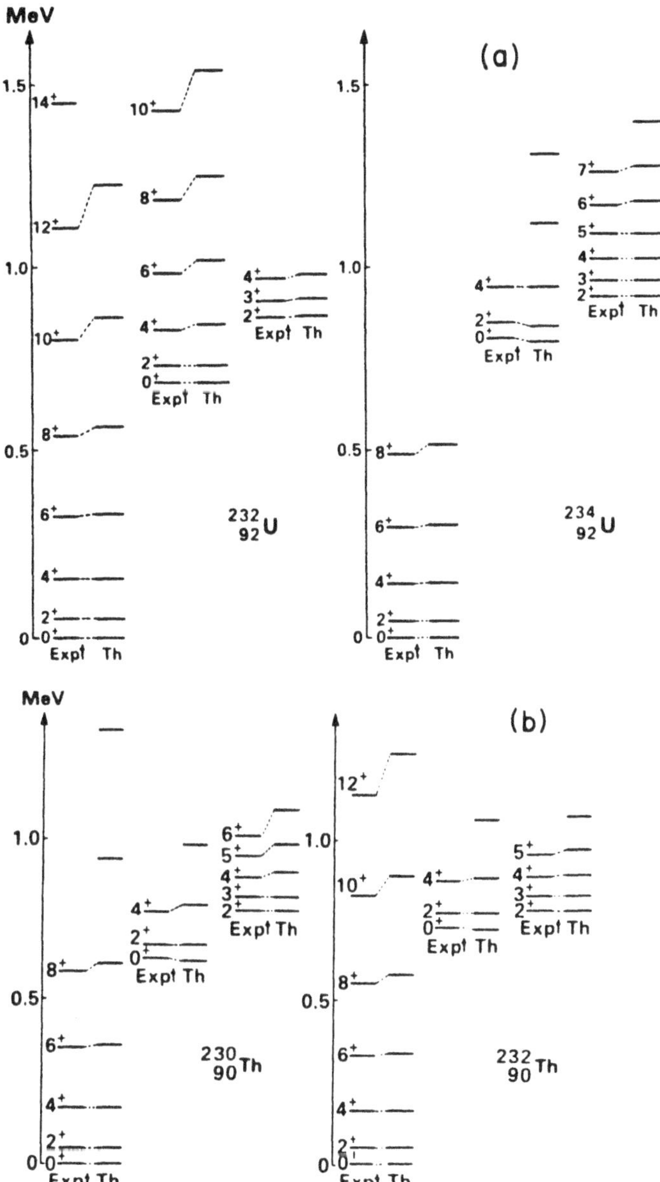

Fig. 1.8 Comparison between the experimental spectra of some Uranium (a) and Thorium (b) isotopes and the predictions of an IBM-1 Hamiltonian with broken SU(3) symmetry. (Taken from Zhang *et al.* (1985)).

1.10.3 Another Hamiltonian with broken symmetry

The Hamiltonian used in the previous subsection gives spectra with structure intermediate between SU(3) and O(6). Another Hamiltonian often used is

$$H = \epsilon_d n_d + \kappa(Q \odot Q), \tag{1.116}$$

where Q has the same form as before. Since the first term is a term occuring in U(5) symmetry, while the second term is the hallmark of SU(3) symmetry (when $\chi = -\sqrt{7}/2$), we expect this Hamiltonian to be able to describe nuclei having spectra intermediate between these two symmetries. If, in addition, one allows χ to vary, the second term has symmetry intermediate between SU(3) and O(6). Thus the overall Hamiltonian should be able to describe nuclei intermediate among the three symmetries U(5), SU(3), and O(6). In fact, it turns out that this Hamiltonian is capable of describing a wide range of nuclei, with the exception of very deformed actinides, which are best described by the Hamiltonian of the previous subsection.

References

Much work has been done on IBM-1, so that any attempt to describe the details of it would have been hopeless. In the extensive list of references given here, in addition to a variety of theoretical considerations, many comparisons of the IBM-1 predictions with experiment have been included. This chapter was mainly based on Iachello (1979, 1980, 1981). Many details of the model can be found in the "classic" papers of Arima and Iachello (1976, 1978b, 1979) and Scholten *et al.* (1978). Review articles have been written by Elliott (1985), Dieperink and Wenes (1985), Arima and Iachello (1984, 1981), and Barrett (1981). Some of the conferences and schools focused totally or partially on IBM have been held at Dubrovnik (1986), Drexel (1984, 1983, 1980), Gull Lake (1984), Erice (1982, 1980, 1978), Granada (1981), and the interested reader can find much material in the corresponding volumes. Compilations of early experimental tests of the model can be found in Casten (1980a, 1980b, 1980c) and Wood (1983).

Al-Janabi, T J, Jafar, J D, Youhana, H M, Demidov,
　　A M, and Govor, L I, 1983. *J. Phys.*, **G9,** 779.
Aprahamian, A, 1986. *Dubrovnik 1986* **2,** 671.
Aprahamian, A, Brenner, D S, Casten, R F, Gill, R L, and
　　Piotrowski, A, 1987. *Phys. Rev. Lett.*, **59,** 535
Arima, A, 1980. *Beijing 1980* 858.

Arima, A, and Iachello, F, 1975. *Phys. Rev. Lett.*, **35**, 1069.

Arima, A, and Iachello, F, 1976. *Ann. Phys.*, **99**, 253.

Arima, A, and Iachello, F, 1977. *Phys. Rev.*, **C16**, 2085.

Arima, A, and Iachello, F, 1978a. *Phys. Rev. Lett.*, **40**, 385.

Arima, A, and Iachello, F, 1978b. *Ann. Phys.*, **111**, 201.

Arima, A, and Iachello, F, 1979. *Ann. Phys.*, **123**, 468.

Arima, A, and Iachello, F, 1981. *Ann. Rev. Nucl. Part. Sci.*, **31**, 75.

Arima, A, and Iachello, F, 1984. In *Advances in Nuclear Physics* (ed. J W Negele and E Vogt) Vol 13 p 139. Plenum, New York.

Bäcklin, A, 1980. *Erice 1980* 41.

Bäcklin, A, *et al.*, 1982. *Nucl. Phys.*, **A380**, 189.

Baker, F T, Grimm, Jr, M A, Scott, A, Styles, R C, Kruse, T H, Jones, K, and Suchannek, R, 1981. *Nucl. Phys.*, **A371**, 68.

Barrett, B R, 1981. *Rev. Mex. Fis.* **27**, 533.

Bauer, R W, Becker, J A, Proctor, I D, Decman, D J, Lanier, R G, and Cizewski, J A, 1986. *Phys. Rev.*, **C34**, 1110.

Betts, R R, and Mortensen, M H, 1979. *Phys. Rev. Lett.*, **43**, 616.

Bijker, R, and Dieperink, A E L, 1982. *Phys. Rev.*, **C26**, 2688.

Bockisch, A, Miller, M, Kleinfeld, A M, Gelberg, A, and Kaup, U, 1979. *Z. Phys.*, **A292**, 265.

Bolotin, H H, Morrison, I, Ryan, C G, and Stuchbery, A E, 1983. *Nucl. Phys.*, **A401**, 175.

Bolotin, H H, Stuchbery, A E, Morrison, I, Kennedy, D L, Ryan, C G, and Sie, S H, 1981. *Nucl. Phys.*, **A370**, 146.

Bonatsos, D, and Klein, A, 1986. *Ann. Phys.*, **169**, 61.

Brink, D M, and Satchler, G R, 1968. *Angular Momentum.* Clarendon, Oxford.

Bucurescu, D, Cata, G, Cutoiu, D, Constantinescu, G, and Ivaşcu, M, 1983. *Nucl. Phys.*, **A401**, 22.

Bucurescu, D, Cata, G, Cutoiu, D, Constantinescu, G, Ivaşcu, M, and Zamfir, N V, 1986. *Z. Phys.*, **A324**, 387.

Bucurescu, D, Cata, G, Cutoiu, D, Constantinescu, G, Ivaşcu, M, and Zamfir, N V, 1987. *Z. Phys.*, **A327**, 241.

Byrne, A P, Stuchbery, A E, Bolotin, H H, Doran, C E, and Lampard, G J, 1987. *Nucl. Phys.*, **A466**, 419.

Castaños, O, Federman, P, and Frank, A, 1980. *Erice 1980* 21.

Castaños, O, Frank, A, and Federman, P, 1979. *Phys. Lett.*, **88B**, 203.

Castaños, O, Chacón, E, Frank, A, and Moshinsky, M, 1979. *J. Math. Phys.*, **20**, 35.

Castaños, O, Federman, P, Frank, A, and Pittel, S, 1982.

Nucl. Phys., **A379,** 61.

Castaños, O, Frank, A, Chacón, E, Hess, P, and Moshinsky,
    M, 1982. *J. Math. Phys.,* **23,** 2537.

Casten, R F, 1978. *Erice 1978* 37.

Casten, R F, 1980a. *Nucl. Phys.,* **A347,** 173.

Casten, R F, 1980b. *Drexel 1980* 369.

Casten, R F, 1980c. *Erice 1980* 3.

Casten, R F, 1984. *Nucl. Phys.,* **A421,** 27c.

Casten, R F, 1985. *Phys. Lett.,* **152B,** 145.

Casten, R F, and Aprahamian, A, 1984. *Phys. Rev.,* **C29,** 1919.

Casten, R F, and Cizewski, J A, 1978a. *Nucl. Phys.,* **A309,** 477.

Casten, R F, and Cizewski, J A, 1978b. *Phys. Lett.,* **79B,** 5.

Casten, R F, and Cizewski, J A, 1984. *Nucl. Phys.,* **A425,** 653.

Casten, R F, and Cizewski, J A, 1987. *Phys. Lett.,* **185B,** 293.

Casten, R F, and von Brentano, P, 1985. *Phys. Lett.,* **152B,** 22.

Casten, R F, and Warner, D D, 1982a. *Phys. Rev. Lett.,* **48,** 666.

Casten, R F, and Warner, D D, 1982b. *Erice 1982* 311.

Casten, R F, and Wolf, A, 1987. *Phys. Rev.,* **C35,** 1156.

Casten, R F, Aprahamian, A, and Warner, D D, 1984. *Phys.
    Rev.,* **C29,** 356.

Casten, R F, Frank, W, and von Brentano, P, 1985. *Nucl.
    Phys.,* **A444,** 133.

Casten, R F, von Brentano, P, and Haque, A M I, 1985. *Phys.
    Rev.,* **C31,** 1991.

Casten, R F, Warner, D D, and Aprahamian, A, 1983. *Phys.
    Rev.,* **C28,** 894.

Chiang, H C, Hsieh, S T, King Yen, M M, and Han, C S, 1985.
    Nucl. Phys., **A435,** 54.

Chuu, D S, Han, C S, Hsieh, S T, and King Yen, M M, 1984.
    Phys. Rev., **C30,** 1300.

Cizewski, J A, 1980. *Erice 1980* 137.

Cizewski, J A, and Dieperink, A E L, 1985. *Phys. Lett.,* **164B,**
    236.

Cizewski, J A, Flynn, E R, Brown, R E, and Sunier, J W,
    1979. *Phys. Lett.,* **88B,** 207.

Cizewski, J A, Flynn, E R, Brown, R E, Hanson, D L, Orbesen,
    S D, and Sunier, J W, 1981. *Phys. Rev.,* **C23,** 1453.

Cizewski, J A, Casten, R F, Smith, G J, Macphail, M R, Stelts,
    M L, Kane, W R, and Börner, H G, 1979. *Nucl. Phys.,*
    A323, 349.

Cizewski, J A, Casten, R F, Smith, G J, Stelts, M L, Kane,
    W R, Börner, H G, and Davidson, W F, 1978. *Phys. Rev. Lett.,*
    40, 167.

Clement, H, 1980. *Erice 1980*, 31.

Cline, D, 1980. *Erice 1980*, 241.

Davidson, W F, Dixon, W R, Burke, D G, and Cizewski, J A, 1983. *Phys. Lett.* **130B,** 161.

Dewald, A, Gast, W, Gelberg, A, Schuh, H W, Zell, K O, and von Brentano, P, 1980. *Strasbourg 1980* 33.

Dieperink, A E L, 1984. *Trieste 1984* **1,** 95.

Dieperink, A E L, and Wenes, G, 1985. *Ann. Rev. Nucl. Part. Sci.* **35,** 77.

Dobaczewski, J, Rohoziński, S G, and Srebrny, J, 1987. *Nucl. Phys.,* **A462,** 72.

du Marchie van Voorthuysen, E H, de Voigt, M J A, Blasi, N, and Jansen, J F W, 1981. *Nucl. Phys.,* **A355,** 93.

Edmonds, A R, 1960. *Angular Momentum in Quantum Mechanics.* Princeton University Press, Princeton.

Eid, E, and Stewart, N M, 1985. *Z. Phys.* **A320,** 495.

El-Daghmah, M S S, and Stewart, N M, 1983. *Z. Phys.* **A309,** 219.

Elliott, J P, 1958a. *Proc. R. Soc. London* Ser A **245,** 128.

Elliott, J P, 1958b. *Proc. R. Soc. London* Ser A **245,** 562.

Elliott, J P, 1985. *Rep. Prog. Phys.,* **48,** 171.

Emling, H, 1978. *Erice 1978* 75.

Federman, P, and Pittel, S, 1979. *Oaxtepec 1979* 55.

Federman, P, Castaños, O, and Frank, A, 1980. *Drexel 1980* 475.

Feng, D H, Gilmore, R, and Deans, S R, 1981. *Phys. Rev.,* **C23,** 1254.

Fernández-Niello, J, Puchta, H, Riess, F, and Trautmann, W, 1982a. *Nucl. Phys.,* **A391,** 221.

Fernández-Niello, J, Puchta, H, Riess, F, and Trautmann, W, 1982b. *Erice 1982* 567.

Fewell, M P, 1986. *Phys. Lett.,* **167B,** 6.

Gelletly, W, Larysz, J R, Börner, H G, Casten, R F, Davidson, W F, Mampe, W, Schreckenbach, K, and Warner, D D, 1985. *J. Phys.,* **G11,** 1055.

Gelletly, W, Larysz, J R, Börner, H G, Casten, R F, Davidson, W F, Mampe, W, Schreckenbach, K, and Warner, D D, 1987. *J. Phys.,* **G13,** 69.

Giannatiempo, A, Nannini, A, Passeri, A, Perego, A, and Sona, P, 1986. *Z. Phys.,* **A325,** 157.

Gilmore, R, and Draayer, J P, 1985. *J. Math. Phys.,* **26,** 3053.

Gilmore, R, and Feng, D H, 1980. *Drexel 1980* 401.

Gilmore, R, and Feng, D H, 1982a. *Erice 1982* 495.

Gilmore, R, and Feng, D H, 1982b. *Phys. Rev.*, **C26,** 766.

Gilmore, R, Feng, D H, Vallières, M, and Wood, J L, 1983. *Drexel 1983* 207.

Gilmore, R, Feng, D H, Vallières, M, and Wood, J L, 1984. *Phys. Rev.*, **C30,** 1100.

Girit, C, Hamilton, W D, and Kalfas, C A, 1983. *J. Phys.*, **G9,** 797.

Girit, C, Hamilton, W D, and Michelakakis, E, 1980. *J. Phys.*, **G6,** 1025.

Goettig, L, Droste, C, Dugo, A, Morek, T, Srebrny, J, Broda, R, Styczeń, J, Hattula, J, Helppi, H, and Jääskeläinen, M, 1981. *Nucl. Phys.*, **A357,** 109.

Gupta, J B, 1986. *Phys. Rev.*, **C33,** 1505.

Hasselgren, L, and Cline, D, 1980. *Erice 1980* 59.

Hsieh, S T, Chiang, H C, King Yen, M M, and Chuu, D S, 1986. *J. Phys.*, **G12,** L167.

Iachello, F, 1978. *Erice 1978* 1.

Iachello, F, 1979. *Gull Lake 1979* 140.

Iachello, F, 1980. *Dronten 1980* 53.

Iachello, F, 1981. *Granada 1981* 1.

Iachello, F, and Arima, A, 1974. *Phys. Lett.*, **53B,** 309.

Iachello, F, and Talmi, I, 1987. *Rev. Mod. Phys.* **59,** 339.

Janssen, D, Jolos, R V, and Dönau, F, 1974. *Nucl. Phys.*, **A224,** 93.

Jolos, R V, Dönau, F, and Janssen, D, 1975. *Sov. J. Nucl. Phys.*, **22,** 503.

Kane, W R, Casten, R F, Warner, D D, Schreckenbach, K, Faust, H R, and Blakeway, S, 1982. *Phys. Lett.*, **117B,** 15.

Kaup, U, and Holzwarth, G, 1985. *Nucl. Phys.*, **A445,** 419.

King Yen, M M, Hsieh, S T, Chiang, H C, and Chuu, D S, 1984. *Phys. Rev.*, **C29,** 688.

Kota, V K B, 1981. *Ann. Phys.*, **134,** 221.

Kracikova, T I, Finger, M, Pavlov, V N, Deryuga, V A, Lipas, P O, Hammarén, E, and Toivonen, P, 1984. *J. Phys.*, **G10,** 571.

Kracikova, T I, Davaa, S, Finger, M, Fominykh, M I, Hamilton, W D, Lipas, P O, Hammarén, E, Toivonen, P, 1984. *J. Phys.*, **G10,** 1115.

Kracikova, T I, Finger, M, Konicek, J, Lebedev, N A, Deryuga, V A, Lipas, P O, Hammarén, E, and Toivonen, P, 1984. *J. Phys.*, **G10,** 667.

Kyrchev, G, 1980. *Nucl. Phys.*, **A349,** 416.

Kyrchev, G, and Paar, V, 1983. *Nucl. Phys.*, **A395,** 61.

Kyrchev, G, and Paar, V, 1986. *Ann. Phys.*, **170,** 257.

Le Blank, R, and Rowe, D J, 1985a. *J. Phys.*, **A18,** 1891.
Le Blank, R, and Rowe, D J, 1985b. *J. Phys.*, **A18,** 1905.
Leviatan, A, Novoselsky, A, and Talmi, I, 1986. *Phys. Lett.*, **172B,** 144.
Lipas, P O, 1982. *Erice 1982* 511.
Lipas, P O, 1984a. *Juväskylä 1984* 305.
Lipas, P O, 1984b. *Gull Lake 1984* 67.
Lipas, P O, and Warner, D D, 1984. *Phys. Rev.*, **C30,** 1098.
Lipas, P O, Hammarén, E, and Toivonen, P, 1984. *Phys. Lett.*, **139B,** 10.
Lipas, P O, Toivonen, P, and Hammarén, E, 1987. *Nucl. Phys.*, **A469,** 348.
Lipas, P O, Toivonen, P, and Warner, D D, 1985. *Phys. Lett.*, **155B,** 295.
Loiselet, M, Holzmann, R, van Hove, M A, and Vervier, J, 1984. *Phys. Lett.*, **146B,** 187.
Lönnroth, T, Vajda, S, Kistner, O C, and Rafailovich, M H, 1984. *Z. Phys.*, **A317,** 215.
Lopac, V, and Paar, V, 1984. *Z. Phys.*, **A319,** 351.
Mardirosian, G, and Stewart, N M, 1984. *Z. Phys.*, **A315,** 213.
Mc Gowan, F K, 1981. *Phys. Rev.*, **C24,** 1803.
Morrison, I, and Smith, R, 1980a. *Nucl. Phys.*, **A350,** 89.
Morrison, I, and Smith, R, 1980b. *Phys. Lett.*, **95B,** 13.
Moshinsky, M, 1979. *Oaxtepec 1979* 255.
Moshinsky, M, 1981. *Granada 1981* 97.
Ower, H, *et al.*, 1980. *Strasbourg 1980* 119.
Ower, H, *et al.*, 1982. *Nucl. Phys.*, **A388,** 421.
Paar, V, 1978. *Erice 1978* 163.
Paar, V, 1979. *Rhodes 1979* 53.
Paar, V, 1981. *Trieste 1981* 91.
Paar, V, Brant, S, Kyrchev, G, and Meyer, R A, 1984. *Z. Phys.*, **A319,** 357.
Panqueva, J, Hellmeister, H P, Bergmeister, F J, and Lieb, K P, 1980. *Strasbourg 1980* 29.
Park, P, Subber, A R H, Hamilton, W D, Elliott, J P, and Kumar, K, 1985. *J. Phys.*, **G11,** L251.
Partensky, A, and Quesne, C, 1981. *Ann. Phys.*, **136,** 340.
Partensky, A, and Quesne, C, 1982. *Phys. Rev.*, **C25,** 2837.
Passoja, A, Kantele, J, Luontama, M, Julin, R, Hammarén, E, Lipas, P O, and Toivonen, P, 1986. *J. Phys.*, **G12,** 1047.
Preston, M A, and Bhaduri, R K, 1975. *Structure of the Nucleus* Chap. 9. Addison-Wesley, Reading.
Quesne, C, 1981. *J. Math. Phys.*, **22,** 1482.

Robinson, S J, Hamilton, W D, and Snelling, D M, 1983a. *J. Phys.*, **G9**, L71.

Robinson, S J, Hamilton, W D, and Snelling, D M, 1983b. *J. Phys.*, **G9**, 961.

Rosensteel, G, and Rowe, D J, 1983. *J. Math. Phys.*, **24**, 2461.

Saha, A, Scholten, O, Hageman, D C J M, and Fortune, H T, 1979. *Phys. Lett.*, **85B**, 215.

Schaaser, H, and Brink, D M, 1984. *Phys. Lett.*, **143B**, 269.

Schaaser, H, and Brink, D M, 1986. *Nucl. Phys.*, **A452**, 1.

Schiffer, K, Dewald, A, Gelberg, A, Reinhardt, R, Zell, K O, X. Sun, and von Brentano, P, 1986. *Nucl. Phys.*, **A458**, 337.

Schneider, E W, Glascock, M D, Walters, W B, and Meyer, R A, 1979. *Phys. Rev.*, **C19**, 1025.

Scholten, O, Iachello, F, and Arima, A, 1978. *Ann. Phys.*, **115**, 325.

Schüler, P, Recht, J, Wilzek, H, Hardt, K, Günther, C, Blume, K P, Euler, K, and Kölschbach, V, 1984. *Z. Phys.*, **A317**, 313.

Schwengner, R, *et al.*, 1987. *Z. Phys.*, **A326**, 287.

Seiwert, M, Maruhn, J A, and Hess, P O, 1984. *Phys. Rev.*, **C30**, 1779.

Simister, D N, Ekström, L P, Jones, G D, Kearns, F, Morrison, T P, Mustaffa, O M, Price, H G, Twin, P J, Wadsworth, R, and Ward, N J, 1980. *J. Phys.*, **G6**, 81.

Snelling, D M, and Hamilton, W D, 1983a. *J. Phys.*, **G9**, 111.

Snelling, D M, and Hamilton, W D, 1983b. *J. Phys.*, **G9**, 763.

Soloviev, V G, 1986. *Z. Phys.*, **A324**, 393.

Stachel, J, van Isacker, P, and Heyde, K, 1982. *Phys. Rev.*, **C25**, 650.

Stuchbery, A E, Morrison, I, and Bolotin, H H, 1984. *Phys. Lett.*, **139B**, 259

Stuchbery, A E, Ryan, C G, Bolotin, H H, and Morrison, I, 1981. *Nucl. Phys.*, **A365**, 317.

Stuchbery, A E, Bolotin, H H, Doran, C E, Morrison, I, Wood, L D, and Yamada, H, 1985. *Z. Phys.*, **A320**, 669.

Stuchbery, A E, Morrison, I, Wood, L D, Bark, R A, Yamada, H, and Bolotin, H H, 1985. *Nucl. Phys.*, **A435**, 635.

Subber, A R H, Robinson, S J, Hungerford, P, Hamilton, W D, van Isacker, P, Kumar, K, Park, P, Schreckenbach, K, and Colvin, G, 1987. *J. Phys.*, **G13**, 807.

Suhonen, J, 1987. *Phys. Lett.*, **193B**, 405.

Suhonen, J, and Lipas, P O, 1985. *Phys. Lett.*, **150B**, 327.

Sujkowski, Z, 1978. *Erice 1978* 55.

Sun, H Z, Zhang, M, and Feng, D H, 1985. *Phys. Lett.*, **163B**,

7.

Szpikowski, S, and Góźdź, A, 1980. *Nucl. Phys.*, **A340, 76**.

Tokunaga, Y, *et al.*, 1985. *Nucl. Phys.*, **A439**, 427.

van Isacker, P, 1983. *Phys. Rev.*, **C27, 2447**.

van Isacker, P, 1987. *Nucl. Phys.*, **A465**, 497.

van Isacker, P, and Lipas, P O, 1985. *Phys. Rev.*, **C31**, 1546.

van Isacker, P, Frank, A, Dukelsky, J, 1985. *Phys. Rev.*, **C31**, 671.

Vergados, J D, 1968. *Nucl. Phys.*, **A111**, 681.

Vergnes, M, 1980. *Erice 1980* 53.

Vervier, J, 1983. *Phys. Lett.*, **133B, 135**.

Ward, N J, Ekström, L P, Jones, G D, Kearns, F, Morrison, T P, Mustaffa, O M, Simister, D N, Twin, P J, and Wadsworth, R, 1981. *J. Phys.*, **G7, 815**.

Warner, D D, 1981. *Phys. Rev. Lett.*, **47**, 1819.

Warner, D D, 1983. *Drexel 1983* 133.

Warner, D D, and Casten, R F, 1982a. *Phys. Rev.*, **C25**, 2019.

Warner, D D, and Casten, R F, 1982b. *Phys. Rev. Lett.*, **48**, 1385.

Warner, D D, and Casten, R F, 1982c. *Phys. Rev.*, **C26**, 2690.

Warner, D D, and Casten, R F, 1983. *Phys. Rev.*, **C28**, 1798.

Warner, D D, Casten, R F, and Davidson, W F, 1980. *Phys. Rev. Lett.*, **45**, 1761.

Warner, D D, Casten, R F, and Davidson, W F, 1981. *Phys. Rev.*, **C24**, 1713.

Wolf, A, and Casten, R F, 1987. *Phys. Rev.*, **C36**, 851.

Wood, J L, 1980. *Drexel 1980* 451.

Wood, J L, 1983. *Nucl. Phys.*, **A396**, 245c.

Wood, L D, and Morrison, I, 1985. *J. Phys.*, **G11**, 501.

Wu, H C, Dieperink, A E L, and Pittel, S, 1986. *Phys. Rev.*, **C34**, 703.

Yadav, H L, Faessler, A, Toki, H, and Castel, B, 1980. *Phys. Lett.*, **89B**, 307.

Yates, S W, and Molnár, G, 1986. *Dubrovnik 1986* **2**, 624.

Zhang, M, Vallières, M, Ling, Y S, and Gilmore, R, 1984. *Gull Lake 1984* 160.

Zhang, M, Vallières, M, Gilmore, R, Feng, D H, Hoff, R W, and Sun, H Z, 1985. *Phys. Rev.*, **C32**, 1076.

Zhang, M, Vallières, M, Gilmore, R, Sun, H Z, Feng, D H, and Hoff, R W, 1984. *Drexel 1984* 234.

2

HIGHER ORDER INTERACTIONS IN IBM-1

As we have already seen, one of the basic assumptions of IBM-1 is that only one- and two-body interactions need to be included in the Hamiltonian. In this chapter we will examine the effects which higher order terms added in the Hamiltonian have on the predicted spectra.

2.1 Symmetry-conserving higher order interaction terms

In Sec. 1.10 some inadequacies of the exact limiting symmetries of IBM-1 in describing the experimental data were identified. They were removed by breaking the symmetries, which had as a consequence that analytical solutions were not possible any more and numerical calculations had instead to be undertaken. In this section an alternative way of avoiding some of these difficulties will be demonstrated.

We will again focus our attention on the SU(3) limit. One of the problems in this limit is that although the γ_1 and β_1 bands are predicted to be degenerate, experimentally in most cases they are not. To avoid this problem without breaking the SU(3) symmetry, one could include higher order terms in the Hamiltonian, with the property of removing the degeneracy without breaking the SU(3) symmetry. Such terms do exist and will be discussed now.

Remember that the IBM-1 Hamiltonian in the SU(3) limit has been written in the form

$$H = \alpha C_{2O3} + \beta C_{2SU3}. \tag{2.1}$$

Both terms are diagonal in the Elliott basis $|(\lambda, \mu)KLM >$, their eigenvalues being

$$< C_{2O3} >= L(L+1), \tag{2.2}$$

$$< C_{2SU3} >= \frac{1}{9}(\lambda^2 + \mu^2 + \lambda\mu + 3(\lambda + \mu)). \tag{2.3}$$

Both these terms are second order terms, i.e. they correspond to two-body interactions. (Remember that one of the basic assumptions of IBM-1 is that only one-body and two-body interactions are included in the Hamiltonian). The first step in removing this limitation will be to include third order terms (three-body interactions)

50

in the Hamiltonian, keeping the SU(3) symmetry intact (Van den Berghe *et al.* 1985, De Meyer *et al.* 1986, Van der Jeugt 1986). One can prove that two terms of this kind exist. One is the third order Casimir invariant of SU(3). In the Elliott basis this operator is diagonal and has eigenvalues

$$< C_{3SU3} >= \frac{1}{162}(\lambda - \mu)(2\lambda + \mu + 3)(\lambda + 2\mu + 3). \qquad (2.4)$$

Inclusion of this term in the Hamiltonian will only result in a rescaling of the β parameter. The reason is that this term, as C_{2SU3}, only affects the relative positions of the various irreps, while preserving the degeneracies within each irrep. Thus this term is of no interest.

However, another third-order operator exists, the so-called O(3) scalar shift operator O_l^0, which here we will call simply Ω. This operator is not diagonal in the Elliott basis. Its eigenvalues have been calculated by several authors, both numerically and analytically. For our purposes the results quoted here suffice:

i) In a $(\lambda, 0)$ irrep one has

$$< \Omega >= \sqrt{6}L(L + 1)(2\lambda + 3). \qquad (2.5)$$

ii) In a $(\lambda, 2)$ irrep one has:

$$< \Omega >= \sqrt{6}[L(L + 1) - 12](2\lambda + 5) \qquad (2.6)$$

for L odd,

$$< \Omega >= \sqrt{6}\{(L - 2)(L + 3)(2\lambda + 5)$$
$$\pm 6\sqrt{L(L + 1)(L - 1)(L + 2) + (2\lambda + 5)^2}\} \qquad (2.7)$$

for L even, while $< \Omega >= 0$ for $L = 0$.

Notice the $\pm$ sign contained in $< \Omega >$ for even L in the $(\lambda, 2)$ irrep. This double sign means that the pairs of degenerate states contained in this irrep are splitted. But this is the irrep containing the γ_1 and β_1 bands ! Thus, if a term proportional to $< \Omega >$ is included in the Hamiltonian, the unwanted degeneracy between the γ_1 and the β_1 bands is removed, while the SU(3) symmetry is preserved. The Hamiltonian in this case will read

$$H = \alpha C_{2O3} + \beta C_{2SU3} + \gamma \Omega. \qquad (2.8)$$

Some results obtained with this Hamiltonian for two Gadolinium isotopes are compared to experiment in Fig. 2.1. We remark that the degeneracy of the β_1 and γ_1 bands, predicted in IBM-1, is broken by the third-order terms, in agreement with experiment. The rotational structure of the ground state band is reproduced by IBM-1, but the rotational constant appears to be somewhat smaller than the one shown by experiment. This shortcoming is not removed by the third-order terms.

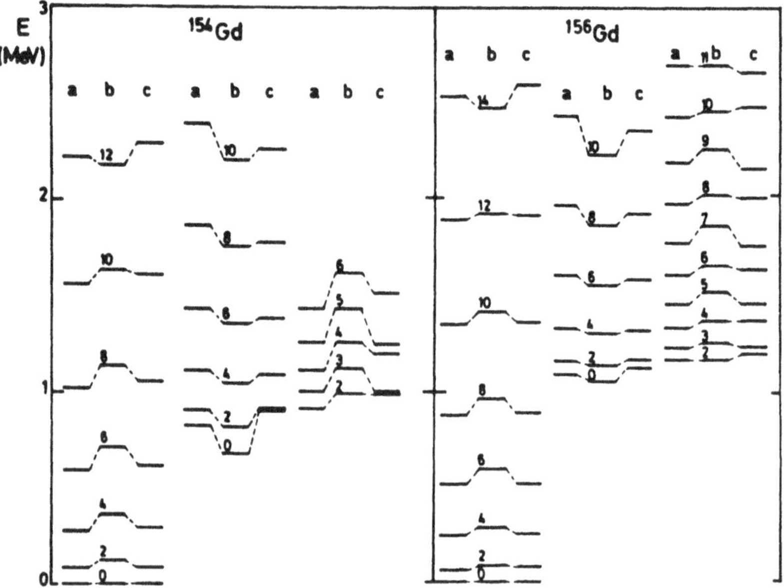

Fig. 2.1 Comparison between experimental and calculated spectra for two Gd isotopes. "a" denotes the predictions of the original IBM-1, "b" represents the experimental data, "c" gives the results obtained from the extended version of IBM-1 including third-order interactions. (Taken from Van den Berghe *et al.* (1985)).

It can be proved that no other linearly independent symmetry-conserving third-order terms exist. One may wonder by now what the effect of fourth-order terms might be. It can be seen that only one linearly independent fourth-order term exists, which we will denote by Λ. As it was the case for Ω, Λ is not diagonal in the Elliott basis. Furthermore, Λ and Ω do not commute. This imposes some extra numerical complications. The most general symmetry-conserving SU(3) Hamiltonian up to fourth order will be

$$H = \alpha C_{2O3} + \beta C_{2SU3} + \gamma \Omega + \delta C_{3SU3}$$

$$+\mu\Lambda + \nu C_{2O3}^2 + \tau C_{2SU3}C_{2O3} + \rho C_{2SU3}^2, \qquad (2.9)$$

since products of lower order operators can also be included. In practice one can put $\delta = \tau = \rho = 0$ without much loss of predictive power. An example of predictions of the remaining five-parameter Hamiltonian is compared with experiment in Fig. 2.2. Not only is the degeneracy between the β_1 and γ_1 levels properly broken by the third- and forth-order terms, but in addition the ground state band is reproduced with the correct rotational constant, removing the shortcoming observed in Fig. 2.1.

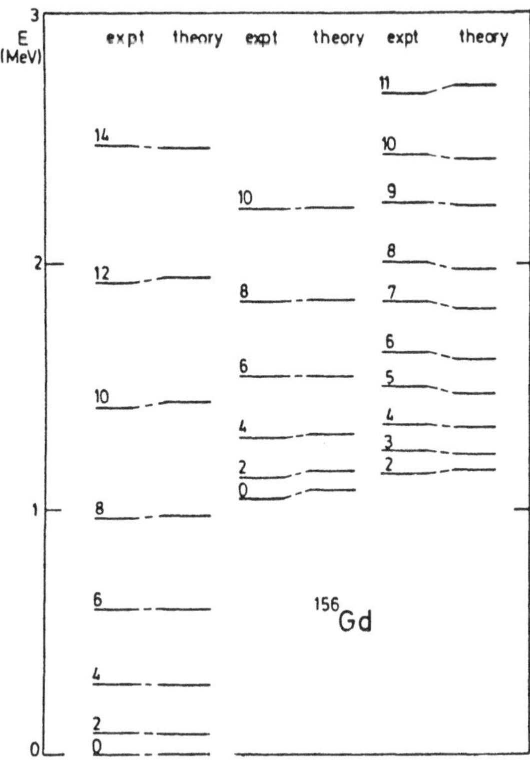

Fig. **2.2** Comparison between the experimental data and the predictions of the extended IBM-1 Hamiltonian involving third- and fourth-order interactions. (Taken from Van den Berghe *et al.* (1985)).

A few comments are appropriate at this point:

i) Since these higher-order terms do conserve the SU(3) symmetry, they do not affect the predictions for the transition rates. Thus the transitions from the γ_1 and β_1 bands to the ground state band remain forbidden. This problem can be avoided only by breaking the SU(3) symmetry.

ii) By studying the classical limit of the higher order terms, one can show that these terms cannot give rise to stable triaxial shapes. Again, triaxiality can be obtained only by breaking the SU(3) symmetry.

iii) Similar studies of higher order symmetry conserving interactions can be undertaken in the other two limits (U(5), O(6)) of IBM-1 as well.

2.2 Symmetry-breaking cubic terms

In Sec. 1.8 the geometric content of the three limiting symmetries of IBM-1 was studied. It was found that the U(5) limit corresponds to a spherical shape (an energy functional which is independent of γ and has a minimum at $\beta = 0$), while the SU(3) limit corresponds to an axially symmetric prolate shape (an energy functional having a minimum at $\gamma = 0$ and $\beta \neq 0$) and the O(6) limit corresponds to a γ-unstable deformed shape (an energy functional which is independent of γ and has a minimum for $\beta \neq 0$). None of the three limits gives rise to a triaxial shape (corresponding to an energy functional having a minimum at $\beta \neq 0$ and $\gamma \neq 0, \pi/3$). However, some nuclei (e.g. ^{104}Ru) have long been thought to have triaxial shapes. Clearly IBM-1 in its original form (only s and d bosons, only one- and two-body interactions in the Hamiltonian) is not able to describe these nuclei. In this section a way of incorporating triaxial shapes in the algebraic framework will be discussed. This can be achieved by including in the Hamiltonian three-body terms (Heyde *et al.* 1984, Casten *et al.* 1985), in addition to the one- and two-body terms of IBM-1. These "cubic" terms read

$$H_c = \sum_L \theta_L [d^+ d^+ d^+]^{(L)} \cdot [\tilde{d}\tilde{d}\tilde{d}]^{(L)}, \quad L = 0, 1, 2, 3, 4. \qquad (2.10)$$

It has been found that only the $L = 3$ term can induce stable triaxial shapes. Thus only this term has been used in applications. In contrast to the higher order terms considered in the previous section, these "cubic" terms do not preserve any of the dynamical symmetries of IBM-1.

It is instructive to consider the geometrical limit of the cubic term with $L = 3$. In terms of geometrical variables one finds

$$E_c = \theta_3 N(N-1)(N-2)\frac{1}{7}\frac{\beta^6}{(1+\beta^3)^2}(-1 + cos^2 3\gamma). \qquad (2.11)$$

In order to study the influence of this "cubic" term, let us consider the O(6) limit. In this case the total energy functional will be

$$E = E^{III} + E_c, \qquad (2.12)$$

where E^{III} is the O(6) energy functional of Sec. 1.8 (eq. 1.90). The total energy functional is not independent of γ any more, and it turns out that it has a minimum at $\gamma = \pi/6$, $\beta \neq 0$, which corresponds to a triaxial shape indeed. Details about the methods used in deriving the classical limit of algebraic Hamiltonians can be found in Chapter 13.

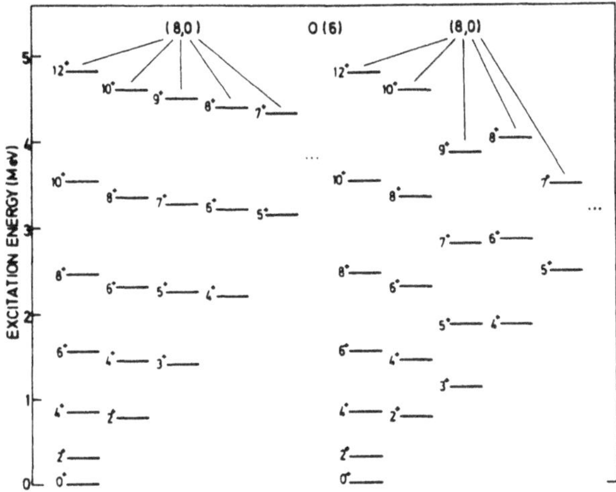

Fig. 2.3 A typical spectrum for a nucleus with $N = 8$ in the O(6) limit of IBM-1 with (right part) and without (left part) the $L = 3$ "cubic" term. (Taken from Heyde *et al.* (1984)).

In Fig. 2.3 a typical IBM-1 spectrum with O(6) symmetry (shown on the left) is compared to the spectrum obtained by adding an $L = 3$ "cubic" term to the O(6) Hamiltonian (shown on the right). We remark that the levels of the ground state band, as well as the even-spin members of the γ band, are hardly influenced by the presence of the "cubic" term, while the odd-spin members of the γ band are appreciably lowered. Notice that in the O(6) limit the odd-spin members of the γ band are strongly displaced upwards relative to their even-spin neighbors. This effect is known as "odd–even staggering" and is indeed observed experimentally in some nuclei, but not to this extent. The main effect of the "cubic" term is to decrease drastically the amount of odd–even staggering in the γ band, which is a result physically desirable.

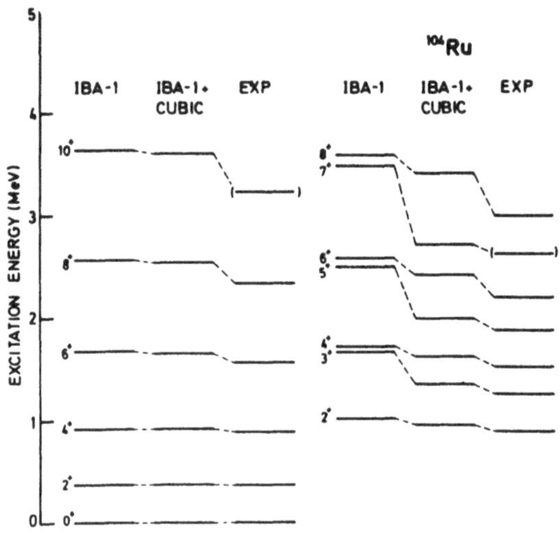

Fig. **2.4** Comparison of the experimental spectrum of ^{104}Ru with the predictions of an IBM-1 Hamiltonian with and without the $L = 3$ "cubic" term. (Taken from Heyde *et al.* (1984)).

In Fig. 2.4 fits of the levels of the triaxial nucleus ^{104}Ru, obtained with an IBM-1 Hamiltonian with and without an $L = 3$ "cubic" term, are compared to experiment. It is clear that while the theoretical predictions for the levels of the ground state band are almost unchanged by the addition of the "cubic" term, the predictions for the γ band are greatly improved. In particular, the excessively strong odd-even staggering predicted by the pure IBM-1 Hamiltonian is largely removed by the "cubic" term, in agreement to the experimental data.

References

Casten, R F, 1984. *Drexel 1984*, 90.
Casten, R F, 1984. *Gull Lake 1984*, 349.
Casten, R F, von Brentano, P, Heyde, K, van Isacker, P, and

Jolie, J, 1985. *Nucl. Phys.*, **A439**, 289.

De Meyer, H, Van den Berghe, G, Van der Jeugt, J, and
Vanthournout, J, 1986. *Dubrovnik 1986*, **2**, 1024.

Heyde, K, van Isacker, P, Waroquier, M, and Moreau, J, 1984.
Phys. Rev. , **C29**, 1420.

Loewenich, K, Zell, K O, Dewald, A, Gast, W, Gelberg,
A, Lieberz, W, von Brentano, P, and van Isacker, P, 1986.
Nucl. Phys., **A460**, 361.

Sun, H Z, Zhang, M, and Feng, D H, 1985. *Phys. Lett.*, **163B**, 7.

Van den Berghe, G, De Meyer, H E, and van Isacker, P, 1985.
Phys. Rev., **C32**, 1049.

Van der Jeugt, J, 1986. *J. Phys.*, **A19**, L463.

3

ALGEBRAIC DESCRIPTION
OF OCTUPOLE VIBRATIONS

3.1 Collective octupole vibrations

Besides the low-lying positive parity collective states which are described in IBM-1 in terms of s and d bosons (of positive parity), in most medium and heavy nuclei negative parity states occur, which are thought to correspond to octupole ($L^\pi = 3^-$) vibrations around a spherical or quadrupoly deformed ground-state shape.

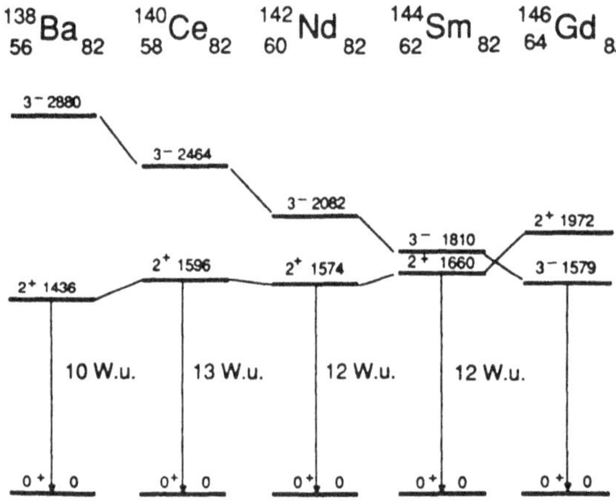

Fig. 3.1 Excitation energies of the lowest 2^+ and 3^- states for a series of $N = 82$ isotones. (Taken from Cottle and Bromley (1987)).

An example is shown in Fig. 3.1, where low-lying 3^- states are shown, along with 2^+ states, for a series of N=82 isotones. It is interesting to observe that in this series of isotones the 2^+ state remains almost constant as one proceeds from one nucleus to another (indicating almost constant quadrupole deformation), while the 3^- state drops dramatically as the number of protons is increased. This drop (i.e. development of collectivity) is thus independent from the quadrupole collectivity. This drop can be explained (Cottle and

58

Bromley 1986, 1987) if one considers the 3^- state as the coherent sum of one-particle–one-hole (1p–1h) excitations between two single particle orbitals (of opposite parity) differing by 3 units of orbital angular momentum. Such $\Delta l = 3$ pairs exist in all major proton (neutron) major shells of interest in medium and heavy nuclei (for more information see Fig. 15.1 and the discussion in Sec. 15.5.2). For example, in the 50–82 proton shell, which is of interest in the case of the isotones of Fig. 3.1, this pair is composed by the unique parity level $1h_{11/2}$ and the normal parity level $1g_{5/2}$. As more protons are added outside the Z=50 closed shell, more protons occupy the lower of the two levels of the $\Delta l = 3$ pair, and additional 1p–1h excitations become available, so that the energy of their coherent sum, which is the collective octupole state, is lowered. Octupole vibrations have been observed in all regions from mass 16 to mass 250. Rich data can be found in Sakai (1984).

3.2 The f boson

As we have already seen, the low-lying octupole vibrational states of negative parity have nothing to do with quadrupole deformation. Thus they cannot be described by the s and d bosons (of positive parity) used in IBM-1 for the description of quadrupole collectivity. Clearly new degrees of freedom have to be included in the algebraic framework in order to allow for the description of octupole states. This is achieved by the inclusion of an f (L=3) boson of negative parity (Arima and Iachello 1976, 1978, Scholten *et al.* 1978, Engel 1986). The total Hamiltonian then takes the form (Barfield *et al.* 1986)

$$H = H_{sd} + H_f + V_{sdf}, \tag{3.1}$$

where H_{sd} is the usual IBM-1 Hamiltonian, H_f is the f-boson Hamiltonian, which in lowest approximation is taken as

$$H_f = \epsilon_f n_f, \tag{3.2}$$

where n_f is the number of f bosons, and V_{sdf} is the interaction between the quadrupole and the octupole degrees of freedom, usually taken as

$$V_{sdf} = A(L_{sd} \cdot L_f) + B(Q_{sd} \cdot Q_f) + C : (E_{df}^+ \cdot E_{sdf}) : . \tag{3.3}$$

In the last equation L_{sd} and Q_{sd} are the usual IBM-1 angular momentum and quadrupole operators, while L_f is the f-boson angular momentum operator

$$L_f = 2\sqrt{7}(f^+ \otimes \tilde{f})^1, \tag{3.4}$$

Q_f is the f-boson quadrupole operator

$$Q_f = -2\sqrt{7}(f^+ \otimes \tilde{f})^2, \tag{3.5}$$

E_{df}^+ is the exchange operator

$$E_{df}^+ = \sqrt{5}(d^+ \otimes \tilde{f})^3, \tag{3.6}$$

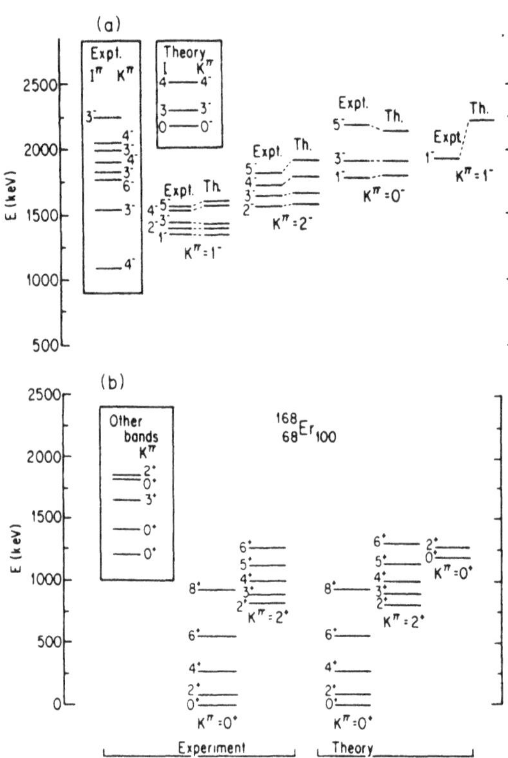

Fig. 3.2 Experimental and theoretical energy spectra for ^{168}Er. Positive (negative) parity states appear in the lower (upper) part. (Taken from Barfield *et al.* (1986)).

and the double dots :: indicate normal ordering of the operators, i.e. that all boson creation operators must appear before any boson annihilation operators. It is found that the first term of V_{sdf}

is of minor importance, the major contributions coming from the quadrupole–quadrupole and the exchange terms.

For calculating electromagnetic transition rates one needs the appropriate transition operators. The most general form of the electric quadrupole transition operator will be

$$T^{(E2)} = e_{sd}Q_{sd} + e_f Q_f. \qquad (3.7)$$

It is assumed, however, that the E2 collectivity is carried by the d boson, so that the second term in this expression can be neglected.

Similarly one can consider E3 transitions. In this case the transition operator is

$$T^{(E3)} = e_3(s^+\tilde{f} + f^+s)^3 + \chi_3(d^+\tilde{f} + f^+\tilde{d})^3. \qquad (3.8)$$

In Fig. 3.2 the results of a calculation of this kind for the spectrum of ^{168}Er are shown. The lower part of the figure contains the positive parity states (the ground state, γ_1 and β_1 bands), while the upper part contains the negative parity ones. The insets in the upper half of the figure contain experimental two-quasiparticle bandheads between 1 and 2 MeV, as well as three calculated bandheads between 2 and 2.5 MeV. The inset in the lower half of the figure contains excited K^+ bandheads. The agreement between the predictions of the model and experiment is quite good.

References

Arima, A, 1984. *Nucl. Phys.*, **A421**, 63c.

Arima, A, and Iachello, F, 1976. *Ann. Phys.*, **99**, 253.

Arima, A, and Iachello, F, 1978. *Ann. Phys.*, **111**, 201.

Barfield, A F, Wood, J L, and Barrett, B R, 1986. *Phys. Rev.*, **C34**, 2001.

Barrett, B R, 1986. *Dubrovnik 1986* **2**, 1010.

Cottle, P D, and Bromley, D A, 1986. *Phys. Lett.*, **182B**, 129.

Cottle, P D, and Bromley, D A, 1987. *Phys. Rev.*, **C35**, 1891.

de Leo, R, Pignanelli, M, Borghols, W T A, Brandenburg, S, Harakeh, M N, Lu, H J, and van der Werf, S Y, 1985. *Phys. Lett.*, **165B**, 30.

Engel, J, 1986. *Phys. Lett.*, **171B**, 148.

Han, C S, Chuu, D S, Hsieh, S T, and Chiang, H C, 1985. *Phys. Lett.*, **163B**, 295.

Hellmeister, H P, Keinonen, J, Lieb, K P, Kaup, U, Rascher, R, Ballini, R, Delaunay, J, and Dumont, H, 1979. *Nucl. Phys.*,

A332, 241.

Konijn, J, de Boer, F W N, van Poelgeest, A, Hesselink, W H A, de Voigt, M J A, Verheul, H, and Scholten, O, 1981. *Nucl. Phys.*, **A352**, 191.

Konijn, J, *et al.*, 1982. *Nucl. Phys.*, **A373**, 397.

Matsuki, S, Sakamoto, N, Ogino, K, Kadota, Y, Tanabe, T, and Okuma, Y, 1981. *Nucl. Phys.*, **A370**, 1.

Matsuki, S, Ogino, K, Kadota, Y, Sakamoto, N, Tanabe, T, Yasue, M, Yokomizo, A, Kubono, S, and Okuma, Y, 1982. *Phys. Lett.*, **113B**, 21.

Sakai, M, 1984. *At. Data Nucl. Data Tables*, **31**, 399.

Scholten, O, Iachello, F, and Arima, A, 1978. *Ann. Phys.*, **115**, 325.

4

THE sdg INTERACTING BOSON MODEL (sdg IBM)

4.1 Some inadequacies of IBM-1

The Interacting Boson Model 1 (IBM-1) gives a reasonable description of the low-lying states of medium and heavy even-even nuclei in terms of s and d bosons only. However, the model has limitations as well. We will discuss some of them here and some additional later on.

The first limitation is that IBM-1 predicts a cut-off for the ground state bands which is too low. Consider for example the nucleus ^{232}Th. This nucleus has 8 valence protons and 16 valence neutrons. As a consequence, $N = N_\pi + N_\nu = 4 + 8 = 12$. Since this is a deformed nucleus, the SU(3) limit is the appropriate IBM-1 limit for its description. According to subsec. 1.7.6, the ground state band of this nucleus will belong to the (24,0) irrep of SU(3), which contains states with angular momenta L=0,2,...,24. However, levels of the ground state band of this nucleus have been identified experimentally up to at least L=30.

Another limitation is that the SU(3) limit of IBM-1 contains only bands of positive parity and K=*even*. However, low lying bands of positive parity and K=*odd* have been identified in several deformed nuclei. For example, in the deformed nucleus ^{176}Hf a $K^\pi = 1^+$ band starting at 1.672 MeV and a $K^\pi = 3^+$ band starting at 1.578 MeV have been identified experimentally. Clearly these bands are not contained in IBM-1.

4.2 The U(15) model

Both these limitations can be overcome if one uses a g-boson (boson with angular momentum 4) in addition to the s and d bosons. The resulting model is called the sdg Interacting Boson Model (sdg IBM) (Wu 1982, Akiyama 1985). This model is built by the boson creation operators s^+, d_μ^+, g_μ^+, and the boson annihilation operators s, d_μ, g_μ, obeying the non-vanishing commutation relations

$$[s, s^+] = 1, \tag{4.1}$$

$$[d_\mu, d_\nu^+] = \delta_{\mu\nu}, \tag{4.2}$$

$$[g_\mu, g_\nu^+] = \delta_{\mu\nu}. \tag{4.3}$$

All other possible commutation relations vanish. Since there are 1+5+9=15 bosons in the model, one can form $15 \times 15 = 225$ generators, in the way demonstrated in detail in Chapter 1. These generators generate the group U(15). Various group chains of U(15) can be studied. Here we will confine ourselves to the SU(3) limit.

4.3 The SU(3) limit

The group chain corresponding to this limit is

$$U(15) \supset SU(3) \supset O(3) \supset O(2). \tag{4.4}$$

The quantum numbers necessary for labelling the states are listed below

$$
\begin{array}{ll}
U(15) & \{N\} \\
SU(3) & (\lambda, \mu) \\
O(3) & L \\
O(2) & M
\end{array}
$$

Furthermore, an additional quantum number χ is needed to fully determine the reduction from SU(3) to O(3), which is not fully decomposable. In the above, N is the total number of bosons (the same for a given nucleus as in IBM-1), while $\{N\}$ denotes the fully symmetric irrep of U(15). The states are then labelled as

$$|\{N\}(\lambda, \mu)\chi LM >. \tag{4.5}$$

The most general Hamiltonian in the SU(3) limit of the sdg IBM has then the same form as in the SU(3) limit of IBM–1. As we already know, its eigenvalues can then be put in the form

$$< H >= (\frac{3}{4}\kappa - \kappa')L(L+1) - \kappa(\lambda^2 + \mu^2 + \lambda\mu + 3(\lambda + \mu)). \tag{4.6}$$

For illustrative purposes we give here the full expressions for the generators of the SU(3) subgroup

$$L_\mu^1 = \sqrt{10}[d^+ \otimes \tilde{d}]_\mu^1 + \sqrt{60}[g^+ \otimes \tilde{g}]_\mu^1, \tag{4.7}$$

$$Q_\mu^2 = 4\sqrt{\frac{7}{15}}([d^+ \otimes \tilde{s}]_\mu^2 + [s^+ \otimes \tilde{d}]_\mu^2) - 11\sqrt{\frac{2}{27}}[d^+ \otimes \tilde{d}]_\mu^2$$

$$+36\sqrt{\frac{1}{105}}(g^+ \otimes \tilde{d}]_\mu^2 + [d^+ \otimes \tilde{g}]_\mu^2) - 2\sqrt{\frac{33}{7}}[g^+ \otimes \tilde{g}]_\mu^2. \qquad (4.8)$$

In this limit the boson creation operators s^+, d_μ^+, and g_μ^+ belong to the (4,0) SU(3) irrep, while the corresponding boson annihilation operators belong to the conjugate irrep, i.e. to (0,4). Thus the most symmetric SU3) irrep contained to the U(15) irrep $\{N\}$ is (4N,0). The values of (λ, μ) contained in each $\{N\}$ irrep of U(15) are

$$\{N\} = (4N,0) \oplus (4N-4,2)$$

$$\oplus(4N-6,3) \oplus (4N-8,4)^2 \oplus (4N-10,5) \oplus \cdots$$

$$\oplus(4N-6,0) \oplus (4N-8,1) \oplus (4N-10,2) \oplus \cdots \qquad (4.9)$$

The values of L contained in each SU(3) irrep are given by the same rules as in IBM-1. As in the case of IBM-1, we are not interested in all the possible SU(3) irreps, but only in these which will lie lowest in energy. It has been argued that the lowest-lying irreps are the irreps with maximum value of the second order Casimir operator of SU(3), C_{2SU3}, since it is in these irreps that the quadrupole–quadrupole interaction is maximized. It can be shown that for $N \geq 4$ the irreps which maximize C_{2SU3} (the so-called "most leading irreps") are

$$(4N,0), (4N-4,2), (4N-6,3), (4N-8,4)^2, (4N-6,0). \quad (4.10)$$

Remember that in the case of IBM–1 the most leading irreps were

$$(2N,0), (2N-4,2), (2N-8,4), (2N-6,0). \qquad (4.11)$$

4.4 Removal of the inadequacies of IBM-1

A comparison between the predictions of the sdg IBM and these of IBM-1 can be made at this point, in order to demonstrate how the inadequacies of IBM-1 mentioned in the beginning of this chapter are removed. In what follows, we will use the symbol $(\lambda, \mu)_0$ to indicate SU(3) irreps of IBM-1 (U(6) structure), while SU(3) irreps of the sdg IBM (U(15) structure) will be denoted by $(\lambda, \mu)_{15}$.

Consider once more the nucleus ^{232}Th, which has N=12. In IBM–1, its ground state belongs to the $(24,0)_6$ irrep, which contains states with angular momentum L=0,2,...,24. In sdg IBM, however, its ground state will belong to the $(48,0)_{15}$ irrep, which contains states with L=0,2,..., 48. Thus the problem of the too low cut-off is removed.

The problem of low-lying bands of positive parity and K=*odd* is also removed in sdg IBM, because the irrep (4N-6,3) appears among the most leading irreps. This irrep contains, according to the SU(3) rules already described, one band with K=1 and one band with K=3. Consider once more the nucleus ^{176}Hf. In that case the most leading irreps are $(64,0)_{15}$, $(60,2)_{15}$, $(58,3)_{15}$, $(56,4)^2_{15}$. The $(58,3)_{15}$ irrep, which has no analog in IBM-1, can easily accomodate the observed $K^\pi = 1^+$ and $K^\pi = 3^+$ bands.

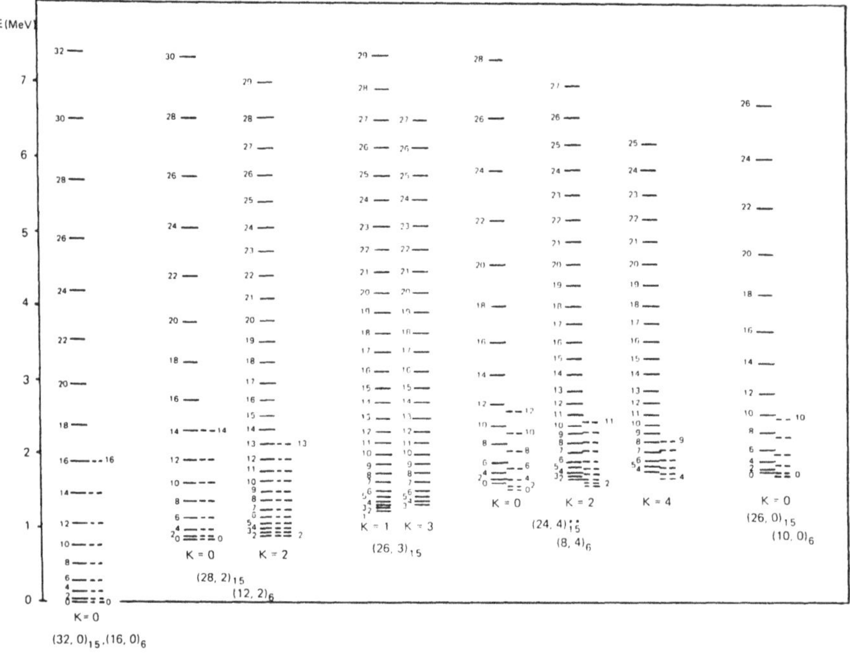

Fig. 4.1 Comparison between the spectra predicted by IBM-1 [U(6)] and sdg IBM [U(15)] for a system of 8 bosons . The predictions of IBM-1 for the energy levels are indicated by broken lines , while those of the sdg IBM are denoted by solid lines . The SU(3) irreps of IBM-1 bear the subscript 6 , while these of the sdg IBM bear the subscript 15 . Notice that no K=1 and K=3 bands are predicted by IBM-1 . (Taken from Wu (1982)) .

The above mentioned differences between IBM-1 and sdg IBM can be seen in Fig. 4.1, where the spectra predicted by the two models for a system of eight bosons are illustrated. It is clear that the cut-off is pushed much higher in sdg IBM. Also notice the presence of K=1 and K=3 bands, absent in IBM-1. The Hamiltonian used is that of Sec. 4.3. Notice that the predictions of the two models for the lowest states of the ground state, β_1 and γ_1 bands are exactly the same, while they are only very slightly different for the lowest states of the higher bands.

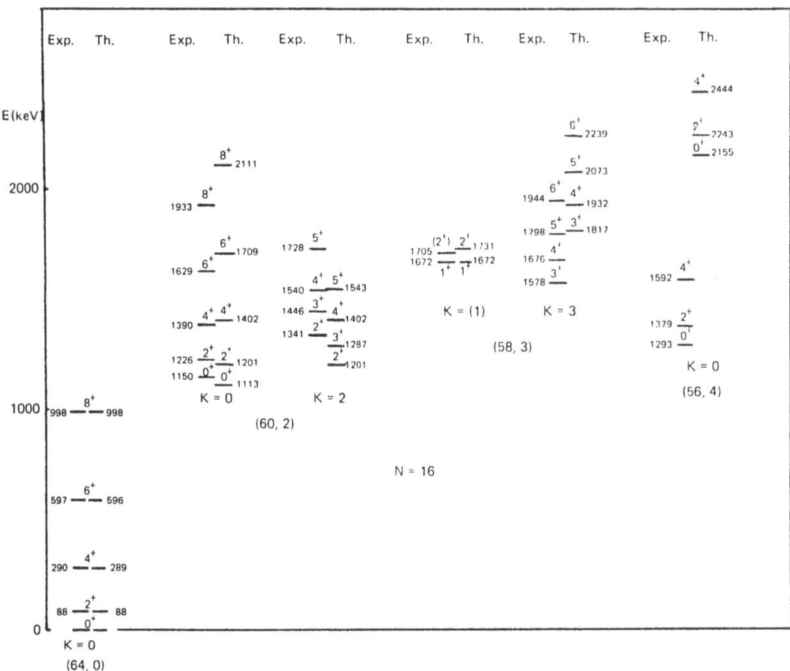

Fig. 4.2 Comparison between the predictions of the sdg IBM and experiment , for the nucleus $^{176}_{72}\text{Hf}_{104}$. (Taken from Wu (1982)) .

A comparison between the predictions of the sdg IBM for the nucleus ^{176}Hf and experiment can be seen in Fig. 4.2. To achieve better agreement with experiment, a term proportional to $(L \odot L)^2$ has been added to the Hamiltonian. Notice that this term is a four-body interaction, thus it is outside the IBM-1 framework, in which only one-body and two-body interactions are included. However, since this term is built exclusively with SU(3) generators, it does not break the SU(3) symmetry, so that analytic solutions can still be found. The energy expression now becomes

$$< H >= aL(L+1)+b(L(L+1))^2+c(\lambda^2+\mu^2+\lambda\mu+3(\lambda+\mu)). \quad (4.12)$$

For the ground state, γ_1 and β_1 bands the agreement between theory and experiment is quite good. The same is true for the K=1 and K=3 bands, while the K=0 band of the (56,4) irrep is predicted much higher than the energy where it is really observed.

4.5 The SU(3) seniority quantum number

The careful reader might have observed that in the SU(3) limit of the sdg IBM the $(4N-8,4)$ irrep occurs twice in a given U(15) irrep. This is a new problem, not encountered in the case of IBM-1. One needs an additional quantum number in order to distinguish these two irreps (and therefore the states belonging to them) from each other. This new quantum number is called the **SU(3) seniority quantum number** (Akiyama 1985) and it is defined as

$$w = \frac{1}{2}(N - v), \quad (4.13)$$

where N is the total number of bosons and v is the usual seniority quantum number (i.e. the number of fermion pairs with angular momentum other than zero). In the case of an SU(3) irrep occurring twice in a given U(15) irrep, the quantum number w takes the values 0 and 1.

In the special case of states belonging to SU(3) irreps for which the relation $\lambda + 2\mu = 4N$ (most of the low-lying states do belong to this category), it can be seen that the Hamiltonian will contain an additional term

$$S = \frac{1}{\sqrt{375}}w(2N - 2w + 3), \quad (4.14)$$

which is called the **SU(3) seniority interaction**. In the case of an SU(3) irrep occurring twice in a given U(15) irrep, it is clear that the states belonging to the $w = 0$ irrep will not be affected by S (since S vanishes for $w = 0$), while the states belonging to the $w = 1$ irrep will be shifted as a whole.

As already mentioned, S is diagonal only in the case of SU(3) irreps satisfying the condition $\lambda + 2\mu = 4N$. In this case S simply splits the multiply occurring irreps by breaking their degeneracy. In the case of SU(3) irreps with $\lambda + 2\mu < 4N$, S is not diagonal any more (actually, it mixes the U(15) irreps).

For practical applications, it is enough to remember that the first SU(3) irrep occuring more than once in a given U(15) irrep is the irrep $(4N - 8, 4)$, occurring twice.

4.6 Effects of bosons higher than g

One may wonder by now what effects the addition of bosons of even higher angular momentum (i-bosons of L=6, k-bosons of L=8, ...) can have. Consider a model including all bosons of even angular momentum up to L, i.e. 0,2,4,...,L. This system has symmetry U((L+1)(L+2)/2). If L is very large, it is impossible to write out the explicit expressions for the group generators. In the case of the rotational limit, however, it is still possible to decompose the $\{N\}$ irrep of U((L+1)(L+2)/2) into SU(3) irreps so as to derive the rotational spectrum. It can be shown that under the conditions $N \geq 4$ and $L \geq 4$ the most leading SU(3) irreps are

$$(NL, 0), (NL - 4, 2), (NL - 6, 3), (NL - 8, 4)^2, (NL - 6, 0). \quad (4.15)$$

The above irreps are very similar to the irreps obtained in the case of the sdg IBM. In both cases the ground state band will belong to the $(NL, 0)$ irrep, the γ_1 and β_1 bands will belong to the $(NL - 4, 2)$ irrep, while the γ_2 and β_2 bands will belong to the $(NL - 8, 4)$ irrep. Finally, the $(NL - 6, 3)$ irrep will contain the lowest $K^\pi = 1^+$ and $K^\pi = 3^+$ bands. As it is seen, the addition of positive parity bosons with angular momenta ≥ 6 to the sdg system does not essentially change the spectrum of the system, while the addition of g-bosons to the sd system resulted in the occurence of positive parity bands with odd K, which do not exist in IBM–1. It is also obvious that the addition of bosons with angular momentum ≥ 6 pushes the cut-offs of the various bands higher and higher.

4.7 Two-nucleon transfer reactions in sdg IBM

4.7.1 The two-nucleon transfer operators

Two-nucleon transfer reactions involving the transfer of a pair of alike nucleons (i.e. two-neutron or two-proton transfer reactions) can be described in the IBM framework in a straightforward manner. In order to do so, one needs the expression for the two-neutron (or two-proton) transfer operators. In lowest order, these will be one-boson operators. In sdg IBM the possible one-boson transfer operators are

$$P^{(0)} = \alpha s^+, \quad (4.16)$$

$$P^{(2)} = \beta d^+, \tag{4.17}$$

$$P^{(3)} = \gamma g^+. \tag{4.18}$$

In IBM-1, of course, only the first two can be present. Furthermore, one needs the states of the initial nucleus and the states of the final nucleus, which will be connected by the transfer operators. These are described in terms of the sdg IBM in the usual way. Here, we will limit ourselves to a qualitative discussion of a specific two-neutron transfer reaction (Akiyama *et al.* 1986).

4.7.2 The $^{166}Er(t,p)^{168}Er$ reaction in the sdg IBM

Consider the reaction ^{166}Er(t,p)^{168}Er. Both the initial nucleus ^{166}Er and the final nucleus ^{168}Er are well-deformed. Thus the SU(3) limit of sdg IBM is the appropriate limit for their description. Suppose that the initial nucleus is in its ground state. Since this nucleus has N=15 bosons, its ground state in sdg IBM will belong to the (60,0) irrep (remember that the s^+, d^+, g^+ bosons belong to the (4,0) SU(3) irrep). The final nucleus has N=16 bosons, thus its ground state band will belong to the (64,0) irrep, while the next most leading irreps will be (60,2), (58,3), $(56,4)^2$, etc. The transition operators given above belong to the (4,0) SU(3) irrep. From the SU(3) outer product

$$(60,0) \otimes (4,0) = (64,0) \oplus (60,2) \oplus (58,3) \oplus (56,4), \tag{4.19}$$

we see that all the most leading irreps of ^{168}Er do appear on the right hand side, thus some states of each of them will be strongly connected to the ground state of ^{166}Er. It is instructive to enumerate them:

i) The transfer operator $P^{(0)}$ can connect the ground state (0_1^+) of ^{166}Er only to states of ^{168}Er with L=0. Such states are contained in K=0 bands only. The lowest L=0 state (0_1^+), which is the ground state of ^{168}Er, belongs to the K=0 band of the (64,0) irrep. The next L=0 state (0_2^+) belongs to the K=0 band of the (60,2) irrep. The next two L=0 states (0_3^+, 0_4^+) will belong to the K=0, w=0 and K=0, w=1 bands of the $(56,4)^{w=0}$ and $(56,4)^{w=1}$ irreps. (Remember that the (56,4) irrep occurs twice, thus the SU(3) seniority quantum number w is used in order to distinguish them).

ii) The transfer operators $P^{(2)}$ can connect the ground state of ^{166}Er only to states of ^{168}Er with L=2. Such states are contained only in K=0, K=1 and K=2 bands. One can easily check that the lowest 2^+ states of ^{168}Er populated by the reaction are the ones listed below (each state is described by the SU(3) irrep and the quantum number K of the band containing it):

$(64,0)$	K=0
$(60,2)$	K=0
$(60,2)$	K=2
$(58,3)$	K=1
$(56,4)^{w=0}$	K=0
$(56,4)^{w=0}$	K=2
$(56,4)^{w=1}$	K=0
$(56,4)^{w=1}$	K=2

iii) The transfer operator $P^{(4)}$ can connect the ground state of ^{166}Er only to states of ^{168}Er with L=4. Such states are contained in bands with K=0, K=1, K=2, K=3, K=4. One can easily check that the populated states of ^{168}Er are:

$(64,0)$	K=0
$(60,2)$	K=0
$(60,2)$	K=2
$(58,3)$	K=1
$(58,3)$	K=3
$(56,4)^{w=0}$	K=0
$(56,4)^{w=0}$	K=2
$(56,4)^{w=0}$	K=4
$(56,4)^{w=1}$	K=0
$(56,4)^{w=1}$	K=2
$(56,4)^{w=1}$	K=4

One can then go on and calculate relative spectroscopic factors, relative (t,p) strengths and other interesting quantities, which we will not consider here. However, some useful conclusions can be drawn if one compares the predictions of the sdg IBM with the predictions of IBM-1 for the same problem. Thus we will analyze the IBM-1 approach to the same reaction now.

4.7.3 The $^{166}Er(t,p)^{168}Er$ reaction in IBM-1

In IBM-1 the s^+ and d^+ bosons belong to the $(2,0)$ irrep of SU(3). Since the nucleus ^{166}Er has N=15 bosons, its ground state band will belong to the $(30,0)$ SU(3) irrep. Only the $P^{(0)}$ and $P^{(2)}$ one-boson transfer operators will be present in this case, belonging to the $(2,0)$ SU(3) irrep. From the SU(3) outer product

$$(30,0) \otimes (2,0) = (32,0) \oplus (28,2) \qquad (4.20)$$

we see that only states in the (32,0) and (28,2) irreps of ^{168}Er can be populated starting from the ground state of ^{166}Er. We enumerate them here:

i) The $P^{(0)}$ transfer operator can only populate L=0 states (starting from an L=0 state). Such states are contained only in K=0 bands. In the present case the 0_1^+ state will belong to the K=0 band of the (32,0) irrep, while the 0_2^+ state will belong to the K=0 band of the (28,2) irrep. Higher 0^+ states, like 0_3^+, 0_4^+, etc, should not be populated according to IBM-1.

ii) The $P^{(2)}$ transfer operator can only populate L=2 levels (starting from an L=0 state). Such states are contained here only in K=0 and K=2 bands (remember that no K=1 bands occur in IBM-1). They are listed below

$$
\begin{array}{ll}
(32,0) & K=0 \\
(28,2) & K=0 \\
(28,2) & K=2
\end{array}
$$

Higher 2^+ states should not be populated according to IBM-1.

Comparing the predictions of IBM-1 and those of the sdg IBM we see the following differences:

i) While only two 0^+ states are predicted to be populated according to IBM-1, four such states are predicted to be populated by the reaction according to the sdg IBM. Experimentally it has been found that the two extra 0^+ states (0_3^+ and 0_4^+) are populated rather strongly, especially 0_3^+.

ii) While only three 2^+ states are predicted to be populated by IBM-1, eight such states are predicted by the sdg IBM. Experimentally it has been found that some of the extra states, especially those belonging to the (56,4) K=0 band and the (56,4) K=2 band, are rather strongly populated.

iii) According to IBM-1, no L=4 states are populated, while in sdg IBM 11 such states are populated.

From the experimental evidence already quoted, it becomes clear that the sdg IBM gives a more complete description of the particular (t,p) reaction than IBM-1 does.

References

Akiyama, Y, 1985. *Nucl. Phys.*, **A433**, 369.

Akiyama, Y, von Brentano, P, and Gelberg, A, 1987. *Z. Phys.*, **A326**, 517.

Akiyama, Y, Heyde, K, Arima, A, and Yoshinaga, N, 1986.

Phys. Lett., **173B**, 1.

Arima, A, 1984. *Nucl. Phys.*, **A421**, 63c.

Baker, F T, 1985. *Phys. Rev.*, **C32**, 1430.

Baker, F T, Sethi, A, Penumetcha, V, Emery, G T, Jones,
W P, Grimm, M A, and Whiten, M L, 1985. *Phys. Rev.*, **C32**,
2212.

Baker, F T, Sethi, A, Penumethcha, V, Scott, A, Emery, G,
Jones, W P, Grimm, M A, and Whiten, M L, 1984. *Gull Lake
1984*, 74.

Barrett, B R, 1986. *Dubrovnik 1986*, **2**, 1010.

Bohr, A, and Mottelson, B R, 1980. *Phys. Scr.*, **22**, 468.

Bohr, A, and Mottelson, B R, 1982. *Phys. Scr.*, **25**, 28.

Chakraborty, M, Kota, V K B, Parikh, J C, 1981. *Phys. Lett.*,
100B, 201.

De Meyer, H, Van der Jeugt, J, Van den Berghe, G, and Kota,
V K B, 1986. *J. Phys.*, **A19**, L565.

Dukelsky, J, Fernández-Niello, J, Sofia, H M, and Perazzo,
R P J, 1983. *Phys. Rev.*, **C28**, 2183.

Goldfarb, L J B, 1981. *Phys. Lett.*, **104B**, 103.

Goldhoorn, P B, Harakeh, M N, Iwasaki, Y, Put, L W, Zwarts,
F, and van Isacker, P, 1981. *Phys. Lett.*, **103B**, 291.

Heyde, K, van Isacker, P, Waroquier, M, Wenes, G, Gigase,
Y, and Stachel, J, 1983. *Nucl. Phys.*, **A398**, 235.

Kota, V K B, 1984. *Gull Lake 1984*, 83.

Kota, V K B, Van der Jeugt, J, De Meyer, H, and Vanden
Berghe, G, 1987. *J. Math. Phys.*, **28**, 1644.

Kuyucak, C, and Morrison, I, 1987. *Phys. Rev. Lett.*, **58**, 315.

Morrison, I, 1986. *Phys. Lett.*, **175B**, 1.

Morrison, I, 1986. *J. Phys.*, **G12**, L201.

Ratna Raju, R D, 1981. *Phys. Rev.*, **C23**, 518.

Ratna Raju, R D, 1982. *J. Phys.*, **G8**, 1663.

Ratna Raju, R D, 1986. *J. Phys.*, **G12**, L279.

van Isacker, P, 1980. *Erice 1980*, 115.

van Isacker, P, Heyde, K, Waroquier, M, and Wenes, G, 1981.
Phys. Lett., **104B**, 5.

van Isacker, P, Heyde, K, Waroquier, M, and Wenes, G, 1982.
Nucl. Phys., **A380**, 383.

Wood, J L, 1984. *Nucl. Phys.*, **A421**, 43c.

Wu, H C, 1982. *Phys. Lett.*, **110B**, 1.

Wu, H C, and Zhou, X Q, 1984. *Nucl. Phys.*, **A417**, 67.

Wu, H C, Dieperink, A E L, and Pittel, S, 1986. *Phys. Rev.*,
C34, 703.

Yoshinaga, N, 1986. *Nucl. Phys.*, **A456**, 21.

Yoshinaga, N, Akiyama, Y, and Arima, A, 1986. *Phys. Rev. Lett.*, **56**, 1116.
Zhou, X Q, and Wu, H C, 1984. *Nucl. Phys.*, **A421**, 159c.

5

THE INTERACTING BOSON MODEL-2 (IBM-2)

5.1 Introduction

In IBM-1 no distinction between protons and neutrons is made. One should expect, however, to obtain a more realistic description of nuclei by treating protons and neutrons as different particles, which they are. This distinction is made by the Interacting Boson Model-2. One introduces proton bosons s_p and d_p, as well as neutron bosons s_n and d_n. The total number of proton bosons introduced equals the number of valence proton pairs (particles or holes, whichever is minimum) in the nucleus under study. Similarly, the number of neutron bosons introduced equals the number of valence neutron pairs (particles or holes). Since in medium and heavy nuclei the valence protons and the valence neutrons occupy different major shells, no proton–neutron pairs are introduced. It is assumed that proton–neutron correlations can be treated as a mean field interaction between the proton bosons and the neutron bosons.

The proton bosons introduced span the dynamic group $U_p(6)$, while the neutron bosons span the dynamic group $U_n(6)$. Addition of the proton degrees of freedom to the neutron degrees of freedom is achieved by taking the **direct product** of the two groups,

$$G = U_p(6) \otimes U_n(6). \tag{5.1}$$

By definition, this group has 72 generators, i.e. the 36 generators of $U_p(6)$ and the 36 generators of $U_n(6)$. The states span the $[N_p] \otimes [N_n]$ irrep of G, where N_p is the number of proton bosons and N_n is the number of neutron bosons.

5.2 Dynamical symmetries in IBM-2

It is of interest to study the various chains of subalgebras of the group G, since they can guide as to dynamic symmetries, allowing for the exact diagonalization of the Hamiltonian. Dynamic symmetries of coupled systems are much more difficult to study than those of single systems. The technique one uses is that of finding all possible **common** subalgebras and combining the appropriate generators for the two (or more) systems.

One class of chains (van Isacker, Heyde, Jolie, and Sevrin 1986) begins with the reduction $U_p(6) \otimes U_n(6) \supset U_{p+n}(6)$, where we denote by $U_{p+n}(6)$ the group generated by the sums of the corresponding generators of the groups $U_p(6)$ and $U_n(6)$. The states are then characterized by the two-row irreps $[N - f, f]$ of $U_{p+n}(6)$, where $N = N_p + N_n$ is the total number of bosons and $f = 0, 1, \ldots, \min(N_p, N_n)$. It is convenient at this point to introduce a quantum number called **F-spin** (Arima, Otsuka, Iachello, and Talmi 1977), defined as

$$F = \frac{N}{2} - f, \qquad (5.2)$$

and its zero-component

$$F_0 = \frac{1}{2}|N_p - N_n|. \qquad (5.3)$$

Then one can characterize the states by the quantum numbers N , F , and F_0. Clearly the totally symmetric states span the one-row irrep of $U_{p+n}(6)$, [N,0]. Thus they are characterized by f=0, which implies that they possess the maximum possible value of F-spin,

$$F_{max} = \frac{N}{2}. \qquad (5.4)$$

It is clear that the states of maximum F-spin are in one to one correspondence with the states of IBM-1. States with F-spin less than the maximum value of N/2 have no counterparts in IBM-1. They have mixed proton–neutron symmetry character, thus they are called **mixed symmetry states** (MISS !) These are of course the states in which the main interest in studying the IBM-2 lies. A more extended discussion of F-spin will be given in Chapter 7.

Example Consider the nuclei $^{124}_{52}\text{Te}_{72}$, $^{128}_{54}\text{Xe}_{74}$, $^{132}_{56}\text{Ba}_{76}$. One can easily check that all of them have the same $N = 6$, while they have different (N_p, N_n) values, $(1,5)$, $(2,4)$, $(3,3)$, respectively. For all of these nuclei the states of maximum F-spin symmetry will have $F = 3$, while the lowest-lying mixed symmetry states will have $F = 2$. Notice that the zero-component of the F-spin has a different value for each of these nuclei, being 2, 1, 0 respectively.

The possible chains beginning with U_{p+n} and containing the angular momentum subalgebra SO(3) are the following

$$U_p(6) \otimes U_n(6) \supset U_{p+n}(6) \supset U_{p+n}(5)$$

$$\supset O_{p+n}(5) \supset O_{p+n}(3) \supset O_{p+n}(2), \tag{5.5}$$

$$U_p(6) \otimes U_n(6) \supset U_{p+n}(6) \supset SU_{p+n}(3) \supset O_{p+n}(3) \supset O_{p+n}(2), \tag{5.6}$$

$$U_p(6) \otimes U_n(6) \supset U_{p+n}(6) \supset O_{p+n}(6)$$

$$\supset O_{p+n}(5) \supset O_{p+n}(3) \supset O_{p+n}(2). \tag{5.7}$$

One thus recovers the symmetries contained in IBM-1. Clearly, these chains remain unchanged if the proton and neutron labels are interchanged. Furthermore, F-spin is a good quantum number for the Hamiltonians corresponding to these chains. This can be seen by remarking that by definition $F = \frac{N}{2} - f$ and both N and f are good quantum numbers, characterizing the irreps of $U_{p+n}(6)$. These chains are called the **F-spin symmetric limits** of IBM-2, and we will refer to them as S–I, S–II, and S–III, respectively.

Besides the chains starting with $U_{p+n}(6)$, other chains of subgroups of G are possible (Iachello 1982). The most interesting ones are these in which the coupling occurs at the level of the first subgroups

$$U_p(6) \otimes U_n(6) \supset U_p(5) \otimes U_n(5) \supset U_{p+n}(5)$$

$$\supset O_{p+n}(5) \supset O_{p+n}(3) \supset O_{p+n}(2), \tag{5.8}$$

$$U_p(6) \otimes U_n(6) \supset SU_p(3) \otimes SU_n(3) \supset SU_{p+n}(3)$$

$$\supset O_{p+n}(3) \supset O_{p+n}(2), \tag{5.9}$$

$$U_p(6) \otimes U_n(6) \supset O_p(6) \otimes O_n(6) \supset O_{p+n}(6)$$

$$\supset O_{p+n}(5) \supset O_{p+n}(3) \supset O_{p+n}(2). \tag{5.10}$$

Clearly, these chains do not remain unchanged under the interchange of the proton and neutron labels. As a result, F-spin is not a good quantum number for the corresponding Hamiltonians. We will refer to them as N–I, N–II, and N–III, respectively.

5.3 The F-spin symmetric limits of IBM-2

5.3.1 The chain S–II

Dynamical symmetries can be exploited in the usual way in these limits. As an example, we give here the results for the chain S-II. The quantum numbers necessary to determine the irreps of each group are listed below.

$$
\begin{array}{cc}
U_p(6) \otimes U_n(6) & [N_p]\,[N_n] \\
U_{p+n}(6) & (N_1, N_2) \\
SU_{p+n}(3) & (\lambda, \mu) \\
O_{p+n}(3) & L \\
O_{p+n}(2) & M
\end{array}
$$

The quantum numbers N_1 and N_2 (which are related to the quantum numbers N and f, used above, by $N_1 = N - f$, $N_2 = f$) are necessary to fully determine the irreps of $U_{p+n}(6)$, since the product $[N_p]\otimes[N_n]$ does not contain only the fully symmetric irreps $[N_p + N_n, 0]$, but also contains states with mixed symmetry $[N_p + N_n - 1, 1]$, ... As mentioned above, the fully symmetric states can be put in one to one correspondence with the states occuring in IBM-1. The mixed symmetry states, however, do not have any counterparts in IBM-1. Thus the main interest in IBM-2 lies in studying the properties of these mixed symmetry states. An additional quantum number K is necessary, since the reduction from SU(3) to O(3) is not fully decomposable. Thus the states are fully determined as

$$
|[N_p][N_n](N_1, N_2)(\lambda, \mu)KLM > . \tag{5.11}
$$

The most general Hamiltonian in this limit can then be written as

$$
H^{S-II} = aC_{1U6} + a'C_{2U6} + \alpha'C_{2O3} + \beta'C_{2SU3}. \tag{5.12}
$$

Its eigenvalues are then

$$
< H^{S-II} >= a(N_1 + N_2) + a'[N_1(N_1 + 5) + N_2(N_2 + 3)] + \alpha'L(L+1)
$$

$$
+\beta'(\lambda^2 + \mu^2 + \lambda\mu + 3(\lambda + \mu)). \tag{5.13}
$$

However, an equivalent formulation, which is physically more transparent, is possible. It has been customary in IBM-2 to include in the Hamiltonian a term describing interaction between proton and neutron bosons, called the **Majorana term**. Its role will be explained later. The Majorana term can be written explicitly in terms of the boson operators as follows

$$
M = \xi_2[(s_n^+ d_p^+ - d_n^+ s_p^+) \odot (s_n \tilde{d}_p - \tilde{d}_n s_p)^2]^0
$$

$$-2 \sum_{k=1,3} \xi_k [(d_n^+ d_p^+)^k \odot (\tilde{d}_n \tilde{d}_p)^k]^0. \tag{5.14}$$

For $\xi_1 = \xi_2 = \xi_3$ the Majorana term is related to the quadratic Casimir operator of $U_{p+n}(6)$ as

$$M = \frac{1}{2}[N(N+5) - C_{2U_{p+n}6}]. \tag{5.15}$$

Then in the S–II limit discussed above, one can use the Hamiltonian

$$H^{S-II} = \alpha' C_{2O3} + \beta' C_{2SU3} + \gamma M, \tag{5.16}$$

with eigenvalues

$$< H^{S-II} >= \alpha' L(L+1) + \beta'(\lambda^2 + \mu^2 + \lambda\mu + 3(\lambda + \mu))$$

$$+\gamma(\frac{N}{2} - F)(\frac{N}{2} + F + 1). \tag{5.17}$$

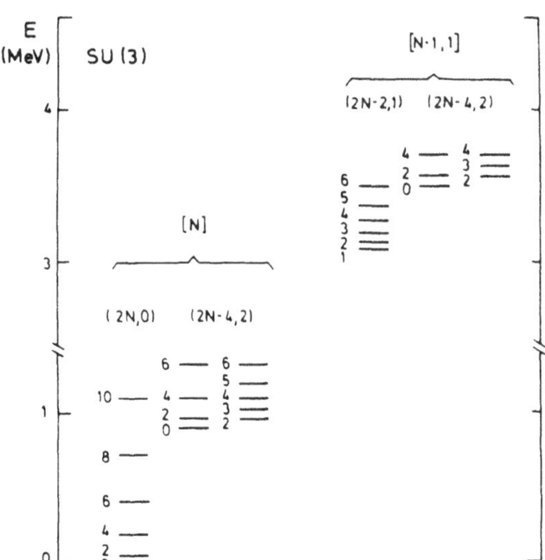

Fig. 5.1 Typical energy spectrum in the F-spin symmetric SU(3) limit of IBM-2. Only the lowest states of the irreps (N) and $(N-1,1)$ are shown. The boson numbers are $N_\pi = 6$ and $N_\nu = 7$. (Taken from van Isacker *et al.* (1986)).

A typical spectrum obtained with this Hamiltonian is shown in Fig. 5.1. The states of the (N) irrep are states which are contained in

IBM-1 as well. The states in the $(N-1,1)$ irrep are mixed symmetry states, appearing only in IBM-2. The rotational character of all bands is clear, as expected since we used the SU(3) limit.

The other two F-spin symmetric dynamical symmetries of IBM-2, can be analyzed in a similar way. Only the final results are quoted here.

5.3.2 The chain S–I

The quantum numbers necessary to label the irreps of each group are

$$
\begin{array}{ll}
U_p(6) \otimes U_n(6) & [N_p][N_n] \\
U_{p+n}(6) & (\text{N-f,f}) \\
U_{p+n}(5) & \{n_1, n_2\} \\
O_{p+n}(5) & (v_1, v_2) \\
O_{p+n}(3) & \text{L} \\
O_{p+n}(2) & \text{M}
\end{array}
$$

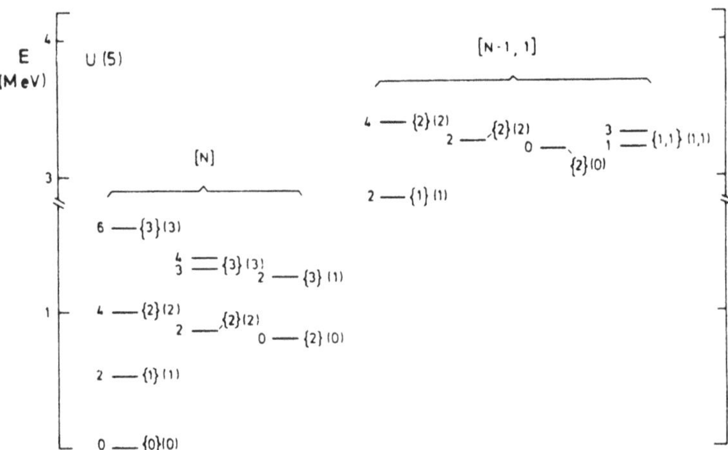

Fig. 5.2 Typical energy spectrum in the F-spin symmetric U(5) limit of IBM-2. Only the lowest states of the irreps (N) and $(N-1,1)$ are shown. The boson numbers are $N_\pi = 1$ and $N_\nu = 5$. (Taken from van Isacker *et al.* (1986)).

In addition, a quantum number α is required to completely specify the reduction $O(5) \supset O(3)$.

The basis is fully determined as

$$|[N_p][N_n](N-f,f)\{n_1, n_2\}(v_1, v_2)\alpha L M > . \qquad (5.18)$$

The Hamiltonian is written as

$$H^{S-I} = A_1 C_{1U_{p+n}5} + A_2 C_{2U_{p+n}5} + B C_{2O_{p+n}5} + C C_{2O_{p+n}3} + aM,$$
(5.19)

with eigenvalues

$$< H^{S-I} >= A_1(n_1 + n_2) + A_2(n_1(n_1 + 4) + n_2(n_2 + 2))$$

$$+B(v_1(v_1+3)+v_2(v_2+1))+CL(L+1)+a(\frac{N}{2}-F)(\frac{N}{2}+F+1). \quad (5.20)$$

A typical spectrum obtained from this Hamiltonian is shown in Fig. 5.2. Again, the (N) states are the ones which occur in IBM-1 as well, while the $(N-1,1)$ states are MISS occurring only in IBM-2. Notice the almost equispaced ground state band, as well as the strong odd–even staggering of the γ_1 band, features already encountered in the U(5) limit of IBM-1.

5.3.3 The chain S–III

The quantum numbers fully determining the states are

$$
\begin{array}{cc}
U_p(6) \otimes U_n(6) & [N_p][N_n] \\
U_{p+n}(6) & (N\text{-}f,f) \\
O_{p+n}(6) & < \sigma_1, \sigma_2 > \\
O_{p+n}(5) & (\tau_1, \tau_2) \\
O_{p+n}(3) & L \\
O_{p+n}(2) & M
\end{array}
$$

An additional quantum number γ is necessary to completely specify the $O(5) \supset O(3)$ reduction.

The states are fully characterized as

$$|[N_p][N_n](N - f, f) < \sigma_1, \sigma_2 > (\tau_1, \tau_2)\gamma LM > . \quad (5.21)$$

The Hamiltonian is written as

$$H^{S-III} = A C_{2O_{p+n}6} + B C_{2O_{p+n}5} + C C_{2O_{p+n}3} + aM, \quad (5.22)$$

with eigenvalues

$$< H^{S-III} >= A(\sigma_1(\sigma_1 + 4) + \sigma_2(\sigma_2 + 2)) + B(\tau_1(\tau_1 + 3) + \tau_2(\tau_2 + 1))$$

$$+CL(L + 1) + a(\frac{N}{2} - F)(\frac{N}{2} + F + 1). \quad (5.23)$$

A typical spectrum obtained from this Hamiltonian is shown in Fig. 5.3. Again, the (N) states are the states which occur in IBM-1 as well, while the $(N-1,1)$ states are MISS occuring only in IBM-2. As in the case of the O(6) limit of IBM-1, strong odd–even staggering is observed in the γ_1 band.

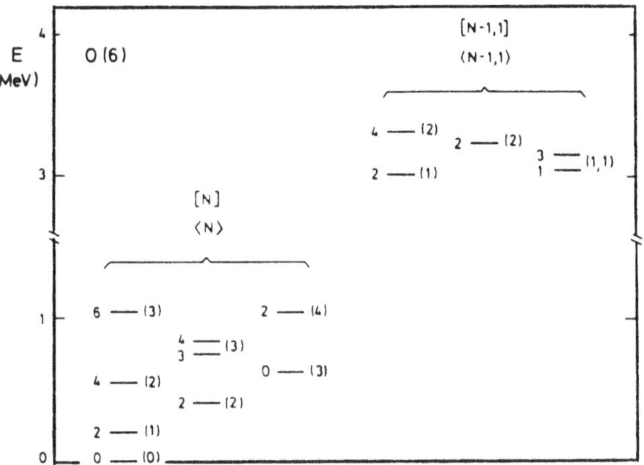

Fig. 5.3 Typical energy spectrum in the F-spin symmetric O(6) limit of IBM-2. Only the lowest states in the irreps (N) and $(N-1,1)$ are shown. The boson numbers are $N_\pi = 2$ and $N_\nu = 4$. (Taken from **van Isacker** *et al.* (1986)).

5.4 F-spin non-symmetric limits of IBM-2

5.4.1 The chain N–II

Consider the group chain N–II. The quantum numbers needed to specify the irreps of the various groups are

$$
\begin{array}{ll}
U_p(6) \otimes U_n(6) & [N_p][N_n] \\
SU_p(3) \otimes SU_n(3) & (\lambda_p, \mu_p)(\lambda_n, \mu_n) \\
SU_{p+n}(3) & (\lambda, \mu) \\
O_{p+n}(3) & \mathrm{L} \\
O_{p+n}(2) & \mathrm{M}
\end{array}
$$

In addition, a quantum number K is needed to fully specify the reduction from SU(3) to O(3).

The states are then specified as

$$|[N_p][N_n](\lambda_p,\mu_p)(\lambda_n,\mu_n)(\lambda,\mu)KLM> .\qquad(5.24)$$

The most general Hamiltonian in this limit is written as

$$H^{N-II} = bC_{2SU_p3} + b'C_{2SU_n3} + \alpha'C_{2O3} + \beta'C_{2SU_{p+n}3},\qquad(5.25)$$

with eigenvalues

$$< H^{N-II} >= b(\lambda_p^2 + \mu_p^2 + \lambda_p\mu_p + 3(\lambda_p + \mu_p))$$

$$+b'(\lambda_n + \mu_n + \lambda_n\mu_n + 3(\lambda_n + \mu_n))$$

$$+\alpha'L(L+1) + \beta'(\lambda^2 + \mu^2 + \lambda\mu + 3(\lambda + \mu)).\qquad(5.26)$$

This spectrum strongly resembles the spectrum obtained in the SU(3) limit of IBM-1. The only difference is that the present spectrum contains more states. To illustrate this point, consider the case where only one proton boson and only one neutron boson are present, i.e. $N_p = 1$, $N_n = 1$. Then $(\lambda_p,\mu_p) = (2,0)$ and $(\lambda_n,\mu_n) = (2,0)$. Using the Young tableaux (described in the Appendices) one can easily see that

$$(2,0) \otimes (2,0) = (4,0) \oplus (2,1) \oplus (0,2).\qquad(5.27)$$

The (4,0) irrep contains a K=0 band with L=0,2,4, while the (0,2) irrep contains another K=0 band with L=0,2. In addition, the (2,1) irrep contains a K=1 band with L=1,3. In the case of IBM-1, the same situation would have been described as N=2. In that case, according to the rules described in subsec. 1.6.2, only the (4,0) and (0,2) irreps would have been present. Thus the K=1 band occurs only in IBM-2 and not in IBM-1.

5.4.2 The N-II′ chain

At this point it is useful to remember that there is a double sign in front of the $[d^+ \otimes \tilde{d}]_\mu^2$ term in the G_μ^2 generator of SU(3). It has been suggested that in IBM-1, the minus sign corresponds to nuclei with prolate deformation, while the plus sign represents nuclei with oblate deformation. In the case of N–II, we assumed that the sign is the same, either positive or negative, in both the $SU_p(3)$ and the $SU_n(3)$ generators. Obviously, another possibility exists: the sign can be positive in the $SU_p(3)$ generator and negative in the $SU_n(3)$ generator, or vice versa. It has been suggested that this situation occurs when the proton bosons are particles and the neutron bosons are holes, or vice versa. In that case a bar is used to indicate the hole nature. For example, the situation where one particle proton boson and one hole neutron boson are present is denoted by $N_p = 1$, $N_n = \bar{1}$. Holes are supposed to be described by the conjugate SU(3) irreps, i.e. one hole boson belongs to the (0,2) $\overline{SU(3)}$ irrep, etc. The $SU_{p+n}(3)$ group obtained from the coupling of one particle $SU_p(3)$ and one hole $\overline{SU_n(3)}$ (or from a hole $\overline{SU_p(3)}$ and a particle $SU_n(3)$) is denoted by $SU^*_{p+n}(3)$ in order to be distinguished from the normal case, although from the mathematical point of view there is no difference. The group chain is then

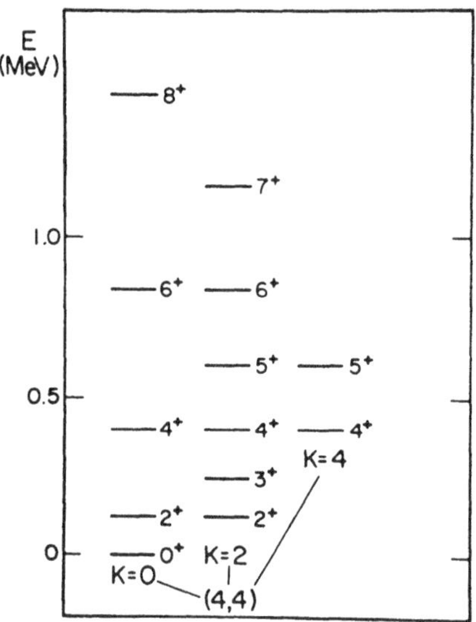

Fig. 5.4 Typical energy spectrum in the $SU(3)^*$ limit of IBM-2. The boson numbers are $N_\pi = 2$ and $N_\nu = 2$. (Taken from Iachello (1983)).

$$U_p(6) \otimes U_n(6) \supset SU_p(3) \otimes \overline{SU_n(3)} \supset SU^*_{p+n}(3)$$
$$\supset O_{p+n}(3) \supset O_{p+n}(2), \tag{5.28}$$

and will be denoted as $N\text{-}II'$.

In the example of $N_p = 1$, $N_n = \bar{1}$ which was considered above, one easily finds that

$$(2,0) \otimes (0,2) = (2,2) \oplus (1,1) \oplus (0,0). \tag{5.29}$$

The (2,2) irrep contains a K=0 band with L=0,2 and a K=2 band with L=2,3,4, while the (1,1) irrep contains a K=1 band with L=1,2 and the (0,0) irrep contains a K=0 band with only L=0. It is clear that this spectrum is very different from the spectrum of $N_p = 1$, $N_n = 1$ studied earlier. While in the particle–particle case the ground state band turns out to belong to the $(N_p + N_n, 0)$ irrep, in the particle–hole case the ground state band turns out to belong to the (N_p, N_n) irrep. It has been suggested that the later case corresponds to nuclei with stable triaxial deformation. As we have already seen, such shapes are not present in the case of IBM-1, but they can be described by the addition of symmetry-breaking "cubic" terms.

A typical spectrum obtained in this limit is shown in Fig. 5.4. This spectrum has been produced with $N_p = 2$, $N_n = 2$. In this case the lowest irrep, coming from the direct product $(4,0) \otimes (0,4)$, is $(4,4)$. Only this lowest irrep is shown in the picture. Notice that the γ_1 band (the K=2 band shown in the figure) occurs very low in energy relative to the ground state band. This is a feature seen in the experimental spectrum of the nucleus ^{104}Ru, which is thought to be triaxial, shown in Fig. 3.4.

5.4.3 Weak coupling group chains

Besides the group chains studied above, many more exist, in which the coupling between protons and neutrons occurs at a later stage. An example is provided by the group chain (Dieperink 1984a)

$$U_p(6) \otimes U_n(6) \supset U_p(5) \otimes U_n(5) \supset O_p(5) \otimes O_n(5)$$

$$\supset O_p(3) \otimes O_n(3) \supset O_{p+n}(3) \supset O_{p+n}(2). \qquad (5.30)$$

The quantum numbers needed to specify the states are listed below.

$$
\begin{array}{cc}
U_p(6) \otimes U_n(6) & [N_p][N_n] \\
U_p(5) \otimes U_n(5) & (n_{d_p})(n_{d_n}) \\
O_p(5) \otimes O_n(5) & \tau_p \tau_n \\
O_p(3) \otimes O_n(3) & L_p \, L_n \\
O_{p+n}(3) & L \\
O_{p+n}(2) & M
\end{array}
$$

This chain is not particularly interesting from the physical point of view. However, this is the group chain used in the computer program NPBOS (T. Otsuka, 1977, University of Tokyo, Japan), which is used to diagonalize numerically IBM–2 Hamiltonians in practical applications.

5.5 Identification and geometric picture of mixed symmetry states

As we have seen, the major difference between IBM-1 and IBM-2 is that the latter contains a whole class of states (Iachello 1984a), the mixed symmetry states (MISS), which are completely MISSing (!) from IBM-1. In the early days of the model the existence of MISS was a puzzle, since they were predicted to occur at rather low energies but no such state had been seen experimentally. It was then argued that "obviously" these states were lying very high in energy, and the coefficient of the Majorana term in the Hamiltonian was made "big", in order to push the MISS far up in the spectrum. Actually, this was the reason the Majorana term was introduced at all.

By now, however, quite a few of these states have been identified (Bohle 1986). They are 1^+ states, seen in rare earth nuclei at an excitation energy around 3 MeV. A list of some of the nuclei where such states have been observed is given below (van Isacker *et al.* 1986), along with the excitation energy of the 1^+ MISS in MeV:

$$\begin{array}{ll}
{}^{154}_{62}\mathrm{Sm}_{92} & 3.2 \\
{}^{156}_{64}\mathrm{Gd}_{92} & 3.075 \\
{}^{158}_{64}\mathrm{Gd}_{94} & 3.2 \\
{}^{164}_{66}\mathrm{Dy}_{98} & 3.11 \\
{}^{168}_{68}\mathrm{Er}_{100} & 3.39 \\
{}^{174}_{70}\mathrm{Yb}_{104} & 3.555
\end{array}$$

Moral: *Nothing is obvious ! (Abraham Klein).*

It is instructive to consider the geometrical shapes associated with F-spin symmetric states and states of mixed symmetry (MISS). They are illustrated in Fig. 5.5. The neutron distribution is indicated by a solid line, while the proton distribution is indicated by a dashed line. If the two distributions coincide, they are indicated by a solid, thick line. In the vibrational (U(5)) limit, the ground state of the nucleus looks spherical. Excited states are created by vibrations. In F-spin symmetric excited states, the proton and neutron distributions vibrate in phase. For MISS the proton and neutron vibrations are out of phase. In the rotational limit (SU(3)), the ground state looks deformed but axially symmetric (the axis of rotational symmetry is indicated in the figure). Its shape can be either prolate (corresponding to the minus sign in the SU(3) quadrupole operator) or oblate (corresponding to the plus sign in the SU(3) quadrupole operator). F-spin symmetric states are obtained in the SU(3) limit when the axes of rotational symmetry of the proton and neutron

distributions coincide. When they do not, MISS are obtained. The
O(6) symmetry corresponds to γ-unstable nuclei. They can be vi-
sualized as very soft nuclei, continuously changing their shape from
prolate to oblate and vice versa.

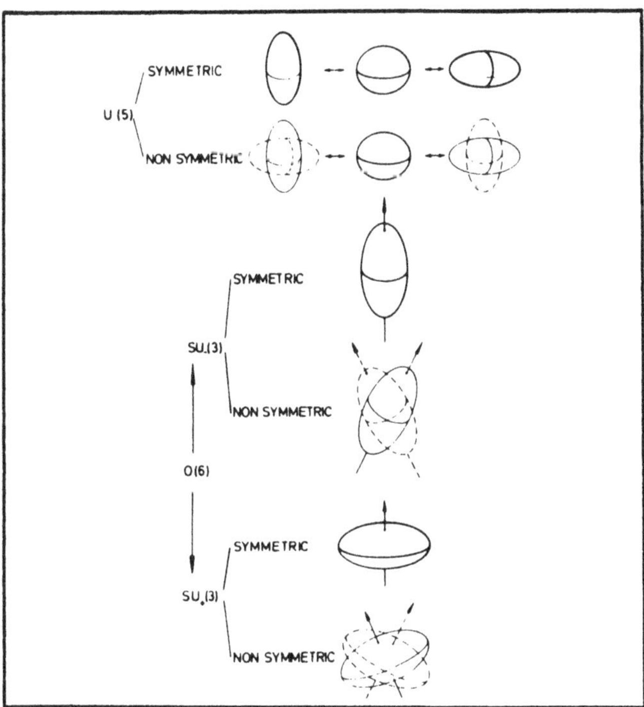

Fig. 5.5 Schematic illustration of the geometric shapes associated with
the F-spin symmetric and the F-spin non-symmetric states in IBM-2. See
the text for discussion. (Taken from van Isacker *et al.* (1986)).

5.6 Numerical calculations in IBM-2

As in the case of IBM-1, few nuclei can be described in terms of
the exact limiting symmetries of IBM-1. In most cases, the symmetries have to be broken, and the Hamiltonian has to be diagonalized
numerically. This is a task much more difficult than in the case of
IBM-1, because the number of states in IBM-2 is much higher than

in IBM-1. As an example, we give here a list (taken from Novoselsky (1985)) of the number of states of a system of 15 bosons (for example, the nucleus ^{178}Hf is such a system) occurring for each value of the angular momentum in each of the models:

J	IBM-1	IBM-2
0	27	580
1	-	1080
2	45	2121
3	19	2438
4	56	3187
5	30	3229
6	61	3629
7	36	3406
8	60	3501

Thus, in the case of J=6, the number of states in IBM-2 is more than 60 times the number of states in IBM-1.

The first program developed for the purpose of performing IBM-2 calculations was NPBOS (T. Otsuka, 1977, University of Tokyo, Japan). It is still widely used. By now, however, another program, called BOSON2 (A. Novoselsky, 1985, The Weizmann Institute of Science, Israel), able to handle the large boson numbers involved in deformed nuclei (like the above mentioned ^{178}Hf) has been developed.

The IBM-2 Hamiltonian usually used in numerical applications has the form (Arima and Iachello 1984)

$$H = \epsilon(n_{d_p} + n_{d_n}) + \kappa(Q_p \cdot Q_\nu) + V_{pp} + V_{nn} + M, \qquad (5.31)$$

where n_{d_p} and n_{d_n} are the number operators for the proton and neutron d bosons, Q_p, and Q_n, are the quadrupole operators for proton and neutron bosons

$$Q_p = (d_p^+ s_p + s_p^+ \tilde{d}_p)^2 + \chi_p(d_p^+ \tilde{d}_p)^2, \qquad (5.32)$$

$$Q_n = (d_n^+ s_n + s_n^+ \tilde{d}_n)^2 + \chi_n(d_n^+ \tilde{d}_n)^2, \qquad (5.33)$$

V_{pp} and V_{nn} are the proton–proton and neutron–neutron interactions

$$V_{pp} = \sum_{L=0,2,4} \frac{1}{2}\sqrt{2L+1}\,c_{L,p}[(d_p^+ d_p^+)^L(\tilde{d}_p\tilde{d}_p)^L]^0, \qquad (5.34)$$

$$V_{nn} = \sum_{L=0,2,4} \frac{1}{2}\sqrt{2L+1}\, c_{L,n}[(d_n^+ d_n^+)^L (\tilde{d}_n \tilde{d}_n)^L]^0, \qquad (5.35)$$

and M is the Majorana term given in Sec. 5.3.

From the IBM-1 experience it is expected that the first term in the Hamiltonian is suitable for describing vibrational features, while the second term decribes rotational characteristics. The proton–proton and neutron–neutron interactions usually have little effect on the structure of the spectra, thus they are usually omitted. The Majorana term, as we have already seen, controls the separation of the F-spin symmetric from the mixed symmetry states.

5.7 Electromagnetic transition operators

Electromagnetic transition operators in IBM-2 have a form more general than IBM-1, given by (Arima and Iachello 1984)

$$T^{(l)} = T_p^{(l)} + T_n^{(l)}, \qquad (5.36)$$

where $T_p^{(l)}$ and $T_n^{(l)}$ are the already known IBM-1 operators with the proton or neutron label attached to them. The most commonly used transition operator is the quadrupole one, which can be put in the form

$$T^{(E2)} = e_p^{(2)} Q_p + e_n^{(2)} Q_n, \qquad (5.37)$$

where the symbols $e_p^{(2)}$ and $e_n^{(2)}$ are the proton and neutron boson effective E2 charges, and Q_p, Q_n are the proton and neutron quadrupole operators given in the previous section. Although in principle the quadrupole operators appearing in the transition operator need not be the same with the quadrupole operators appearing in the Hamiltonian, they are usually taken as such in numerical calculations. The proton (neutron) boson effective charge is assumed to depend only on the number of protons (neutrons). In calculations they are usually kept constant. In calculations of B(E2;$0^+ \rightarrow 2^+$) values for various Xe, Ba, and Ce isotopes, for example, they both have been kept constant and equal to 0.12 eb.

Magnetic dipole transitions are also specially interesting. The relevant transition operator is

$$T^{(M1)} = g_p L_p + g_n L_n, \qquad (5.38)$$

where L_p, L_n are are the proton and neutron angular momentum operators

$$L_p = \sqrt{10}(d_p^+ \tilde{d}_p)^1, \qquad (5.39)$$

$$L_n = \sqrt{10}(d_n^+ \tilde{d}_n)^1, \tag{5.40}$$

and g_p (g_n) are the proton (neutron) g factors, which are assumed to depend only on the proton (neutron) boson number, respectively. Notice that while in IBM-1 all M1 transitions were forbidden if only lowest order terms were included in the transition operator, here this is no longer the case if $g_p \neq g_n$. Thus the need to include higher order terms in the M1 transition operator in IBM-1 in order to allow for M1 transitions to occur can be seen as a way to simulate the proton–neutron degree of freedom in the IBM-1 framework.

5.8 The quartet model

In this section we will discuss a model which uses exactly the same formalism as IBM-2, but from a different physical point of view.

The presence of quartets (a pair of protons and a pair of neutrons coupled together) has been discussed several times in nuclear physics. In particular, the systematics of the binding energies of heavy nuclei have been discussed in such a framework. In the quartet model (Daley *et al.* 1986), though, one does not deal with binding energies but with the systematics of the low-lying levels of the ground state band in series of even nuclei of the type

$$(Z, N), (Z + 2, N + 2), (Z + 4, N + 4), (Z + 6, N + 6), \ldots. \tag{5.41}$$

Thus the nuclei in each series differ from their closest neighbors by one α-particle.

It has been argued that for deformed nuclei the proton–neutron correlations are strong, so that the formation of quartets is favored. Thus, after counting the number of proton bosons (proton valence pairs) and the number of neutron bosons (neutron valence pairs), one couples together as many proton pairs and neutron pairs as possible, forming quartets. The bosons participating in the formation of quartets are called **quartetted** bosons, and denoted by the subscript q. The "left-over" proton bosons or neutron bosons (if $N_\pi > N_\nu$ the left-over bosons will be proton bosons, if $N_\pi < N_\nu$ the left-over boson are neutron bosons) are called **unquartetted** bosons and they are denoted by the subscript u.

Example 1 Consider the nucleus $^{154}_{62}\text{Sm}_{92}$. This nucleus has 6 proton bosons ($N_\pi = 6$) and 5 neutron bosons ($N_\nu = 5$). One can then couple together the 5 neutron bosons with 5 of the proton bosons to form quartets. Thus one has 5+5=10 quartetted bosons ($N_q = 10$). The left-over one proton boson is counted as one unquartetted boson ($N_u = 1$). Notice that the N_q, N_u numbers (10, 1) are very different from the N_π, N_ν numbers (6, 5).

Example 2 Consider the nucleus $^{164}_{70}$Yb$_{94}$. For this nucleus $N_\pi = 6$ and $N_\nu = 6$. Thus $N_q = 12$ and $N_u = 0$. No unquartetted bosons are obtained in this case.

Example 3 Consider the nucleus $^{174}_{72}$Hf$_{102}$. This nucleus has $N_\pi = 5$ and $N_\nu = 10$. By coupling the 5 proton bosons to five of the neutron bosons one can form 10 quartetted bosons ($N_q = 10$). The left-over 5 neutron bosons are counted as 5 unquartetted bosons ($N_u = 5$). Notice that in this case the N_q, N_u numbers (10, 5) are the same as the N_ν, N_π numbers.

It has been argued that the systematics of low-lying members of the ground state bands of deformed even nuclei in the rare earth region are described better if one uses in the IBM-2 Hamiltonian the number of quartetted bosons and the number of unquartetted bosons instead of the number of proton bosons and the number of neutron bosons. The Hamiltonian used was

$$H = \alpha(Q \odot Q) + \beta(Q_u \odot Q_u) + \gamma(Q_q \odot Q_u), \qquad (5.42)$$

where $Q = Q_q + Q_u$. By replacing the subscripts q and u by π and ν, one obtains the corresponding IBM-2 Hamiltonian. The operators Q, Q_π, Q_ν are the well known quadrupole operators, defined as the generators of the appropriate SU(3) algebra as usually. Since this Hamiltonian emphasizes the role of the quadrupole–quadrupole interaction, it is expected to work well only in the rotational (SU(3)) region, which turns out to be correct. A more symmetric form of the Hamiltonian can be trivially obtained through use of the $Q = Q_q + Q_u$ relation

$$H = \alpha(Q_q \odot Q_q) + (\alpha + \beta)(Q_u \odot Q_u) + (2\alpha + \gamma)(Q_q \odot Q_u). \quad (5.43)$$

This Hamiltonian has been diagonalized numerically. The NPBOS computer program is used, with N_q in the place of N_π and N_u in the place of N_ν.

So far the quartet model has only been checked versus the experimental data for low-lying members of the ground state band. Its predictive power for higher bands is currently under investigation. Of course, the nuclei mostly interesting in attempting a comparison between the quartet model and IBM-2 are the nuclei where the N_q, N_u numbers are very different from the N_π, N_ν numbers, as in the case of the nuclei ^{154}Sm and ^{164}Yb mentioned above. Nuclei like ^{174}Hf, where these pairs of numbers are the same, are identical from the numerical point of view and thus can offer no help in checking the differences between the two models.

It must be mentioned here that a different quartet model had been introduced before the present one. In the early quartet model, quartet bosons were considered, corresponding to two valence protons plus two valence neutrons. Thus in that model the nucleus $^{164}_{70}\mathrm{Yb}_{94}$, which has boson numbers $N_\pi = 6$ and $N_\nu = 6$, was considered as a system of 6 quartet bosons. The problem of that model was the small total number of bosons. In the present quartet model, this problem is avoided, since the total number of bosons in the model equals the total number of bosons in IBM-2.

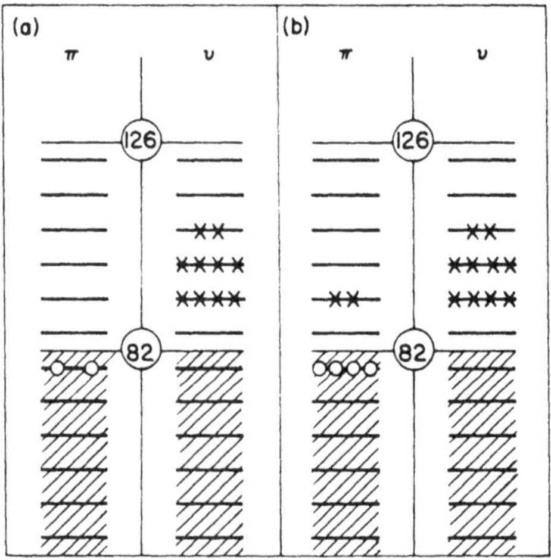

Fig. 5.6 (a) The single-particle proton and neutron configurations for an even–even Hg isotope. Particles are denoted as x's, holes as circles. (b) The single-particle proton and neutron configurations for the same even–even Hg isotope when a proton pair has been excited across the 82 shell gap. (Taken from Duval and Barrett (1981a)).

5.9 Algebraic description of intruder states

Besides the "usual" collective states described in Chapter 0, one encounters in some nuclei additional low-lying states, which cannot be described in the framework of IBM-1 or IBM-2. These are called

intruder states and can be divided into different classes, according to their shell-model structure (van Isacker *et al.* 1982b). In this section we will first discuss briefly the various classes of intruder states, and we will subsequently see how IBM can be generalized in order to be able to describe them.

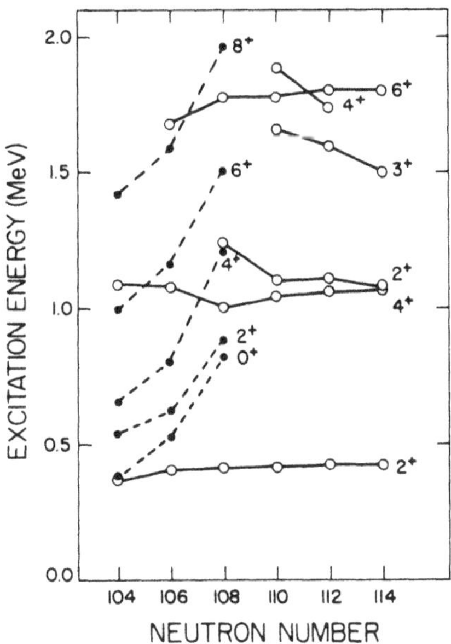

Fig. 5.7 Experimental energy levels for a series of Hg isotopes. Vibrational (rotational) states are connected by solid (dashed) lines. (Taken from Duval and Barrett (1981a)).

5.9.1 First class of intruder states

The first class of intruder states arises in even–even nuclei which have proton (neutron) number nearly equal to a magic proton (neutron) number, i.e. their protons (neutrons) are near a closed shell configuration. It turns out that in these nuclei it is quite easy for a pair of protons (neutrons) below the closed shell to be excited into the next major shell, leaving a pair of proton (neutron) holes in the major shell it was previously occupying. An example is shown in Fig. 5.6, where the single-particle proton and neutron configurations of an even–even $_{80}$Hg isotope are shown. Mercury isotopes have two proton holes in the 50–82 proton major shell. It is then easy for a pair of protons to be excited into the 82–126 major shell, so that in the new configuration one has four proton holes in the 50–82 proton major shell and two proton particles in the 82–126 proton major shell.

Excitations of this kind (referred to as two-particle two-hole (2p-2h) excitations) are thought to be responsible for the appearance of intruder 0^+ and 2^+ states in $_{48}$Cd isotopes, as well as for the appearance of intruder $K^\pi = 0^+$ rotational bands (with a bandhead around 2 MeV) in $_{50}$Sn isotopes. In the case of the mercury isotopes considered here, 2p-2h excitations give rise to $K^\pi = 0^+$ rotational bands, as it can be seen in Fig. 5.7, where the experimental energy levels of a series of Hg isotopes are shown. Two kinds of states are seen in the figure:

i) Collective vibrational states, connected by solid lines. (Since Hg isotopes have only two valence proton holes, they are expected to be vibrational in character). These remain almost constant when moving from one isotope to another. Of the expected two-phonon triplet $(0^+, 2^+, 4^+)$, only the 2^+ and 4^+ members are seen. Of the expected three-phonon quintet $(0^+, 2^+, 3^+, 4^+, 6^+)$, only the 3^+, 4^+, 6^+ members are seen.

ii) Intruder rotational states, connected by dashed lines. They come lower in energy as the number of valence neutron holes in the isotope increases. These bands cannot be described in the framework of IBM-1 or IBM-2 according to what we have already seen.

5.9.2 Second class of intruder states

As we have already seen, the first class of intruder states arises in nuclei near a major *shell* closure. A second class of intruder states is thought to arise in nuclei near *subshell* closures, like the one at Z=64. The intruder $K^\pi = 0^+$ bands observed in $_{64}$Gd isotopes are thought to arise because of the Z=64 subshell closure. As an example, the spectrum of ^{156}Gd is shown in Fig. 5.8 (solid lines). These bands cannot be described in the framework of IBM-1 or IBM-2 according to what we have already discussed.

5.9.3 Third class of intruder states

In Fig. 5.8 we observe in the spectrum of ^{156}Gd a low-lying $K^\pi = 4^+$ intruder band. Similar bands are observed in other Gd isotopes as well. These intruder bands are thought to arise from the g-boson degree of freedom, already studied in Chapter 4. The same is true for $K^\pi = 3^+$ intruder bands, observed for example in ^{168}Er and ^{176}Hf, as we have already seen in Chapter 4 (see Fig. 4.2).

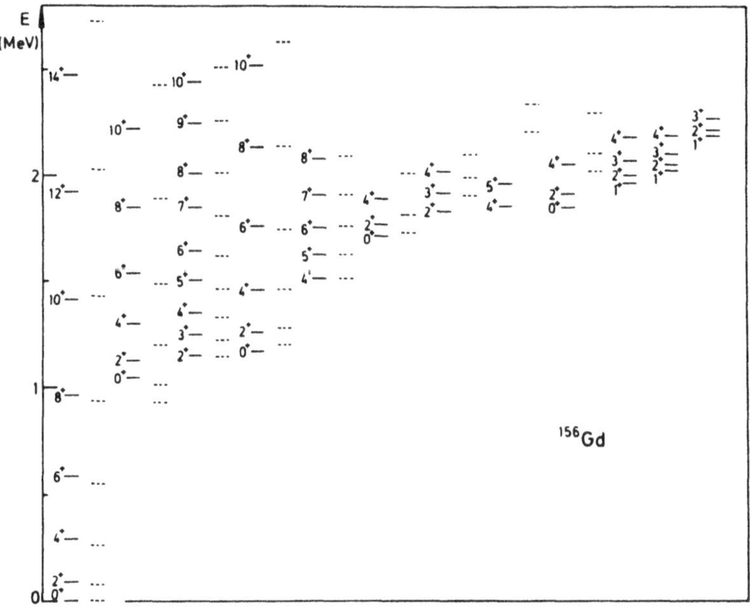

Fig. 5.8 Energy spectrum of ^{156}Gd. Experimental levels are indicated by solid lines, while dashed lines represent the results of the calculations described in the text. (Taken from van Isacker *et al.* (1982b)).

5.9.4 Configuration mixing in IBM-2

A description of the first class of intruder states can be given in IBM-2, if one considers **configuration mixing** (Duval and Barrett 1981a). To clarify the method, we consider the above mentioned example of the Hg isotopes, $^{182}_{80}$Hg$_{102}$ in particular. In the usual IBM-2 framework this nucleus is described as a system of one proton boson (of hole character) and 10 neutron bosons (of particle character). Thus the basic configuration is ($N_\pi = 1, N_\nu = 10$). Since this configuration contains only one valence proton boson, it gives rise to a vibrational spectrum. If a 2p-2h proton excitation occurs, the final nucleus has 10 neutron bosons of particle character, two proton bosons of hole character (in the 50-82 major shell) and one proton boson of particle character (in the major shell 82-126). In lowest approximation it turns out that it is legitimate to ignore the difference between particle bosons and hole bosons and consider the new configuration as ($N_\pi = 3, N_\nu = 10$). Since this configuration contains 3 proton bosons, it can give rise to rotational spectra, which are desirable in order to get a description of the rotational intruder states.

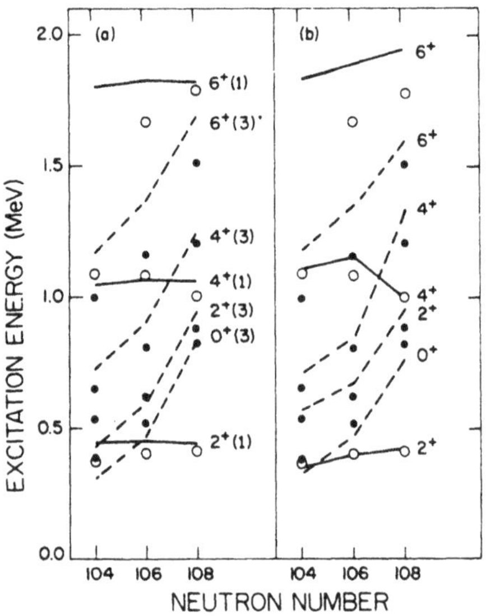

Fig. 5.9 Comparison between theoretical and experimental energy levels of some Hg isotopes (a) for no configuration mixing and (b) for configuration mixing. Experimental vibrational (rotational) levels are indicated by open (solid) circles. Solid lines connect the theoretical predictions for states in the ($N_\pi = 1$) configuration (labelled as (1)), while dashed lines connect the theoretical predictions for states in the ($N_\pi = 3$) configuration (labelled as (3)). (Taken from Duval and Barrett (1981a)).

For practical applications in the Hg isotopes one can use the IBM-2 Hamiltonian of Sec. 5.6. One can make a separate calculation for the collective vibrational states, using for each nucleus the configuration involving $N_\pi = 1$, and a separate calculation for the intruder rotational states, using the configuration involving $N_\pi = 3$. As seen in Fig. 5.9 (left part), agreement with experiment is quite good. The results can be improved if one allows for mixing of the two configurations to take place. In the specific case considered here, the most general one-body mixing boson Hamiltonian will be

$$H_{mix} = \alpha(s_\pi^+ s_\pi^+ + s_\pi s_\pi)^0 + \beta(d_\pi^+ d_\pi^+ + \tilde{d}_\pi \tilde{d}_\pi)^0. \qquad (5.44)$$

This mixing Hamiltonian does not conserve the proton boson number, as expected. In practical applications, one first makes two separate IBM-2 calculations, each using one of the two different configurations, and then he mixes the results using the above given mixing Hamiltonian. Results of a mixing calculation of this kind for the Hg isotopes are shown in Fig. 5.9 (right part). Considerable improvement over the results of the calculations without any mixing (left part of the same figure) is observed.

5.9.5 Addition of s' and d' bosons

We have seen that the first class of intruder states can be explained by considering configuration mixing in IBM-2. However, a different approach is necessary in order to explain the second class of intruder states, since these states occur in nuclei far away from closed shells, in regions where 2p-2h excitations are highly unlikely to occur. One possible approach is to consider that subshell closures (like the one at $Z=64$) give rise to s' and d' bosons (van Isacker *et al.* 1982b), which are correlated $J=0$ and $J=2$ pairs with structure different from the usual s and d bosons and lie higher in energy. For simplicity, no distinction between protons and neutrons will be made here. As we know, the usual IBM-1 basis contains states of the type $|(sd)^N >$. In addition to them, here we will also consider states of the form $|(s'd')(sd)^{N-1} >$, i.e. states in which one "primed" boson is present. States with more "primed" bosons cannot be excluded *a priori*, but here we will neglect them in a lowest order approximation, since these states are expected to lie higher in energy. In constructing the Hamiltonian we should bear in mind that it has to conserve the total number of bosons, since no excitations across major shells are considered here. The total Hamiltonian will be

$$H = H_{sd} + H_{s'd'} + H_{mix}, \qquad (5.45)$$

where H_{sd} is the usual IBM-1 Hamiltonian, $H_{s'd'}$ is the Hamiltonian involving the primed bosons, and H_{mix} is the Hamiltonian mixing primed and unprimed bosons.

Confining ourselves to lowest order, we assume that only one s' boson can be present. Then

$$H_{s'd'} = \epsilon_{s'}(s')^+ s', \qquad (5.46)$$

where $\epsilon_{s'}$ is the unperturbed energy of the s' boson. For the mixing Hamiltonian in the same lowest approximation we have

$$H_{mix} = m_1(s^+s^+ss' + (s')^+s^+ss) + m_2[(d^+d^+)^0ss' + (s')^+s^+(\tilde{d}\tilde{d})^0]$$

$$+m_3[(d^+d^+)^2(\tilde{d}s')^2 + ((s')^+d^+)^2(\tilde{d}\tilde{d})^2]^0. \qquad (5.47)$$

The results of applying such a Hamiltonian (involving only one s' boson) to ^{156}Gd are shown as dashed lines in Fig. 5.8. Comparing the theoretical predictions to the experimental data we see that the intruder $K^\pi = 0^+$ bands are well accounted for. Remember that these intruder bands are supposed to arise from effects occurring in one and the same major shell, in contrast to the intruder states of class one, which arise from excitations across major shells. The fact that the intruder $K^\pi = 4^+$ band is also predicted at the right energy is due to the fact that an admixture of a g boson has been used in the calculation.

5.10 High spin states

In this section the high spin effect of backbending will be discussed and two methods for its algebraic description will be given.

5.10.1 Backbending

The study of high spin states is a very active field of nuclear physics (de Voigt, Dudek, and Szymanski 1983). Before proceeding to the discussion of high spin effects in the algebraic framework, we need some terminology. **Yrast band** is the band composed by considering for each spin the level lying lowest in energy, **Yrare band** is the band composed by considering for each spin the second lowest lying level. At low spin the Yrast band coincides with the ground state band. At higher spins this is not necessarily the case, since other bands, lying above the ground state band at low spin, can become the lowest at higher spin.

One of the most interesting phenomena occuring at high spin is **backbending** (Lieder and Ryde 1978), illustrated in Fig. 5.10. If one makes a plot of the moment of inertia of a nucleus in its Yrast band versus the square of its angular velocity, in several cases one gets a smoothly rising curve, as in the case of ^{174}Yb in Fig. 5.10. But in some nuclei, like ^{164}Yb and ^{166}Yb in Fig. 5.10, at some value of the angular velocity the curve shows a very rapid increase and it even "bends back". This backbending effect has been observed in several nuclei, and is thought to be due to the crossing of the ground state band with another band, called the **superband**, which is thought to be a two-quasiparticle band. The two-quasiparticle band is thought to arise from the breaking at high spin and the subsequent alignment of a neutron or proton pair. The breaking of the pair is thought to be due to the **Coriolis Antipairing** effect (CAP), while the subsequent alignment is thought to be the consequence of the **Rotation Alignment** effect (RAL). Both of these effects increase with increasing spin, thus in a given major shell one expects the first pair to be broken as spin rises to be a pair belonging to the unique parity orbital, since this is the orbital with the highest spin. (The reader who is not familiar with unique parity orbitals can find more information in Sec. 15.5.2). In the 50–82 major shell the unique parity orbital is the $1h_{11/2}$ orbital, while in the 82–126 major shell it is the $1i_{13/2}$ one. Notice that backbending is a property of the Yrast band, **not** of the ground state band !

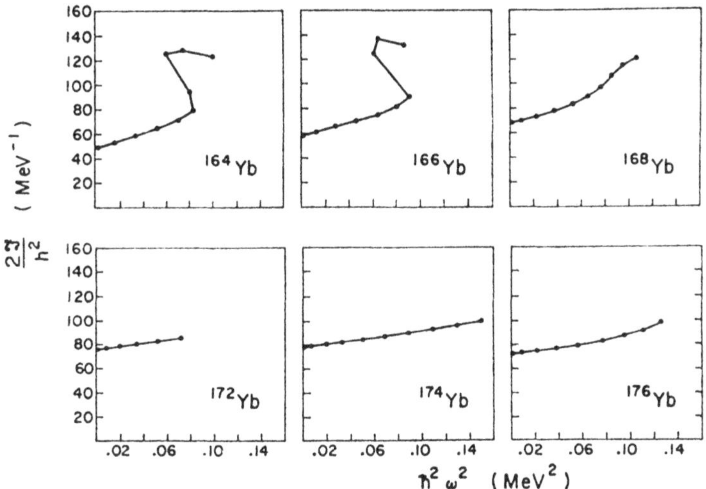

Fig. 5.10 Standard backbending plots for the Yrast bands of various Yb isotopes. (Taken from Ward *et al.* (1976)).

Backbending has also been observed in γ and β bands. Also notice that in order to check if a nucleus shows backbending one has to plot the moment of inertia versus the square of the angular velocity. By simply looking at plots of energy versus I(I+1) one can hardly notice any effect.

5.10.2 IBM-2 plus two quasiparticles

In order to be able to describe the backbending effect in the IBM framework, one has to take into account the two-quasiparticle band. In the most popular approach (Yoshida, Arima, and Otsuka 1982), one describes the low-lying states by IBM-2, and assumes that the two-quasiparticle band can be obtained by changing one of the bosons into a pair of nucleons in the relevant unique parity orbital with spin j. In order to avoid double counting, it is assumed that the angular momentum of the nucleon pair takes the values $4, 6, \ldots, 2j - 1$, i.e. the pairs of spin 0 and 2 are excluded, since they are assumed to be already included in the s and d bosons. The Hamiltonian takes the form

$$H = H^B + H^F + H^{BF}, \tag{5.48}$$

where H^B is the usual IBM-2 Hamiltonian, H^F is the Hamiltonian describing the two quasiparticles, and H^{BF} indicates the interaction between bosons and quasiparticles (fermions). The nucleon Hamiltonian can be written in the form

$$H^F = \epsilon_\pi n_\pi + \epsilon_\nu n_\nu + V_\pi^F + V_\nu^F, \tag{5.49}$$

where ϵ_π (ϵ_ν) is the single-particle proton (neutron) energy, n_π (n_ν) is the number operator of protons (neutrons), and V_π^F (V_ν^F) is the two-body interaction between protons (neutrons), which is usually assumed to be the delta interaction. For the boson–fermion interaction, one can assume that it takes the quadrupole–quadrupole form

$$V^{BF} = \kappa(Q_\pi \cdot Q_\nu) - \kappa(Q_\pi^B \cdot Q_\nu^B), \tag{5.50}$$

where

$$Q_\sigma = Q_\sigma^B + \alpha_\sigma(a_\sigma^+ \tilde{a}_\sigma)^2 + \beta_\sigma[(a_\sigma^+ a_\sigma^+)^4 \cdot \tilde{d}_\sigma]^2 - \beta_\sigma[d_\sigma^+ \cdot (\tilde{a}_\sigma \tilde{a}_\sigma)^4]^2,$$

$$\sigma = \pi, \nu. \tag{5.51}$$

In the above a_{jm}^+ are nucleon creation operators in the unique parity level j, while Q_σ^B are the usual boson quadrupole operators. Notice that the quadrupole–quadrupole interaction between bosons is subtracted from V^{BF}, since it is already included in H^B.

To apply the above formalism, one has in principle to consider three different kinds of states: i) the usual IBM-2 space with N_π proton bosons and N_ν neutron bosons, ii) the space with two proton quasiparticles, $N_\pi - 1$ proton bosons and N_ν neutron bosons, and

iii) the space with two neutron quasiparticles, N_π proton bosons and $N_\nu - 1$ neutron bosons. In most applications only i) and ii) or i) and iii) need to be considered, since it is quite clear if the first pair to be broken as spin rises will be a proton or a neutron pair.

As an **example** of application of the method, consider the nucleus $^{130}_{58}\text{Ce}_{72}$, which has 4 proton bosons (of particle nature) and 5 neutron bosons (of hole nature), both in the 50–82 major shell. The relevant unique parity orbital is $1h_{11/2}$. Having a look at the Nilsson diagrams (easily accessible at Lederer and Shirley (1978)) for this major shell, one sees that it is quite unlikely for valence protons, located at the beginning of the major shell, to reach the unique parity orbital, while valence neutron holes can easily get into it. It is thus expected that in this nucleus the two-quasiparticle pair will be a neutron pair, not a proton pair, so that one has to consider only the (N_π, N_ν) and $(N_\pi, N_\nu - 1)$ boson systems. In the actual calculation one first diagonalizes H^B for each of these boson systems separately, and for each of them he stores the lowest few eigenvectors of each spin. Then, matrix elements of the total Hamiltonian are calculated, using the stored vectors. Results of such calculations for various Ba and Ce isotopes are shown in Fig. 5.11. Good agreement with experiment is observed.

A method similar to the one described in this subsection, involving the coupling of two quasiparticles to an IBM-1 core, has been developed by Faessler *et al.* (Morrison *et al.* 1981, Kuyucak *et al.* 1984, Faessler *et al.* 1985, 1986).

5.10.3 Core excitations

An alternative way (Heyde, Jolie, van Isacker, Moreau, and Waroquier 1984) of describing high spin phenomena in the algebraic framework will be discused here. This method is a generalization of the configuration mixing method described in the previous section. One assumes that for deformed nuclei the concept of a closed shell is not rigidly valid, and as a consequence one must allow for core excitations, which in even–even nuclei can be of 2p-2h, 4p-4h, ..., type. If N_0 is the usual number of bosons, in addition to the usual IBM-1 states $|(sd)^{N_0}>$ one has to consider 2p-2h excited states $|(sd)^{N_0+2}>$, 4p-4h excited states $|(sd)^{N_0+4}>$, ... In practical calculations one has to truncate the space and stop at a cut-off boson number N_{max}. The Hamiltonian will be

$$H = H^B + \Delta(N - N_0), \tag{5.52}$$

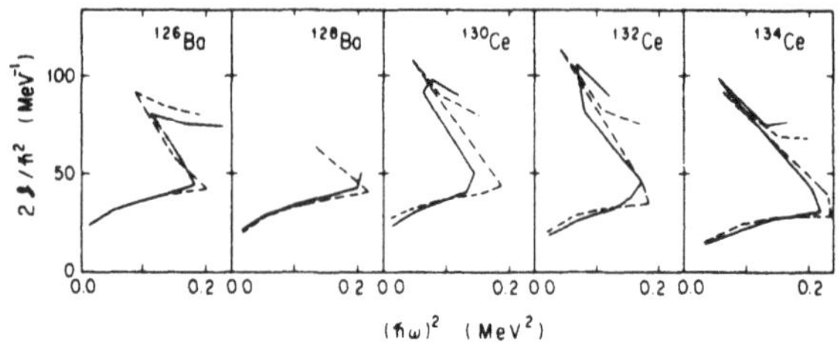

Fig. 5.11 Standard backbending plots for the Yrast bands of some Ba and Ce isotopes. Solid curves connect experimental points, while broken curves represent the results of the calculation described in the text. (Taken from Yoshida *et al.* (1982)).

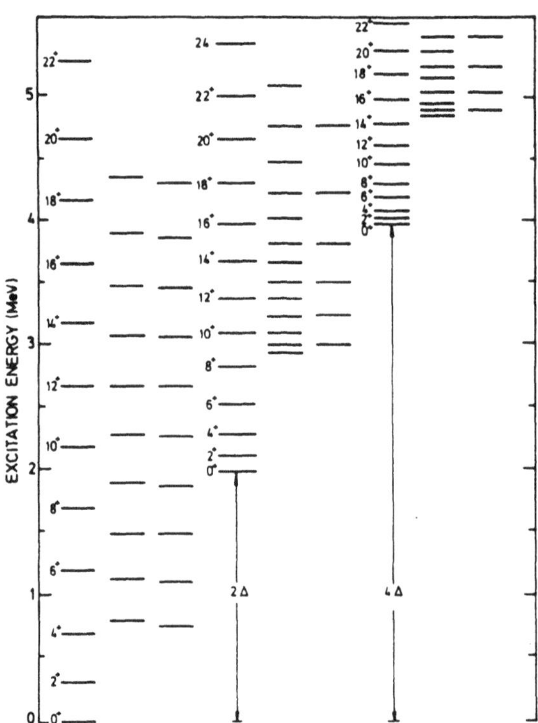

Fig. 5.12 Typical spectrum given by the Hamiltonian (5.52), which allows for core excitations to take place. (Taken from Heyde, Jolie, van Isacker, Moreau, and Waroquier (1984)).

where H^B is the usual IBM-1 Hamiltonian and N the total number
of bosons. The physical meaning of the parameter Δ is clear: 2Δ
is the energy required to create a 2p-2h excitation. In addition, one
must consider the mixing between adjacent configurations, i.e. the
mixing of the ground state with the 2p-2h excitations, the mixing of
the 2p-2h excitations with the 4p-4h excitations, etc. The simplest
one-body boson Hamiltonian will be

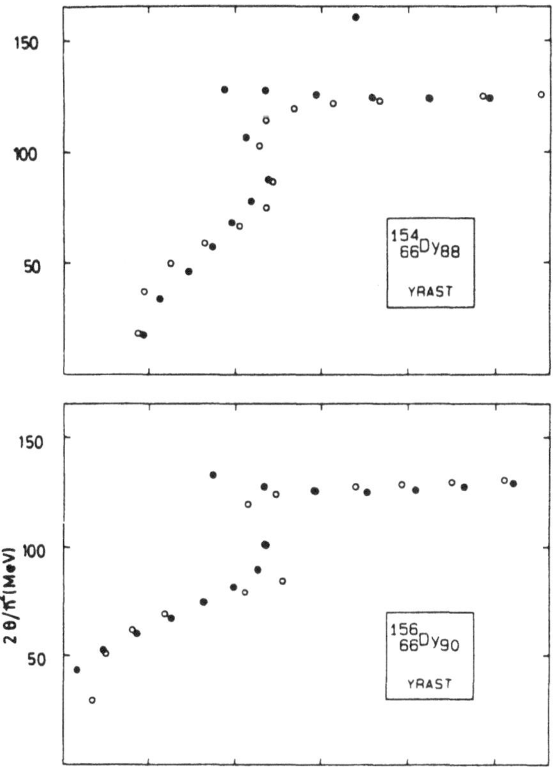

Fig. 5.13 Standard backbending curves for the Yrast bands of ^{154}Dy
and ^{156}Dy. Experimental points are denoted by filled circles, while the
theoretical predictions are shown as open circles. (Taken from Heyde,
Jolie, van Isacker, Moreau, and Waroquier (1984)).

$$H_{mix} = \alpha(s^+ s^+ + ss)^0 + \beta(d^+ d^+ + \tilde{d}\tilde{d})^0, \qquad (5.53)$$

as in eq. (5.44). Depending on the values of α, β, and Δ, backbending can occur.

In Fig. 5.12 typical results of the present method are shown. We remark that the lowest 2p-2h excited state is shifted above the ground state by 2Δ, while the lowest 4p-4h excited state is shifted above the ground state by 4Δ. Because of the mixing Hamiltonian, the Yrast band becomes a mixture of N_0, $N_0 + 2$, $N_0 + 4$, ..., components. In Fig. 5.12 notice that the levels of the lowest K=0 band of the $N_0 + 2$ configuration are consistently higher than the corresponding levels of the ground state band (the lowest K=0 band of the N_0 configuration) up to spin 18, the spin 20 levels are lying at almost the same energy, while above spin 20 the $N_0 + 2$ levels are lying below the corresponding N_0 levels. Thus the N_0 band is Yrast up to spin 20, while the $N_0 + 2$ band is becoming Yrast above this spin. In Fig. 5.13 results of numerical calculations of backbending curves using the present core-excitation method are shown for two Dy isotopes. In these calculations N_{max} was taken to be $N_0 + 6$, while Δ turned out to be around 1 MeV. We remark that the pronounced backbending in the Yrast band of ^{156}Dy is well reproduced. The upbending in the Yrast band of ^{158}Dy is also described satisfactorily.

References

Much work has been done on IBM-2, so that any attempt to go into details in describing it will be hopeless. In the extensive list of references given here one can find a variety of theoretical considerations as well as many experimental tests of the theory. The present discussion of the F-spin symmetric limits of IBM-2 has been based on van Isacker, Heyde, Jolie and Sevrin (1986), while the discussion of the F-spin non-symmetric limits has been based on Iachello (1982). Relevant review articles have been written by Elliott (1985), Dieperink and Wenes (1985), Arima and Iachello (1984), and Barrett (1981). Extensive accounts of IBM-2 can be found also in Dieperink (1986b, 1985, 1984a) and Iachello (1979). The properties of the collective 1^+ mode have been recently reviewed by Faessler and Nojarov (1987). Compilations of early experimental tests of IBM-2 can be found in Casten (1980a, 1980b, 1980c) and Wood (1983b). A recent collection of papers on mixed symmetry states can be found in Dubrovnik (1986).

Alonso, C E, Arias, J M, and Lozano, M, 1986. *Phys. Lett.*, **177B**, 130.
Aprahamian, A, Brenner, D S, Casten, R F, and Heyde, K, 1984. *Gull Lake 1984*, 78.
Aprahamian, A, Brenner, D S, Casten, R F, Gill, R L,

Piotrowski, A, and Heyde, K, 1984. *Phys. Lett.*, **140B**, 22.

Arima, A, 1984a. *Osaka 1984*, 550.

Arima, A, 1984b. *Nucl. Phys.*, **A421**, 63c.

Arima, A, 1984c. *Nucl. Phys.* , **A421**, 419c.

Arima, A, 1986. *Beijing 1986*, 3.

Arima, A, and Iachello, F, 1984. In *Advances in Nuclear Physics* (ed. J W Negele and E Vogt) Vol 13, p 139. Plenum, New York.

Arima, A, Otsuka, T, Iachello, F, and Talmi, I, 1977. *Phys. Lett.*, **66B**, 205.

Balantekin, A B, and Barrett, B R, 1987. *Phys. Rev.*, **C35**, 1878.

Barclay, M E, *et al.*, 1986. *J. Phys.*, **G12**, L295.

Barfield, A F, and Barrett, B R, 1984. *Phys. Lett.*, **149B**, 277.

Barfield, A F, Barrett, B R, Sage, K A, and Duval, P D, 1983. *Z. Phys.*, **A311**, 205.

Barrett, B R, 1981. *Rev. Mex. Fis.*, **27**, 533.

Barrett, B R, and Duval, P D, 1980. *Erice 1980*, 47.

Barrett, B R, and Halse, P, 1984. *Drexel 1984*, 621.

Barrett, B R, and Halse, P, 1985. *Phys. Lett.*, **155B**, 133.

Barrett, B R, and Sage, K A, 1980. *Erice 1980*, 123.

Berg, U E P, *et al.*, 1984. *Phys. Lett.*, **149B**, 59.

Bijker, R, Dieperink, A E L, Scholten, O, and Spanhoff, R, 1980. *Nucl. Phys.*, **A344**, 207.

Bohle, D, 1986. *Dubrovnik 1986*, **1**, 350.

Bohle, D, Küchler, G, Richter, A, and Steffen, W, 1984. *Phys. Lett.*, **148B**, 260.

Bohle, D, Richter, A, Heyde, K, van Isacker, P, Moreau, J, and Sevrin, A, 1985. *Phys. Rev. Lett.*, **55**, 1661.

Bohle, D, Richter, A, Steffen, W, Dieperink, A E L, Lo Iudice, N, Palumbo, F, and Scholten, O, 1984. *Phys. Lett.*, **137B**, 27.

Casten, R F, 1980a. *Erice 1980*, 3.

Casten, R F, 1980b. *Drexel 1980*, 369.

Casten, R F, 1980c. *Nucl. Phys.* , **A347**, 173.

Chiang, H C, Hsieh, S T, King Yen, M M, and Han, C S, 1985. *Nucl. Phys.*, **A435**, 54.

Colvin, G G, Hoyler, F, and Robinson, S J, 1987. *J. Phys.*, **G13**, 191.

Daley, H, Nagarajan, M A, Rowley, N, Morrison, D, and May, A D, 1986. *Phys. Rev. Lett.* , **57**, 198.

de Voigt, M J A, Dudek, J, and Szymanski, Z, 1983. *Rev. Mod. Phys.*, **55**, 949.

Dewald, A, Kaup, U, Gast, W, Gelberg, A, Schuh, H W, Zell, K O, and von Brentano, P, 1982. *Phys. Rev.* , **C25**, 226.

Dieperink, A E L, 1984a. *Trieste 1984*, **1**, 95.
Dieperink, A E L, 1984b. *Drexel 1984*, 271.
Dieperink, A E L, 1984c. *Nucl. Phys.*, **A421**, 189c.
Dieperink, A E L, 1985. *La Rábida 1985*, 205.
Dieperink, A E L, 1986a. *Dubrovnik 1986*, **1**, 322.
Dieperink, A E L, 1986b. *Harrogate 1986*, **2**, 139.
Dieperink, A E L, and Bijker, R, 1982. *Phys. Lett.* , **116B**, 77.
Dieperink, A E L, and Talmi, I, 1983. *Phys. Lett.*, **131B**, 1.
Dieperink, A E L, and Wenes, G, 1985. *Ann. Rev. Nucl. Part. Sci.*, **35**, 77.
Dieperink, A E L, Scholten, O, and Warner, D D, 1987. *Nucl. Phys.*, **A469**, 173.
Dobes, J, 1987. *Nucl. Phys.*, **A469**, 424.
Druce, C H, McCullen, J D, Duval, P D, and Barrett, B R, 1982. *J. Phys.*, **G8**, 1565.
Dukelsky, J, Federman, P, Perazzo, R P J, and Sofia, H M, 1982. *Phys. Lett.*, **115B**, 359.
Duval, P D, and Barrett, B R, 1981a. *Phys. Lett.*, **100B**, 223.
Duval, P D, and Barrett, B R, 1981b. *Phys. Rev.* , **C23**, 492.
Duval, P D, and Barrett, B R, 1982. *Nucl. Phys.*, **A376**, 213.
Duval, P D, Goutte, D, and Vergnes, M, 1983. *Phys. Lett.*, **124B**, 297.
Duval, P D, Pittel, S, Barrett, B R, and Druce, C H, 1983. *Phys. Lett.*, **129B**, 289.
Eid, S A A, Hamilton, W D, and Elliott, J P, 1986. *Phys. Lett.*, **166B**, 267.
Elliott, J P, 1985. *Rep. Prog. Phys.*, **48**, 171.
Elliott, J P, Evans, J A, and Williams, A P, 1987. *Nucl. Phys.*, **A469**, 51.
Faessler, A, 1982. *Rep. Prog. Phys.*, **45**, 653.
Faessler, A, and Nojarov, R, 1987. In *Progress in Particle and Nuclear Physics* (ed. A. Faessler) Vol 19, p 167. Pergamon, Oxford.
Faessler, A, Kuyucak, S, and Wakai, M, 1986. *Nucl. Phys.*, **A458**, 381.
Faessler, A, Kuyucak, S, Petrovici, A, and Petersen, L, 1985. *Nucl. Phys.*, **A438**, 78.
Fettweis, P, and Dehaes, J C, 1983. *Z. Phys.*, **A314**, 159.
Fields, C A, de Boer, F W N, Sugarbaker, E, and Walker, P M, 1981. *Nucl. Phys.*, **A363**, 352.
Fortune, H T, 1985. *J. Phys.* , **G11**, 1305.
Fortune, H T, 1987. *Phys. Rev.* , **C35**, 2318.
Frank, A, 1986. *Phys. Rev.* , **C34**, 351.

Frank, A, and van Isacker, P, 1985. *Phys. Rev.*, **C32**, 1770.

Garrett, C, Rastikerdar, S, Gelletly, W, and Warner, D D, 1982. *Phys. Lett.*, **118B**, 292.

Gelberg, A, and Kaup, U, 1978. *Erice 1978*, 59.

Gelberg, A, and Zemel, A, 1980a. *Erice 1980*, 129.

Gelberg, A, and Zemel, A, 1980b. *Strasbourg 1980*, 45.

Gelberg, A, and Zemel, A, 1980c. *Phys. Rev.*, **C22**, 937.

Gelberg, A, von Brentano, P, Dewald, A, Hanewinkel, H, Harter, H, Kaup, U, Reinhardt, R, Schiffer, K, Schmittgen, K P, Xiangfu, S, and Zell, K O, 1984. *Osaka 1984*, 87.

Gelletly, W, van Isacker, P, Warner, D D, Colvin, G, and Schreckenbach, K, 1987. *Phys. Lett.*, **191B**, 240.

Gill, R L, Casten, R F, Warner, D D, Brenner, D S, and Walters, W B, 1982. *Phys. Lett.*, **118B**, 251.

Gilmore, R, and Draayer, J P, 1985. *J. Math. Phys.*, **26**, 3053.

Ginocchio, J N, 1984. *Drexel 1984*, 220.

Ginocchio, J N, and van Isacker, P, 1986. *Phys. Rev.*, **C33**, 365.

Hamilton, W D, 1986. *Dubrovnik 1986*, **1**, 338.

Hamilton, W D, Irbäck, A, and Elliott, J P, 1984. *Phys. Rev. Lett.*, **53**, 2469.

Hanewinkel, H, Gast, W, Kaup, U, Harter, H, Dewald, A, Gelberg, A, Reinhardt, R, von Brentano, P, Zemel, A, Alonso, C E, and Arias, A, 1983. *Phys. Lett.* , **133B**, 9.

Harter, H, 1986. *Dubrovnik 1986*, **1**, 374.

Harter, H, Gelberg, A, and von Brentano, P, 1985. *Phys. Lett.* **157B** 1.

Harter, H, von Brentano, P, and Gelberg, A, 1986. *Phys. Rev.* , **C34**, 1472.

Harter, H, Gelberg, A, Kaup, U, and von Brentano, P, 1984. *Debrecen 1984*, **1**, 341.

Harter, H, von Brentano, P, Gelberg, A, and Casten, R F, 1985. *Phys. Rev.*, **C32**, 631.

Harter, H, von Brentano, P, Gelberg, A, and Otsuka, T, 1987. *Phys. Lett.*, **188B**, 295.

Hartmann, U, Bohle, D, Guhr, T, Hummel, K D, Kilgus, G, Milkau, U, and Richter, A, 1987. *Nucl. Phys.*, **A465**, 25.

Hellmeister, H P, Lieb, K P, and Panqueva, J, 1980. *Erice 1980*, 65.

Hellmeister, H P, Kaup, U, Keinonen, J, Lieb, K P, Rascher, R, Ballini, R, Delaunay, J, and Dumont, H, 1979a. *Phys. Lett.*, **85B**, 34.

Hellmeister, H P, Kaup, U, Keinonen, J, Lieb, K P, Rascher, R, Ballini, R, Delaunay, J, and Dumont, H, 1979b. *Rhodes*

1979, 233.

Heyde, K, 1986. *Dubrovnik 1986*, **1**, 288.

Heyde, K, and Sau, J, 1984. *Phys. Rev.*, **C30**, 1355.

Heyde, K, Moreau, J, and Waroquier, M, 1984. *Phys. Rev.* , **C29**, 1859.

Heyde, K, van Isacker, P, Moreau, J, and Waroquier, M, 1984. *Debrecen 1984*, **1**, 151.

Heyde, K, Jolie, J, van Isacker, P, Moreau, J, and Waroquier, M, 1984. *Phys. Rev.*, **C29**, 1428.

Heyde, K, van Isacker, P, Jolie, J, Moreau, J, and Waroquier, M, 1983. *Phys. Lett.*, **132B**, 15.

Heyde, K, van Isacker, P, Waroquier, M, Wenes, G, and Sambataro, M, 1982. *Phys. Rev.* , **C25**, 3160.

Higo, T, Matsuki, S, and Yanabu, T, 1983. *Nucl. Phys.* , **A393**, 224.

Iachello, F, 1979. *Rhodes 1979*, 161.

Iachello, F, 1982. *Erice 1982*, 5.

Iachello, F, 1983. *Drexel 1983*, 3.

Iachello, F, 1984a. *Drexel 1984*, 279.

Iachello, F, 1984b. *Phys. Rev. Lett.* , **53**, 1427.

Iachello, F, Puddu, G, Scholten, O, Arima, A, and Otsuka, T, 1979. *Phys. Lett.*, **89B**, 1.

Jänecke, J, Becchetti, F D, Overway, D, Cossairt, J D, and Spross, R L, 1981. *Phys. Rev.*, **C23**, 101.

Kaup, U, and Gelberg, A, 1979. *Z. Phys.*, **A293**, 311.

Kaup, U, Mönkemeyer, C, and von Brentano, P, 1983. *Z. Phys.*, **A310**, 129.

Kern, B D, Sistemich, K, Lauppe, W D, and Lawin, H, 1982. *Z. Phys.*, **A306**, 161.

Kleppinger, E W, and Yates, S W, 1983. *Phys. Rev.*, **C27**, 2608.

Kneissl, U, 1986. *Dubrovnik 1986*, **1**, 362.

Kusakari, H, and Sugawara, M, 1984. *Z. Phys.*, **A317**, 287.

Kuyucak, S, Faessler, A, and Wakai, M, 1984. *Nucl. Phys.*, **A420**, 83.

Lederer, C M, and Shirley, V S, 1978. *Tables of Isotopes*, (7th edition). Wiley, New York.

Lieder, R M, and Ryde, H, 1978. In *Advances in Nuclear Physics*, (ed. M Baranger and E Vogt) Vol 10, p 1. Plenum New York.

Lipas, P O, and Helimäki, K, 1985. *Phys. Lett.*, **165B**, 244.

Lo Iudice, N, 1985. *La Rábida 1985*, 190.

Lo Iudice, N, and Palumbo, F, 1985. *Legnaro 1985*, 167.

Luontama, M, Julin, R, Kantele, J, Passoja, A, Trzaska, W, Bäcklin, A, Jonsson, N G, and Westerberg, L, 1986. *Z. Phys.*,

A324, 317.

Mamane, G, Cheifetz, E, Dafni, E, Zemel, A, and Wilhelmy, J B, 1986. *Nucl. Phys.*, **A454**, 213.

Matsuaki, T, and Taketani, H, 1982. *Nucl. Phys.* , **A390**, 413.

Meyer, R A, Henry, E A, Mann, L G, and Heyde, K, 1986. *Phys. Lett.*, **177B**, 271.

Meyer, R A, Henry, E A, Molnár, G, Yates, S W, and Sistemich, K, 1984. *Drexel 1984*, 435.

Meyer, R A, Wild, J F, Eskola, K, Leino, M E, Väisälä, S, Forssten, K, Kaup, U, and Gelberg, A, 1983. *Phys. Rev.* , **C27**, 2217.

Mheemeed, A, *et al.*, 1984. *Nucl. Phys.*, **A412**, 113.

Molnár, G, Diószegi, I, Veres, Á, and Sambataro, M, 1983. *Nucl. Phys.*, **A403**, 342.

Morrison, I, 1981. *Phys. Rev.*, **C23**, 1831.

Morrison, I, and Spear, R H, 1981. *Phys. Rev.*, **C23**, 932.

Morrison, I, Faessler, A, and Lima, C, 1981. *Nucl. Phys.* , **A372**, 13.

Mundy, S J, Gelletly, W, Lukasiak, J, Phillips, W R, and Varley, B J, 1985. *Nucl. Phys.*, **A441**, 534.

Novoselsky, A, 1985. *Phys. Lett.*, **155B**, 299.

Novoselsky, A, and Talmi, I, 1985. *Phys. Lett.*, **160B**, 13.

Novoselsky, A, and Talmi, I, 1986. *Phys. Lett.*, **172B**, 139.

Otsuka, T, and Ginocchio, J N, 1985a. *Phys. Rev. Lett.*, **54**, 777.

Otsuka, T, and Ginocchio, J N, 1985b. *Phys. Rev. Lett.* , **55**, 276.

Park, P, and Elliott, J P, 1986. *Nucl. Phys.* , **A448**, 381.

Pfeiffer, B, and Kratz, K L, 1987. *Z. Phys.*, **A327**, 163.

Pignanelli, M, 1986. *Dubrovnik 1986*, **1**, 368.

Puddu, G, Scholten, O, and Otsuka, T, 1980. *Nucl. Phys.* , **A348**, 109.

Rastikerdar, S, Garrett, C, Foote, G S, and Gelletly, W, 1983. *J. Phys.*, **G9**, 555.

Richter, A, 1983. *Florence 1983*, 189.

Robinson, S J, Hamilton, W D, Hungerford, P, Pfeiffer, B, Jung, G, and Snelling, D M, 1986. *J. Phys.*, **G12**, 903.

Rohoziński, S G, and Greiner, W, 1985. *Z. Phys.*, **A322**, 271.

Sage, K A, and Barrett, B R, 1980a. *Drexel 1980*, 495.

Sage, K A, and Barrett, B R, 1980b. *Phys. Rev.*, **C22**, 1765.

Saha, A, 1984. *Gull Lake 1984*, 217.

Saha, A, Seth, K K, Casey, L, Godman, D, Kielczewska, D, Seth, R, Stuart, J, and Scholten, O, 1983. *Phys. Lett.*, **132B**, 51.

Sala, P, Gelberg, A, and von Brentano, P, 1986. *Z. Phys.*, **A323**, 281.

Sala,P, von Brentano, P, Harter, H, Barrett, B R, and Casten, R F, 1986. *Nucl. Phys.*, **A456**, 269.

Sambataro, M, 1982. *Nucl. Phys.* , **A380**, 365.

Sambataro, M, and Dieperink, A E L, 1981. *Phys. Lett.* , **107B**, 249.

Sambataro, M, and Molnár, G, 1982. *Nucl. Phys.* , **A376**, 201.

Sambataro, M, Scholten, O, Dieperink, A E L, and Piccitto, G, 1984a. *Debrecen 1984*, 749.

Sambataro, M, Scholten, O, Dieperink, A E L, and Piccitto, G, 1984b. *Nucl. Phys.*, **A423**, 333.

Schaaser, H, and Brink, D M, 1986. *Nucl. Phys.* , **A452**, 1.

Schiffer, K, Dewald, A, Gelberg, A, Reinhardt, R, Zell, K O, von Brentano, P, and Sun, X, 1983. *Z. Phys.* , **A313**, 245.

Scholten, O, 1978. *Erice 1978*, 17.

Scholten, O, 1986. *Dubrovnik 1986*, **1**, 315.

Scholten, O, Heyde, K, and van Isacker, P, 1985. *Phys. Rev. Lett.*, **55**, 1866.

Scholten, O, Dieperink, A E L, Heyde, K, and van Isacker, P, 1984. *Phys. Lett.*, **149B**, 279.

Scholten, O, Heyde, K, van Isacker, P, and Otsuka, T, 1985. *Phys. Rev.* , **C32**, 1729.

Scholten, O, Heyde, K, van Isacker, P, Jolie, J, Moreau, J, Waroquier, M, and Sau, J, 1985. *Nucl. Phys.* , **A438**, 41.

Scholten, O, *et al.*, 1986. *Phys. Rev.*, **C34**, 1962.

Scott, S M, Hamilton, W D, Hungerford, P, Warner, D D, Jung, G, Wünsch, K D, and Pfeiffer, B, 1980. *J. Phys.*, **G6**, 1291.

Semmes, P B, Leander, G A, and Wood, L D, 1984. *Gull Lake 1984*, 209.

Semmes, P B, Barfield, A F, Barrett, B R, and Wood, J L, 1987. *Phys. Rev.*, **C35**, 844.

Skarnemark, G, Brodén, K, Kaffrell, N, Prussin, S G, Trautmann, N, Rengan, K, Eriksen, D, Kusnezov, D F, and Meyer, R A, 1986. *Z. Phys.*, **A323**, 407.

Solari, H G, Gilmore, R, and Vallières, M, 1986. *Dubrovnik 1986*, **1**, 280.

Solari, H G, Gilmore, R, and Vallières, M, 1987. *Phys. Rev.* , **C35**, 320.

Sorensen, R A, and Fowler, K, 1986. *Dubrovnik 1986*, **2**, 1036.

Stachel, J, Kaffrell, N, Trautmann, N, Brodén, K, Skarnemark, G, and Eriksen, D, 1984. *Z. Phys.*, **A316**, 105.

Stuchbery, A E, Morrison, I, and Bolotin, H H, 1984. *Phys. Lett.*, **139B**, 159.

Stuchbery, A E, Bolotin, H H, Doran, C E, Morrison, I, Wood, L D, and Yamada, H, 1985. *Z. Phys.*, **A320**, 669.

Stuchbery, A E, Morrison, I, Wood, L D, Bark, R A, Yamada, H, and Bolotin, H H, 1985. *Nucl. Phys.*, **A435**, 635.

Subber, A R H, Park, P, Hamilton, W D, Kumar, K, Schreckenbach, K, and Colvin, G, 1986. *J. Phys.*, **G12**, 881.

Theuerkauf, J, Harter, H, von Brentano, P, and Casten, R F, 1987. *Z. Phys.* , **A326**, 65.

Tokunaga, Y, Seyfarth, H, Schult, O W B, Börner, H G, Hofmeyr C, Barreau, G, Brissot, R, Kaup, U, and Mönkemeyer, C, 1983. *Nucl. Phys.* , **A411**, 209.

van Isacker, P, 1980. *Erice 1980*, 115.

van Isacker, P, 1987. *Oaxtepec 1987*, 321.

van Isacker, P, and Puddu, G, 1980. *Nucl. Phys.* , **A348**, 125.

van Isacker, P, Harter, H, Gelberg, A, and von Brentano, P, 1987. *Phys. Rev.* , **C36**, 441.

van Isacker, P, Heyde, K, Jolie, J, and Sevrin, A, 1986. *Ann. Phys.*, **171**, 253.

van Isacker, P, Heyde, K, Jolie, J, and Scholten, O, 1984. *Gull Lake 1984*, 334.

van Isacker, P, Heyde, K, Waroquier, M, and Wenes, G, 1982a. *Oaxtepec 1982*, 314.

van Isacker, P, Heyde, K, Waroquier, M, and Wenes, G, 1982b. *Nucl. Phys.*, **A380**, 383.

van Isacker, P, Heyde, K, Jolie, J, Waroquier, M, Moreau, J, and Scholten, O, 1984. *Phys. Lett.*, **144B**, 1.

van Ruyven, J J, Hesselink, W H A, Akkermans, J, van Nes, P, and Verheul, H, 1982. *Nucl. Phys.*, **A380**, 125.

Veskovic, M, Harder, M, Kumar, K, and Hamilton, W D, 1987. *J. Phys.*, **G13**, L155.

von Brentano, P, Gelberg, A, Harter, H, and Sala, P, 1985. *J. Phys.*, **G11**, L85.

Wadsworth, R, Cohler, M D, Lane, S M, Smithson, M J, and Watson, D L, 1985. *J. Phys.* , **G11**, 1045.

Ward, D, Colombani, P, Lee, I Y, Butler, P A, Simon, R S, Diamond, R M, and Stephens, F S, 1976. *Nucl. Phys.* **A266,** 194.

Warner, D D, 1983. *Drexel 1983*, 133.

Warner, D D, 1986. *Phys. Rev.* , **C34**, 1131.

Wesselborg, C, *et al.*, 1986. *Z. Phys.*, **A323**, 485.

Wolf, A, Warner, D D, and Benczer-Koller, N, 1985. *Phys. Lett.*, **158B**, 7.

Wolf, A, Berant, Z, Warner, D D, Gill, R L, Shmid, M,
    Chrien, R E, Peaslee, G, Yamamoto, H, Hill, J C, Wohn,
    F K, Chung, C, Walters, W B, 1983. *Phys. Lett.*, **123B**, 165.
Wood, J L, 1983a. *Drexel 1983*, 19.
Wood, J L, 1983b. *Nucl. Phys.* , **A396**, 245c.
Wu, H C, Dieperink, A E L, and Pittel, S, 1986. *Phys. Rev.* ,
    C34, 703.
Wu, H C, Dieperink, A E L, and Scholten, O, 1987. *Phys. Lett.*,
    187B, 205.
Yoshida, N, and Arima, A, 1985. *Phys. Lett.*, **164B**, 231.
Yoshida, N, Arima, A, and Otsuka, T, 1982. *Phys. Lett.*, **114B**,
    86.
Zemel, A, 1983. *Phys. Lett.* , **126B**, 145.
Zemel, A, and Dobes, J, 1983. *Phys. Rev.* , **C27**, 2311.
Zimmermann, M, and Dobeš, J, 1985. *Phys. Lett.*, **156B**, 7.

6

RADIAL DEGREES OF FREEDOM

6.1 Spatial dependence of electromagnetic transition operators

The Interacting Boson Model can be used in the analyis of elastic and inelastic electron (proton) scattering experiments in medium and heavy nuclei. These experiments are capable of exploring the spatial dependence of the nuclear degrees of freedom, which had been ignored so far since it does not influence the static properties we had been studying. The relevant IBM operators are the electromagnetic transition operators of multipolarity l. In the case of IBM-2, where one distinguishes between protons (π) and neutrons (ν), these have the general form

$$T^{(l)} = T_\pi^{(l)} + T_\nu^{(l)}. \tag{6.1}$$

These operators describe the coupling of the protons (neutrons) to the external electromagnetic field. In previous chapters, where only static properties were considered, the coupling constants were taken as numbers. When studying electron (proton) scattering, these numbers must be replaced by **boson form factors**, describing the spatial dependence of the coupling.

In lowest approximation we assume that only one-body terms need to be present in the electromagnetic transition operators. Then their most general form is (Iachello 1981)

$$T_\sigma^{(l)}(r) = \delta_{l0}(\tilde{\eta}_\sigma^{(0)}(r) + \tilde{\gamma}_\sigma^{(0)}(r)(s^+s)_\sigma^0)$$

$$+\delta_{l2}\tilde{\alpha}_\sigma^{(2)}(r)(d^+s + s^+\tilde{d})_\sigma^2 + \tilde{\beta}_\sigma^{(l)}(r)(d^+\tilde{d})_\sigma^l, \tag{6.2}$$

where $\sigma = \pi, \nu$. If one restricts himself in s and d bosons only and in one-body terms only, it is clear that no multipole higher than $l = 4$ is possible. For the wave functions of the IBM-2 Hamiltonian we will use the symbol $|N_\pi N_\nu J_i >$.

6.2 Scalar densities in IBM-2

When $l = 0$ only the $(s^+s)^0$, $(d^+\tilde{d})^0$ and a constant term occur in the transition operator. Since

$$(s^+s)^0 = n_s, \tag{6.3}$$

113

$$(d^+\tilde{d})^0 = \frac{n_d}{\sqrt{5}}, \qquad (6.4)$$

and

$$N = n_s + n_d, \qquad (6.5)$$

one can eliminate n_s from the expression of the transition operator. Thus for the $l = 0$ proton transition operator one obtains

$$T_\pi^{(0)}(r) = \eta_\pi^{(0)}(r) + \gamma_{\pi\pi}^{(0)}(r)N_\pi$$

$$+\gamma_{\pi\nu}^{(0)}(r)N_\nu + \beta_{\pi\pi}^{(0)}(r)n_{d_\pi} + \beta_{\pi\nu}^{(0)}n_{d_\nu}, \qquad (6.6)$$

while for the corresponding neutron operator one has

$$T_\nu^{(0)}(r) = \eta_\nu^{(0)}(r) + \gamma_{\nu\pi}^{(0)}(r)N_\pi$$

$$+\gamma_{\nu\nu}^{(0)}N_\nu + \beta_{\nu\pi}^{(0)}(r)n_{d_\pi} + \beta_{\nu\nu}^{(0)}(r)n_{d_\nu}. \qquad (6.7)$$

In the above the coefficients $\eta^{(0)}(r)$, $\gamma^{(0)}(r)$, $\beta^{(0)}(r)$ are simply related to $\tilde{\eta}^{(0)}(r)$, $\tilde{\gamma}^{(0)}(r)$, $\tilde{\beta}^{(0)}(r)$, as one can easily see.

Taking matrix elements of the above scalar transition operators between eigenstates $|N_\pi N_\nu 0_i^+ >$ of the IBM-2 Hamiltonian one obtains the diagonal densities

$$\rho_{\pi,0_i^+ \to 0_i^+}^{(0)}(r) = \eta_\pi^{(0)}(r) + \gamma_{\pi\pi}^{(0)}(r)N_\pi$$

$$+\gamma_{\pi\nu}^{(0)}(r)N_\nu + \beta_{\pi\pi}^{(0)}(r)B_{\pi,ii}^0 + \beta_{\pi\nu}^{(0)}(r)B_{\nu,ii}^0, \qquad (6.8)$$

$$\rho_{\nu,0_i^+ \to 0_i^+}^{(0)}(r) = \eta_\nu^{(0)}(r) + \gamma_{\nu\pi}^{(0)}(r)N_\pi$$

$$+\gamma_{\nu\nu}^{(0)}(r)N_\nu + \beta_{\nu\pi}^{(0)}(r)B_{\pi,ii}^{(0)} + \beta_{\nu\nu}^{(0)}(r)B_{\nu,ii}^{(0)}, \qquad (6.9)$$

and the transition densities

$$\rho_{\pi,0_j^+ \to 0_i^+}^{(0)} = \beta_{\pi\pi}^{(0)}(r)B_{\pi,ji}^{(0)} + \beta_{\pi\nu}^{(0)}(r)B_{\nu,ji}^{(0)}, \qquad (6.10)$$

$$\rho_{\nu,0_j^+ \to 0_i^+}^{(0)}(r) = \beta_{\nu\pi}^{(0)}(r)B_{\pi,ji}^{(0)} + \beta_{\nu\nu}^{(0)}(r)B_{\nu,ji}^{(0)}. \qquad (6.11)$$

In the above expressions the notation

$$B_{\pi,ji}^{(0)} =< N_\pi N_\nu 0_i^+ |n_{d_\pi}| N_\pi N_\nu 0_j^+ >, \qquad (6.12)$$

$$B_{\nu,ji}^{(0)} = <N_\pi N_\nu 0_i^+ |n_{d_\nu}| N_\pi N_\nu O_j^+ >, \qquad (6.13)$$

has been used. The matrix elements B contain the nuclear structure information and can be calculated numerically, using the already mentioned in Chapter 5 computer programs NPBOS or BOSON2.

The diagonal and transition densities found above can be used to calculate electron scattering cross sections. They can also be used for the calculation of static properties, as we will see below.

6.3 Mean-square radii in IBM-2

If the diagonal proton density found above is normalized as

$$\int_0^\infty 4\pi r^2 \rho_{\pi,0_1^+ \to 0_1^+}^{(0)}(r)dr = Z, \qquad (6.14)$$

the proton mean-square radius is

$$<r_\pi^2>_{0_1^+} = \frac{4\pi}{Z}\int_0^\infty r^4 \rho_{\pi,0_1^+ \to 0_1^+}^{(0)} dr. \qquad (6.15)$$

Then one can write for the proton mean-square radius the expression

$$<r_\pi^2>_{0_1^+} = <r_\pi^2>^{(0,0)} + \gamma_{\pi\pi}^{(0)} N_\pi + \gamma_{\pi\nu}^{(0)} N_\nu$$

$$+ \beta_{\pi\pi}^{(0)} B_{\pi,11}^{(0)} + \beta_{\pi\nu}^{(0)} B_{\nu,11}^{(0)}, \qquad (6.16)$$

where $<r_\pi^2>^{(0,0)}$ is the proton mean-square radius of the closed shell and $\gamma_{\pi\pi}^{(0)}$, $\gamma_{\pi\nu}^{(0)}$, $\beta_{\pi\pi}^{(0)}$, $\beta_{\pi\nu}^{(0)}$ are integrals over the corresponding quantities $\gamma_{\pi\pi}^{(0)}(r)$, $\gamma_{\pi\nu}^{(0)}(r)$, $\beta_{\pi\pi}^{(0)}(r)$, $\beta_{\pi\nu}^{(0)}(r)$. A similar expression holds for the neutron mean-square radius.

6.4 Isotope shifts in IBM-2

Using the expression for the proton mean-square radius obtained above, one can calculate isotope shifts. These are given by

$$\Delta <r_\pi^2> = <r_\pi^2>_{0_1^+}^{(N_\pi,N_\nu+1)} - <r_\pi^2>_{0_1^+}^{(N_\pi,N_\nu)}$$

$$= \gamma_{\pi\nu}^{(0)} + \beta_{\pi\pi}^{(0)}(B_{\pi,11}^{(0)(N_\pi,N_\nu+1)} - B_{\pi,11}^{(0)(N_\pi,N_\nu)})$$

$$+\beta_{\pi\nu}^{(0)}(B_{\nu,11}^{(0)(N_\pi,N_\nu+1)} - B_{\nu,11}^{(0)(N_\pi,N_\nu)}), \tag{6.17}$$

where the superscripts show the values of N_π and N_ν for which the mean-square radii are calculated.

6.5 Isomer shifts in IBM-2

Following a similar procedure one can construct the diagonal densities for the states $|N_\pi N_\nu 2_i^+ >$. Then one can calculate the isomer shift, which is given by

$$\delta < r_\pi^2 > = < r_\pi^2 >_{2_1^+}^{(N_\pi,N_\nu)} - < r_\pi^2 >_{0_1^+}^{(N_\pi,N_\nu)}$$

$$= \beta_{\pi\pi}^{(0)}(B_{\pi,11^*}^{(0)(N_\pi,N_\nu)} - B_{\pi,11}^{(0)(N_\pi,N_\nu)})$$

$$+\beta_{\pi\nu}^{(0)}(B_{\nu,11^*}^{(0)(N_\pi,N_\nu)} - B_{\nu,11}^{(0)(N_\pi,N_\nu)}), \tag{6.18}$$

where

$$B_{\pi,11^*}^{(0)(N_\pi,N_\nu)} = < N_\pi N_\nu 2_1^+ |n_{d_\pi}| N_\pi N_\nu 2_1^+ >, \tag{6.19}$$

$$B_{\nu,11^*}^{(0)(N_\pi,N_\nu)} = < N_\pi N_\nu 2_1^+ |n_{d_\nu}| N_\pi N_\nu 2_1^+ > . \tag{6.20}$$

6.6 Boson densities from elastic and inelastic cross sections

In the last three subsections we saw how information on the radial integrals of the boson densities can be obtained. Here we will see how information on the boson densities themselves can be obtained.

Measuring elastic cross sections in a series of isotopes one gets information on the quantity

$$\Delta\rho_{\pi,0_1^+ \to 0_1^+}^{(0)}(r) = \rho_{\pi,0_1^+ \to 0_1^+}^{(0)(N_\pi,N_\nu+1)}(r) - \rho_{\pi,0_1^+ \to 0_1^+}^{(0)(N_\pi,N_\nu)}(r)$$

$$= \gamma_{\pi\nu}^{(0)}(r) + \beta_{\pi\pi}^{(0)}(r)(B_{\pi,11}^{(0)(N_\pi,N_\nu+1)} - B_{\pi,11}^{(0)(N_\pi,N_\nu)})$$

$$+\beta_{\pi\nu}^{(0)}(r)(B_{\nu,11}^{(0)(N_\pi,N_\nu+1)} - B_{\pi,11}^{(0)(N_\pi,N_\nu)}). \tag{6.21}$$

Similarly, measuring the inelastic cross section for the excitation of the state 0_2^+ one gets information on the quantity

$$\rho_{\pi,0_1^+ \to 0_2^+}^{(0)} = \beta_{\pi\pi}^{(0)}(r)B_{\pi,12}^{(0)} + \beta_{\pi\nu}^{(0)}(r)B_{\nu,12}^{(0)}. \tag{6.22}$$

Since only three independent quantities $(\gamma_{\pi\nu}^{(0)}(r),\ \beta_{\pi\pi}^{(0)}(r),\ \beta_{\pi\nu}^{(0)}(r))$ are involved in these two equations, a systematic study of elastic and inelastic scattering cross sections can provide first a determination of these boson densities and subsequently a test of the model.

6.7 Scalar densities in IBM-1

In the case of IBM-1, where no distinction between protons and neutrons is made, the scalar transition operator takes the form

$$T^{(0)}(r) = \eta^{(0)}(r) + \gamma^{(0)}(r)N + \beta^{(0)}(r)n_d. \qquad (6.23)$$

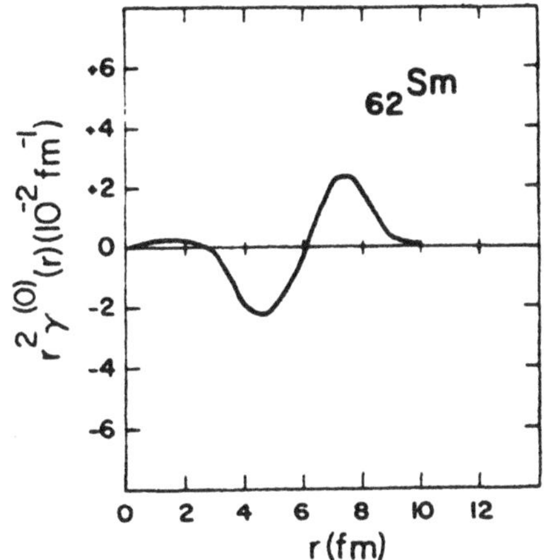

Fig. 6.1 The boson scalar density $r^2\gamma^{(0)}(r)$ as extracted from a fit to the ground state densities of a series of Sm isotopes, obtained from electron scattering experiments. (Taken from Iachello (1981)).

The eigenstates of the Hamiltonian are now $|NJ_i^+>$. Then the diagonal and transition densities take the form

$$\rho_{0_i^+\to 0_i^+}^{(0)}(r) = \eta^{(0)}(r) + \gamma^{(0)}(r)N + \beta^{(0)}(r)B_{ii}^{(0)}, \quad i\neq j, \qquad (6.24)$$

$$\rho_{0_i^+\to 0_j^+}^{(0)}(r) = \beta^{(0)}(r)B_{ij}^{(0)}, \qquad (6.25)$$

where

$$B_{ii}^{(0)} = < N0_i^+ |n_d| N0_i^+ >, \qquad (6.26)$$

$$B_{ij}^{(0)} = < N0_i^+ |n_d| N0_j^+ > . \qquad (6.27)$$

The matrix elements $B_{ii}^{(0)}$, $B_{ij}^{(0)}$ contain the nuclear structure information and must be calculated using the computer program **PHINT**, mentioned in Chapter 1. The nuclear mean-square radius now has the form

$$< r^2 >_{0_1^+} = < r^2 >^{(0)} + \gamma^{(0)} N + \beta^{(0)} B_{11}^0. \qquad (6.28)$$

From this the isotope shift can be obtained as

$$\Delta < r^2 >_{0_1^+}^{(N)} = \gamma^{(0)} + \beta^{(0)} (B_{11}^{(0)(N+1)} - B_{11}^{(0)(N)}). \qquad (6.29)$$

In Fig. 6.1 the boson scalar density $r^2 \gamma^{(0)}(r)$ is shown, as extracted from a fit to the ground state densities of Sm isotopes.

6.8 Quadrupole densities in IBM-2

In the case of $l = 2$ the transition operators are

$$T_\pi^{(2)}(r) = \alpha_\pi^{(2)}(r)[d^+ s + s^+ \tilde{d}]_\pi^2 + \beta_\pi^{(2)}(r)[d^+ \tilde{d}]_\pi^2, \qquad (6.30)$$

$$T_\nu^{(2)}(r) = \alpha_\nu^{(2)}(r)[d^+ s + s^+ \tilde{d}]_\nu^2 + \beta_\nu^{(2)}(r)[d^+ \tilde{d}]_\nu^2. \qquad (6.31)$$

The interesting transition densities are

$$\rho_{\pi,0_1^+ \to 2_1^+}^{(2)}(r) = \alpha_\pi^{(2)}(r) A_{\pi,1i}^{(2)} + \beta_\pi^{(2)}(r) B_{\pi,1i}^{(2)}, \qquad (6.32)$$

$$\rho_{\nu,0_1^+ \to 2_1^+}^{(2)}(r) = \alpha_{nu}^{(2)}(r) A_{\nu,1i}^{(2)} + \beta_\nu^{(2)}(r) B_{\nu,1i}^{(2)}, \qquad (6.33)$$

where

$$A_{\sigma,1i}^{(2)} = < N_\pi N_\nu 2_i^+ ||[d^+ s + s^+ \tilde{d}]_\sigma^2|| N_\pi N_\nu 0_1^+ >, \quad \sigma = \pi, \nu, \quad (6.34)$$

$$B_{\sigma,1i}^{(2)} = < N_\pi N_\nu 2_i^+ ||[d^+ \tilde{d}]_\sigma^2|| N_\pi N_\nu 0_1^+ >, \quad \sigma = \pi, \nu. \qquad (6.35)$$

The reduced matrix elements A and B contain the nuclear structure information and can be obtained by using the computer programs

NPBOS or BOSON2, mentioned in Chapter 5. The total transition
density has the form

$$\rho^{(2)}_{0^+_1 \to 2^+_i}(r) = e^{(2)}_\pi \rho^{(2)}_{\pi,0^+_1 \to 2^+_i}(r) + e^{(2)}_\nu \rho^{(2)}_{\nu,0^+_1 \to 2^+_i}, \qquad (6.36)$$

where $e^{(2)}_\pi$ and $e^{(2)}_\nu$ are the **boson effective quadrupole charges**.

From the total transition density one can obtain the corresponding B(E2) value. First, one has to calculate the integral

$$\int_0^\infty r^4 \rho^{(2)}_{0^+_1 \to 2^+_i}(r)dr = e^2_\pi(\alpha^2_\pi A^2_{\pi,1i} + \beta^2_\pi B^2_{\pi,1i})$$

$$+ e^2_\nu(\alpha^2_\nu A^2_{\nu,1i} + \beta^2_\nu B^2_{\nu,1i}). \qquad (6.37)$$

Then the B(E2) value is obtained as the square of this integral

$$B(E2; 0^+_1 \to 2^+_i) = \left(\int_0^\infty r^4 \rho^{(2)}_{0^+ \to 2^+_i}(r)dr \right)^2. \qquad (6.38)$$

6.9 Quadrupole densities in IBM-1

In IBM-1 one ignores the distinction between protons and neutrons.
Then the quadrupole transition operator becomes

$$T^{(2)}(r) = \alpha^{(2)}(r)[d^+ s + s^+ \tilde{d}]^2 + \beta^{(2)}(r)[d^+ \tilde{d}]^2. \qquad (6.39)$$

Then the transition density is

$$\rho^{(2)}_{0^+_1 \to 2^+_1}(r) = e^{(2)}(\alpha^{(2)}(r)A^{(2)}_{1i} + \beta^{(2)}(r)B^{(2)}_{1i}), \qquad (6.40)$$

where the reduced matrix elements A and B can be calculated using
the computer program PHINT. To obtain B(E2) values one needs
the integral

$$\int_0^\infty r^4 \rho^{(2)}_{0^+_1 \to 2^+_i}(r)dr = e^{(2)}(\alpha^{(2)} A^{(2)}_{1i} + \beta^{(2)} B^{(2)}_{1i}). \qquad (6.41)$$

Then

$$B(E2; 0^+_1 \to 2^+_i) = \left(\int_0^\infty r^4 \rho^{(2)}_{0^+_1 \to 2^+_i}(r)dr \right)^2. \qquad (6.42)$$

In Fig. 6.2 the boson quadrupole densities $\alpha^{(2)}(r)$ and $\beta^{(2)}(r)$ are shown, as extracted from a fit to the $0_1^+ \to 2_1^+$ and $0_1^+ \to 2_3^+$ transition densities in $^{150}_{60}\mathrm{Nd}_{90}$.

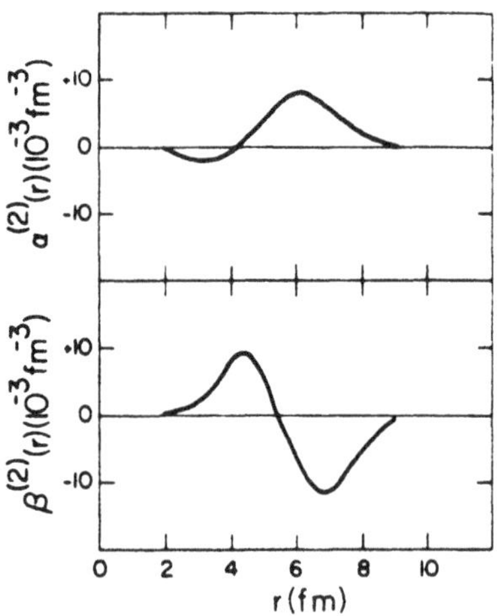

Fig. 6.2 The boson quadrupole densities $\alpha^{(2)}(r)$ and $\beta^{(2)}(r)$ as extracted from a fit to the $0_1^+ \to 2_1^+$ and $0_1^+ \to 2_3^+$ transition densities in $^{150}_{60}\mathrm{Nd}_{90}$, obtained from electron scattering experiments. (Taken from Iachello (1981)).

Similar calculations can be performed for hexadecapole ($l = 4$) densities, as well as for magnetic dipole ($l = 1$) and magnetic octupole ($l = 3$) densities.

References

This chapter has been based on Iachello (1981). See de Jager (1986) and Goutte (1986) for recent applications in electron scattering, as well as Riech *et al.* (1986a) for a recent application in proton inelastic scattering. Recent calculations of isotope shifts can be found in Alonso *et al.* (1986).

Alonso, C E, and Arias, J M, 1985. *La Rábida 1985*, 611.
Alonso, C E, Arias, J M, and Iachello, F, 1985. *Phys. Lett.*,
     164B, 241.
Alonso, C E, Arias, J M, and Lozano, M, 1986. *Dubrovnik 1986,*

2, 1048.

Bazantay, J P, *et al.*, 1985. *Phys. Rev. Lett.* , **54**, 643.

Borghols, W T A, Blasi, N, Bijker, R, Harakeh, M N, de Jager, C W, van der Laan, J B, de Vries, H, and van der Werf, S Y, 1985. *Phys. Lett.*, **152B**, 330.

Cereda, E, Pignanelli, M, Micheletti, S, von Geramb, H V, Harakeh, M N, De Leo, R, D' Erasmo, G, and Pantaleo, A, 1983. *Nucl. Phys.*, **A396**, 281c.

Dasso, C H, and Vitturi, A, 1987. *Nucl. Phys.*, **A469**, 437.

de Jager, C W, 1984a. *Drexel 1984*, 142.

de Jager, C W, 1984b. *Gull Lake 1984*, 225.

de Jager, C W, 1986. *Dubrovnik 1986*, **1**, 404.

de Leo, R, *et al.*, 1985. *Phys. Lett.*, **162B**, 1.

Dieperink, A E L, 1978. *Erice 1978*, 129.

Dieperink, A E L, 1981. *Nucl. Phys.*, **A358**, 189c.

Dieperink, A E L, Iachello, F, Rinat, A, and Creswell, C, 1978. *Phys. Lett.*, **76B**, 135.

Goutte, D, 1984. *Gull Lake 1984*, 241.

Goutte, D, 1986. *Dubrovnik 1986*, **1**, 411.

Hersman, F W, *et al.*, 1983. *Phys. Lett.* , **132B**, 47.

Iachello, F, 1981. *Nucl. Phys.* , **A358**, 89c.

Jänecke, J, and Becchetti, F D, 1983. *Phys. Rev.*, **C27**, 2282.

Jänecke, J, and Scholten, O, 1984. *Phys. Lett.*, **141B**, 281.

Lee, T S H, 1980. *Erice 1980*, 143.

Moinester, M A, Alster, J, Azuelos, G, and Dieperink, A E L, 1982. *Nucl. Phys.*, **A383**, 264.

Paar, V, and Kyrchev, G, 1984. *Phys. Lett.*, **148B**, 251.

Pignanelli, M, Micheletti, S, Cereda, E, Harakeh, M N, van der Werf, S Y, and de Leo, R, 1984. *Phys. Rev.*, **C29**, 434.

Riech, V, Fretwurst, E, Lindström, G, von Reden, K F, Scherwinski, R, and Blok, H P, 1986a. *Dubrovnik 1986*, **1**, 424.

Riech, V, Fretwurst, E, Lindström, G, von Reden, K F, Scherwinski, R, and Blok, H P, 1986b. *Phys. Lett.*, **178B**, 10.

van der Laan, J B, Burghardt, A J C, de Jager, C W, and de Vries, H, 1985. *Phys. Lett.*, **153B**, 130.

7

THE INTERACTING BOSON MODELS 3 AND 4

7.1 Introduction

As we have already seen, IBM-1 makes no distinction between protons and neutrons, while in IBM-2 one considers both proton bosons and neutron bosons, microscopically corresponding to **correlated** proton pairs and neutron pairs, respectively. Proton–neutron pairs are ignored in both models. This is maybe justified in medium and heavy nuclei, where the valence protons and the valence neutrons occupy different major shells and can belong to very different orbitals. In s-d shell nuclei, however, both protons and neutrons occupy the same major shell, so that the omission of proton–neutron pairs is not justified any more. IBM-3 (introduced by Elliott and White (1980)) and IBM-4 (introduced by Elliott and Evans (1981)) are extensions of IBM-2 taking into account the existence of proton-neutron pairs. The difference between IBM-3 and IBM-4 will be explained later.

The study of sd shell nuclei in the IBM framework is interesting because of the following reasons:

i) There is good experimental data to test the theory against, even for nuclei described by only one or two bosons.

ii) Detailed shell-model calculations exist throughout the region, i.e. the nuclei in this region are well-understood theoretically.

iii) The region contains some nuclei with clearly rotational character. A few examples are ^{20}Ne (the lightest deformed nucleus), ^{24}Mg (the most deformed nucleus in this region), ^{32}Si.

iv) As mentioned above, valence protons and valence neutrons occupy the same major shell, in contrast to the heavy nuclei, where IBM-1 and IBM-2 have been exclusively applied.

v) The level density for single-nucleon states is low and the shell is small, so that the variation in nuclear properties from nucleus to nucleus will be greater than in the collective regions of heavy nuclei.

We now proceed to the study of the group theoretical structure of the various IBM models (Halse, Elliott, and Evans 1984). A brief review of the group theoretical structure of IBM-1 and IBM-2 (already studied in Chapter 1 and Chapter 5 respectively) will be given before we proceed to the description of the more general models IBM-3 and IBM-4.

7.2 Group theoretical structure of IBM-1 and IBM-2. The concept of F-spin

In IBM-1 the complete set of states for a system of N bosons is given by the totally symmetric states. These are contained in the $U_{s-d}(6)$ irrep represented by a Young tableau of one line with N boxes, as we have already seen.

In IBM–2 the bosons have, in addition to the s-d space, a two-dimensional space corresponding to the choice between the neutron bosons (ν) and the proton bosons (π). Mathematically this space can be described by assigning a "spin" label $\pm 1/2$ to the ν and π bosons, respectively. As it has been mentioned in Chapter 5, this label has been called **F-spin**. Each boson has

$$F = 1/2, \tag{7.1}$$

with

$$F_0 = +1/2 \tag{7.2}$$

for neutron bosons and

$$F_0 = -1/2 \tag{7.3}$$

for proton bosons. The complete set of F-spin states of a system of N bosons is represented by Young tableaux $[f_1, f_2]$ which have two lines, the first one with f_1 boxes, the second with f_2 boxes. The F-spin of a system of N bosons is then given by

$$F = \frac{N}{2} - f_2 = \frac{f_1 - f_2}{2}, \tag{7.4}$$

since

$$N = f_1 + f_2. \tag{7.5}$$

Thus the possible values of F-spin are

$$F = \frac{N}{2}, \frac{N}{2} - 1, \frac{N}{2} - 2, \ldots, \frac{1}{2} \quad or \quad 0. \tag{7.6}$$

The total F-spin of a system of bosons describes the symmetry in the F-spin space only. (Like the total isospin of a system of nucleons describes the symmetry in the isospin space only). The bosons of IBM-2, however, have angular momentum (L=0 if they are s bosons or L=2 if they are d bosons) in addition to F-spin. Furthermore, the states of any system of bosons have to be overall symmetric. This means that for each boson state its symmetry character in the

s-d space must be the same as its symmetry character in the F-spin
space. Thus the F-spin also labels the symmetry in the s-d space.
The same Young diagram $[f_1, f_2]$ will apply to both spaces.

The maximum value of F-spin, N/2, corresponds to total symmetry in both spaces. It has been suggested that the states with
maximum F-spin lie lowest in energy. These are the states occurring
in IBM-1. In fact for such states it is possible to transform an IBM-2
Hamiltonian into an effective IBM-1 Hamiltonian, although this task
will not be undertaken here. Thus it is clear that IBM-2 augments
IBM-1, since it contains, in addition to the fully symmetric states of
maximum F-spin, states of lower values of F-spin, which correspond
to mixed symmetry in the s-d space. These are the mixed symmetry
states (MISS !), encountered in Chapter 5, which are found to lie
higher in the spectrum.

From a group-theoretical point of view, there are now twelve
states for a single boson. The F-spin classification corresponds to
the subgroup $SU_F(2) \otimes U_{sd}(6)$, with F-spin labelling the irreps of
both SU(2) and U(6). These are the same concepts described in
Chapter 5 in a somewhat different language. It should be noted here
that a realistic boson Hamiltonian cannot be F-spin invariant, since
that would demand the same properties for nuclei having the same
F-spin but different values of F_0.

Example Consider the s-d shell nuclei ^{20}O and ^{20}Ne. These
have the same maximum F-spin, 1, but different values of F_0 (+1 and
0, respectively), as it can easily be seen. If the boson Hamiltonian
were F-spin invariant, these two nuclei would have had the same
properties (which is not observed experimentally).

7.3 Group theoretical structure of IBM-3

Although IBM-2 distinguishes neutron and proton bosons, it has in-
sufficient flexibility to give states of good isospin. The reason is clear:
the neutron and proton bosons provide only two of the components
$(T_z = \pm 1)$ of a $T = 1$ triplet. This defect may be unimportant in
heavy nuclei, where neutrons and protons are filling very different
single-particle orbits. However, in s-d shell nuclei, where protons
and neutrons are filling the same shell, it is known from shell-model
calculations that large errors can result from a failure to take into
account the isospin symmetry.

For such nuclei the boson model can be put into an isospin invariant form by the inclusion of a third type of boson, with $T = 1$
and $T_z = 0$, to complete the isospin $T = 1$ triplet. The new boson is imagined to be constructed from correlated proton–neutron

pairs. The resulting model is called **IBM-3**. The bosons have now, in addition to the s-d space, a three-dimensional isospin space. The states can be classified by the same procedure used above, i.e. by their symmetry in the two separate spaces. The appropriate subgroup now is $SU_T(3) \otimes U_{sd}(6)$, instead of $SU_F(2) \otimes U_{sd}(6)$. Thus the F-spin label is replaced by the labels of an $SU_T(3)$ irrep, (λ, μ). If we again restrict attention to states which are totally symmetric in the sd space, the $SU_T(3)$ label is $(N,0)$, where N is the total number of bosons. The isospin group $SU_T(2)$ is now a subgroup of $SU_T(3)$. Its irreps are labelled by the isospin quantum number T. The values of T contained in $(N,0)$ are

$$T = N, N - 2, \ldots, 1 \quad or \quad 0. \tag{7.7}$$

These are precisely the values corresponding to the even–even nuclei.

Example Consider the nuclei with mass 20. They have 4 valence nucleons, so that they are described in terms of two bosons ($N = 2$). In this case the possible values of T are 2 and 0, corresponding to ^{20}O and ^{20}Ne, respectively. Thus these two nuclei are properly distinguished by their isospin, although they lie in the same $SU_T(3)$ irrep. Thus the Hamiltonian can be isospin invariant.

7.4 Group theoretical structure of IBM-4

Once a proton–neutron boson of $T = 1$ has been introduced, one immediately asks why a proton–neutron boson with $T = 0$ is not included. Experimentally it is known that the nucleus ^{18}F, which has one valence proton and one valence neutron in the s-d shell, has both $T = 1$ and $T = 0$ states. In fact the ground state has $T = 0$. It is indeed possible to include $T = 0$ bosons into the model, thus obtaining **IBM-4**. However, a comparison with the two-nucleon system suggests that such a boson should carry an intrinsic spin of $S = 1$, in contrast to the $T = 1$ bosons which have intrinsic spin $S = 0$. This is so because if two nucleons forming a pair occupy the same point in space, their orbital wave function must be symmetric under particle exchange. Since the overall wave function has to be antisymmetric, in order to satisfy the Pauli principle, their total spin and isospin have to be in the combinations $T = 1$, $S = 0$ or $T = 0$, $S = 1$. The properties of nuclei such as the deuteron and ^{18}F (which we mentioned above) show that both combinations have roughly the same energy, so that there is no reason for excluding either.

There are now 36 types of bosons in total. The 18 $T = 1$ bosons are the same ones used in IBM-3. For the 18 $T = 0$ bosons, which possess both orbital angular momentum and intrinsic spin, the total angular momentum is given by vector coupling, using the notation $^{2S+1}l_j$. Thus the $T = 0$ bosons are denoted $^3s_1$, $^3d_1$, $^3d_2$, $^3d_3$. It is important to realize at this point that, although in the qualitative argument above the boson "intrinsic spin" has been associated with the sum of intrinsic spins of the two nucleons, this precise relationship is *not* assumed. The conventional d-boson represents a 2^+ state of a correlated pair of nucleons. Although the $L = 2$ for a d-boson is regarded as a boson orbital angular momentum, the angular momentum of the 2^+ state will be derived from both orbital and spin angular momenta of the nucleons. Thus the assumption of L–S coupling among bosons does *not* imply an L–S coupling assumption in the underlying shell-model. Actually, it can be the case that the L–S coupling in the boson model is a better approximation than the L–S coupling in the shell-model, because part of the departure from the shell-model L–S coupling limit (caused by the strong spin-orbit interaction, as we will see in more detail in Sec. 15.5.2) can be incorporated in the L–S coupling picture in the boson space.

7.5 Classification scheme for IBM-4 states

We now proceed to the construction of a classification scheme for the states of IBM-4. The appropriate group structure here is $SU_{ST}(6) \otimes U_{sd}(6)$, where $U_{sd}(6)$ refers to the s-d space, as before, while $SU_{ST}(6)$ describes the T and S degrees of freedom. One way to get the T- and S-labels is to reduce the $SU_{ST}(6)$ group to its subgroup $SU_S(2) \otimes SU_T(2)$ of the isospin and spin groups, i.e.

$$SU_{ST}(6) \otimes U_{sd}(6) \supset SU_S(2) \otimes SU_T(2) \otimes U_{sd}(6). \qquad (7.8)$$

However, we can get the T- and S-labels in a different way, using as an intermediate stage the group SU(4), the Wigner supermultiplet group for bosons, studied in the Appendices. Then

$$SU_{ST}(6) \otimes U_{sd}(6) \supset SU(4) \otimes U_{sd}(6). \qquad (7.9)$$

The presence of this group is expected because the 6 types of bosons $T = 1$, $S = 0$ and $T = 0$, $S = 1$ belong to a single representation (supermultiplet) of the group SU(4).

In order to proceed, we need some notation for the irreps of SU(4). One possibility is to use as labels the number of boxes in each row of the corresponding Young diagram. Since the maximum number of rows for SU(4) is 3, the irreps can be denoted as $[f_1, f_2, f_3]$, where f_1 is the number of boxes in the first row, etc. An alternative notation is (λ, μ, ν), where λ, μ, ν are the differences between the lengths of successive rows in the Young diagram, i.e.

$$\lambda = f_1 - f_2, \tag{7.10}$$

$$\mu = f_2 - f_3, \tag{7.11}$$

$$\nu = f_3, \tag{7.12}$$

We are going to use the latter case here, which is obviously a generalization of the Elliott notation (λ, μ) for the irreps of SU(3), used earlier.

It turns out that the 6 types of bosons $T = 1$, $S = 0$ and $T - 0$, $S = 1$ belong to the 6-dimensional SU(4) irrep (0,1,0). For a system of N bosons only states which are totally symmetric in the sd space will be taken into account, as usual. Then

A) The SU(6) irrep is [N]=(N,0,0,0,0,0).

B) The above mentioned SU(6) irrep contains the SU(4) irreps $(0, \sigma, 0)$, with

$$\sigma = N, N - 2, \ldots, 1 \quad or \quad 0. \tag{7.13}$$

C) The T and S values occuring in a representation $(0, \sigma, 0)$ are defined simply by the condition

$$T + S = \sigma, \sigma - 2, \ldots, 1 \quad or \quad 0. \tag{7.14}$$

For a given nucleus we know N and T, but (with the exception of the case $N = T$, when $\sigma = N$ is the only possibilty) we do not know the value of σ. Therefore we assume that the irreps $(0, \sigma, 0)$ are ordered as they would be ordered in the shell-model, where we know that small σ lie lowest in energy. This tells us that $\sigma = T$ and $S = 0$. (Notice that σ cannot be less than T because of eq. (7.14)). This assumption is a reasonable one because of the similarity between the boson SU(4) and the Wigner SU(4). As we have already seen, the boson SU(4) contains both isospin and intrinsic spin. There is no question that the boson isospin is the real isospin. However, the intrinsic spin of the bosons is not the same as the intrinsic spin of the nucleons (recall the earlier discussion about the distinction between the l=2 boson angular momentum and the angular momentum of the 2^+ state). In light nuclei, shell-model calculations indicate that the Wigner SU(4) is a reasonable approximation but it is mixed by the spin–orbit interaction. As we have already seen, it can be the case that the boson SU(4) is more valid than the Wigner SU(4), because the definition of boson intrinsic spin has incorporated some part of the effect of the nucleon spin–orbit force. However, it is interesting to notice that although the individual bosons can carry spin in IBM-4, the lowest states of even–even nuclei have total $S = 0$, as found above. However, the rest of the states are composed by a mixture of some $S = 0$ bosons and some $S = 1$ bosons.

7.6 Dynamical symmetries in IBM-4

The concept of dynamical symmetries has been proven very useful in IBM-1 and IBM-2. Their use is possible in IBM-3 and IBM-4, too. In this section the dynamical symmetries of IBM-4 (Han, Sun, and Li 1987) will be discussed in general, while a particular example will be described in more detail.

7.6.1 Definitions of operators

In IBM-4, as we have already seen, pairs of valence nucleons are assumed to behave like bosons with angular momentum $l = 0, 2$, spin s and isospin t in the combinations $t = 1$, $s = 0$ or $t = 0$, $s = 1$. To avoid the physically transparent but rather cumbersome notation explained in the previous section, we introduce boson creation and annihilation operators $b^+_{lmsm_stm_t}$ and $b_{lmsm_stm_t}$, where $l = 0, 2$, and $t = 1$, $s = 0$ or $t = 0$, $s = 1$. The values of the rest of the indices are

$$m = -l, -l+1, \ldots, l, \tag{7.15}$$

$$m_s = -s, -s+1, \ldots, s, \tag{7.16}$$

$$m_t = -t, -t+1, \ldots, t. \tag{7.17}$$

There are 36 creation operators and 36 annihilation operators in total. If we assume boson number conservation, as we did in the previous models, the largest symmetry group of this system is U(36), generated by $b^+_{lmsm_stm_t} b_{l'm's'm'_st'm'_t}$. We introduce convenient forms of the generators, coupled to good angular momentum, spin and isospin

$$B(lstl's't')^{kST}_{qM_SM_T} = [b^+_{lst} \otimes \tilde{b}_{l's't'}]^{kST}_{qM_SM_T}$$

$$= \sum_{m,m_s,m_t,m',m'_s,m'_t} (lml'm'|kq)(sm_ss'm'_s|SM_S)(tm_tt'm'_t|TM_T)$$

$$b^+_{lmsm_stm_t} \tilde{b}_{l'm's'm'_st'm'_t}, \tag{7.18}$$

where

$$\tilde{b}_{lmsm_s tm_t} = (-1)^{l+m+s+m_s+t+m_t} b_{l,-m,s,-m_s,t,-m_t}. \qquad (7.19)$$

It is also convenient to introduce the operators

$$B(ll')_q^k = \sum_{s,t} \sqrt{(2s+1)(2t+1)} B(lstl'st)_{q00}^{k00}$$

$$= \sum_{m,m',s,t,m_s,m_t} (-1)^{l'+m'} (lml'm'|kq) b_{lmsm_s tm_t}^{+} b_{l',-m',s,m_s,t,m_t},$$

$$(7.20)$$

$$B(sts't')_{M_S M_T}^{ST} = \sum_l \sqrt{2l+1} B(lstl's't')_{0M_S M_T}^{0ST}$$

$$= \sum_{lmm_s m_t m_s' m_t'} (-1)^{s'+m_s'+t'+m_t'} (sm_s s'm_s'|SM_S)(tm_t t'm_t'|TM_T)$$

$$b_{lmsm_s tm_t}^{+} b_{l,m,s',-m_s',t',-m_t'}. \qquad (7.21)$$

For illustrative purposes, we give here the expressions for the total boson number

$$N = \sum_{lstmm_s m_t} b_{lmsm_s tm_t}^{+} b_{lmsm_s tm_t}, \qquad (7.22)$$

the generators of the orbital angular momentum subgroup $O_d(3)$

$$L_q^1 = \sqrt{10} B(22)_q^1, \qquad (7.23)$$

the generators of the spin subgroup $O_S(3)$

$$S_q^1 = \sqrt{2} B(1010)_{q0}^{10}, \qquad (7.24)$$

and the generators of the isospin subgroup $O_T(3)$

$$T_q^1 = \sqrt{2} B(0101)_{0q}^{01}. \qquad (7.25)$$

The total angular momentum operator is $J_q^1 = L_q^1 + S_q^1$ and generates the $O_J(3)$ subgroup. The operators $B(ll')_q^k$ generate the group $U_{sd}(6)$, which is the group of maximum symmetry in IBM-1.

7.6.2 Classification of states

Taking into account the rotational invariance in space and isospace, the states are classified according to group chains involving $O_J(3) \otimes O_T(3)$ as a subgroup of U(36), in the same way that in IBM-1 by taking into account the rotational invariance in space the states were classified according to the chains of U(6) containing O(3) as a subgroup. However, since in IBM-4 six types of s and d bosons are present (with $t = 1$, $s = 0$ or $t = 0$, $s = 1$), six-rowed irreps of $U_{sd}(6)$ are allowed, labelled as $[f_1, f_2, f_3, f_4, f_5, f_6]$, where f_i is the number of boxes in the ith row of the corresponding Young tableau. Remember that in IBM-1 only one kind of s and d bosons existed, so that only the most symmetric one-rowed U(6) irrep, denoted by $[f_1]$, occured. However, if one deals with low-lying states, it is reasonable to assume that only the most symmetric irrep $[f_1]$ and the next lower symmetric irrep $[f_1, 1]$ are enough.

7.6.3 Subgroups of U(36)

Before proceeding to study the possible subgroup chains of U(36), we give an account of its possible subgroups. Since $U_{sd}(6)$ is a subgroup of U(36), it follows that all the $U_{sd}(6)$ subgroups studied in IBM-1 are also present here. In addition, subgroups associated with the spin and isospin degrees of freedom are present (like $O_S(3)$ and $O_T(3)$ studied above), as well as subgroups associated with mixtures of sd and S degrees of freedom (like the earlier studied $O_J(3)$). The subgroups not studied yet are listed below, followed by their generators

$$
\begin{array}{ll}
U_S(3) & B(1010)^{S0}_{M_S 0} \\[4pt]
U_T(3) & B(0101)^{0T}_{0 M_T} \\[4pt]
U_{ST}(6) & B(sts't')^{ST}_{M_S M_T} \\[4pt]
O_{ST}(6) & B(1010)^{10}_{M_S 0} \,,\; B(0101)^{01}_{0 M_T} \,, \\[4pt]
& B(1001)^{11}_{M_S M_T} \,,\; B(0110)^{11}_{M_S M_T} \\[4pt]
SU_{sdS}(6) & J_q \,,\; Q_q + \sqrt{\tfrac{3}{4}} \, B(1010)^{20}_{q0}
\end{array}
$$

where Q_q denotes the familiar quadrupole operator of subsec. 1.7.2.

7.6.4 Subgroup chains of U(36)

It can be proved that seven different group chains containing $O_J(3) \otimes O_T(3)$ can be constructed starting from U(36). These are

   i) The strong-coupling SU(3) chain

$$
U(36) \supset U_{sd}(6) \otimes U_{ST}(6) \supset SU_{sd}(3) \otimes U_S(3) \otimes U_T(3)
$$

$$\supset SU_{sdS}(3) \otimes O_T(3) \supset O_J(3) \otimes O_T(3). \qquad (7.26)$$

ii) The medium-coupling U(5), O(6), and SU(3) chains

$$U(36) \supset U_{sd}(6) \otimes U_{ST}(6) \supset U_d(5) \otimes O_{ST}(6) \supset O_d(5) \otimes O_S(3) \otimes O_T(3)$$

$$\supset O_d(3) \otimes O_S(3) \otimes O_T(3) \supset O_J(3) \otimes O_T(3), \qquad (7.27)$$

$$U(36) \supset U_{sd}(6) \otimes U_{ST}(6) \supset O_{sd}(6) \otimes O_{ST}(6) \supset O_d(5) \otimes O_S(3) \otimes O_T(3)$$

$$\supset O_d(3) \otimes O_S(3) \otimes O_T(3) \supset O_J(3) \otimes O_T(3), \qquad (7.28)$$

$$U(36) \supset U_{sd}(6) \otimes U_{ST}(6) \supset SU_{sd}(3) \otimes O_{ST}(6)$$

$$\supset O_d(3) \otimes O_S(3) \otimes O_T(3) \supset O_J(3) \otimes O_T(3). \qquad (7.29)$$

It should be noted here that the group $O_{ST}(6)$, which distinguishes these chains from the rest, is isomorphic to $SU_{ST}(4)$. As discussed in the previous section, this is not the Wigner SU(4), but the boson SU(4).

iii) The weak-coupling U(5), O(6), and SU(3) chains, which can be obtained from the corresponding medium-coupling chains if we replace the subgroup $O_{ST}(6)$ by $U_S(3) \otimes U_T(3)$.

In the above chains the subscripts s an d were used to denote the subgroups of IBM-1.

7.6.5 An exactly soluble IBM-4 Hamiltonian

In the case of the medium-coupling O(6) dynamical symmetry, the most general Hamiltonian including only one-body and two-body boson–boson interactions can be written as

$$H = E_0 + EC_{1U_{sd}6} + AC_{2U_{sd}6} + BC_{2O_{sd}6} + CC_{2O_d5}$$

$$+DC_{2O_d3} + \alpha C_{2O_{ST}6} + \beta C_{2O_T3} + \delta C_{2O_S3} + \gamma C_{2O_J3}. \qquad (7.30)$$

With the further assumptions that only the totally symmetric irreps of the subgroups $U_{sd}(6)$ and $U_{ST}(6)$ are important, and that all the low-lying states belong to the most symmetric irrep of $O_{ST}(6)$, the eigenvalues of the Hamiltonian are given by

$$< H >= E_0' + Cv(v+3) + DL(L+1) + \beta T(T+1) + \delta S(S+1) + \gamma J(J+1), \qquad (7.31)$$

where v is the seniority quantum number, labelling the irreps of $O_d(5)$.

The predictions of this Hamiltonian for the spectrum of the nucleus $^{30}_{14}\mathrm{Si}_{16}$ are compared to experiment in Fig. 7.1. We remark that

there is an one-to-one correspondence between experimental and predicted $(ST) = (01)$ levels. Five of the predicted $(ST) = (21)$ levels
have experimental counterparts, while the predicted 1^+ (6.52 MeV)
and 0^+ (6.22 MeV) levels with $(ST) = (21)$, as well as the predicted
1^+ (6.84 MeV) level with $(ST) = (12)$, have not been seen experimentally. In addition, the predicted position of the 1^+ state with
$(ST) = (21)$ is much higher than in the experimental data.

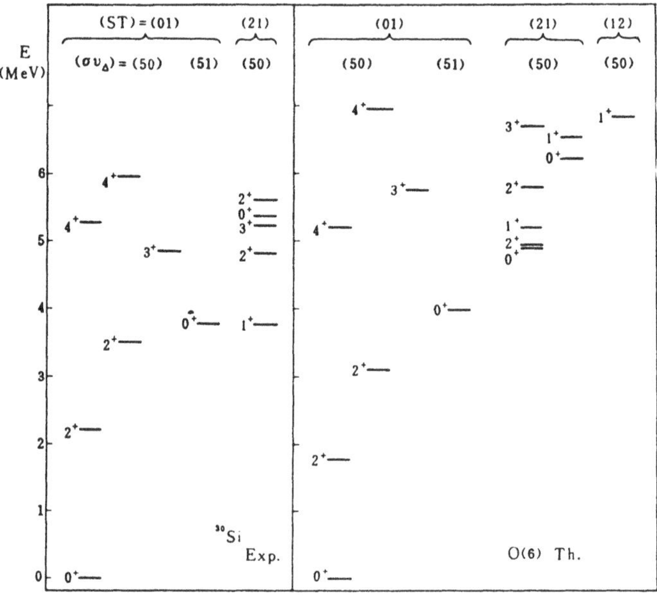

Fig. 7.1 Comparison between the predictions of IBM-4 in the O(6) limit
and experiment for the spectrum of the nucleus ^{30}Si. The upper set of
parentheses indicates the (ST) quantum numbers, while the lower set of
parentheses contains the $(\sigma\nu_\Delta)$ quantum numbers. (Taken from Han *et
al.* (1987)).

The above mentioned nucleus $^{30}_{14}\text{Si}_{16}$ has 6 valence protons and
4 valence neutron holes. Thus it has been described as a system of
$(6 + 4)/2 = 5$ bosons. Consider now the odd–odd nucleus $^{30}_{15}\text{P}_{15}$,
which has 5 valence proton holes and 5 valence neutron holes. In
IBM-4 odd–odd nuclei can be described in the same framework as
even–even nuclei, thus ^{30}P can be considered as a system of 5 bosons,
the same number of bosons as in ^{30}Si. One can then use the same
Hamiltonian used for fitting ^{30}Si in order to fit the spectrum of ^{30}P.

The same parameters should be used in the Hamiltonian, with the exception of E_0', which is associated to binding energies and will be different for each nucleus. The results of such a calculation for ^{30}P are shown in Fig. 7.2. We remark that there is an one-to-one correspondence between all predicted levels and experimental states, although some ordering problems appear in the second band with $(ST) = (10)$.

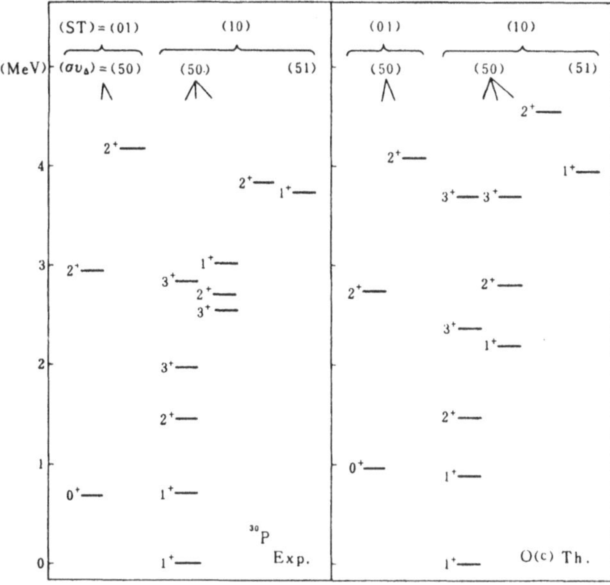

Fig. 7.2 Comparison between the predictions of IBM-4 in the O(6) limit and experiment for the spectrum of the nucleus ^{30}P. The same notation as in Fig. 7.1 has been used. (Taken from Han *et al.* (1987)).

References

The discussion of the group theoretical structure of IBM-2, IBM-3, and IBM-4 has been based on Halse, Elliott, and Evans (1984), while the account of dynamical symmetries in IBM-4 has been based on Han, Sun, and Li (1987).

Elliott, J P, 1982. *Erice 1982*, 101.
Elliott, J P, 1983. *Florence 1983*, 101.

Elliott, J P, 1985. *Rep. Prog. Phys.*, **48**, 171.
Elliott, J P, and Evans, J A, 1981. *Phys. Lett.*, **101B**, 216.
Elliott, J P, and White, A P, 1980. *Phys. Lett.*, **97B**, 169.
Elliott, J P, Evans, J A, and Williams, A P, 1987. *Nucl. Phys.*, **A469**, 51.
Evans, J A, Elliott, J P, and Szpikowski, S, 1985. *Nucl. Phys.*, **A435**, 317.
Halse, P, 1985. *Nucl. Phys.*, **A445**, 93.
Halse, P, Elliott, J P, and Evans, J A, 1984. *Nucl. Phys.*, **A417**, 301.
Han, Q Z, Sun, H Z, and Li, G H, 1987. *Phys. Rev.*, **C35**, 786.

8

THE NUCLEAR VIBRON MODEL

8.1 Introduction

Algebraic techniques can be used for the description of systems composed of two separate fragments. In such cases the collective degree of freedom is the distance separating the two fragments (dipole collectivity). It has been suggested that an algebraic description can be given in terms of four boson operators, a scalar boson of angular momentum 0 and positive parity, s, and a vector boson of angular momentum 1 and negative parity, p_μ, with $\mu = 0, \pm 1$. The p bosons are taken to have negative parity because they correspond to an odd degree of freedom, the separation between the two fragments.

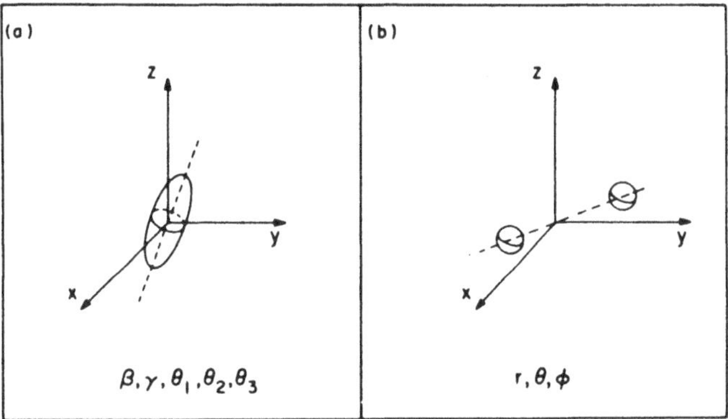

Fig. 8.1 (a) A deformed body with quadrupole deformation. (b) A configuration of two clusters. (Taken from Iachello (1984b)).

A schematic illustration is given in Fig. 8.1. In part (a) a deformed body is shown. As we have already seen, this body has 5 degrees of freedom and can be described geometrically by 5 parameters: the parameters β and γ determining its shape, and the three angles $\theta_1, \theta_2, \theta_3$ determining its orientation in space. A nucleus having such a shape has been described in IBM-1 in terms of s and d bosons. In

135

part (b) a system composed of two fragments is shown. This can be considered as a diatomic molecule, or as a nucleus composed of two clusters (of course in this case the two fragments should be very close to each other). Geometrically this system can be described by 3 degrees of freedom, the separation r of the two fragments and the angles θ, ϕ giving its orientation in space. In the vibron model, this system is described in terms of s and p bosons.

The vibron model approach was first used for the description of diatomic molecules. Subsequently it has been used to describe nuclei composed of two clusters. In this chapter both applications will be discussed.

8.2 The algebra U(4) and its chains of subalgebras

8.2.1 The algebra U(4)

Using the boson creation (annihilation) operators s^+, p_μ^+ (s, p_μ) (Iachello 1985a) we can form the following pairs

$$G_\mu^2(pp) = [p^+ \otimes \tilde{p}]_\mu^2, \tag{8.1}$$

$$G_\mu^1(pp) = [p^+ \otimes \tilde{p}]_\mu^1, \tag{8.2}$$

$$G_0^0(pp) = [p^+ \otimes \tilde{p}]_0^0, \tag{8.3}$$

$$G_\mu^1(ps) = [p^+ \otimes \tilde{s}]_\mu^1, \tag{8.4}$$

$$G_\mu^1(sp) = [s^+ \otimes \tilde{p}]_\mu^1, \tag{8.5}$$

$$G_0^0(ss) = [s^+ \otimes \tilde{s}]_0^0. \tag{8.6}$$

These are the 16 generators of the group U(4).

We wish to find all possible chains of subalgebras of U(4) containing the rotation algebra SO(3) as a subgroup.

8.2.2 Chain I

A) From the generators listed above, we first delete all the generators containing the s boson. The remaining 9 operators (containing only p bosons) are the generators of the group U(3).

B) From the remaining 9 operators, we keep only the 3 operators $G_\mu^1(pp)$. These generate the group SO(3).

C) The single operator $G_0^1(pp)$ generates the group SO(2).

Thus a complete chain of groups is

$$U(4) \supset U(3) \supset SO(3) \supset SO(2).\qquad(8.7)$$

8.2.3 Chain II

A) The 6 operators $G^1_\mu(pp)$ and $G^1_\mu(ps) + G^1_\mu(sp)$ generate the group SO(4).

B) The three operators $G^1_\mu(pp)$ generate the group SO(3).

C) The single operator $G^1_0(pp)$ generates the group SO(2).

Thus a complete chain of groups is

$$U(4) \supset SO(4) \supset SO(3) \supset SO(2).\qquad(8.8)$$

It is possible to show that these are the only two possible chains containing SO(3).

8.3 Classification of states of U(4)

Each group chain can be used to construct a basis in which the Hamiltonian can be diagonalized. The states are labelled by the quantum numbers characterizing the irreps of the various groups which appear in the chain.

8.3.1 Chain I

A) The irreps of U(4) are labelled by the total number of bosons M, and are denoted by [M].

B) The irreps of U(3) are labelled by the number of p-bosons, n_p. They are denoted by (n_p). The values of n_p allowed for a given value of M are

$$n_p = 0, 1, \ldots, M.\qquad(8.9)$$

C) The irreps of SO(3) are labelled by the angular momentum quantum number L. The values of L contained in a given (n_p) are

$$L = n_p, n_p - 2, \ldots, 1 \quad or \quad 0.\qquad(8.10)$$

D) The irreps of SO(2) are labelled by the z-component of angular momentum, M. The values of M allowed for a given value of L are well known to be

$$-L \leq M \leq L.\qquad(8.11)$$

Thus the complete classification scheme for chain I is

$$|[M](n_p)LM > .\qquad(8.12)$$

8.3.2 Chain II

A) The irreps of U(4) are labelled by the total number of bosons M, and are denoted by [M].

B) The irreps of SO(4) are labelled by the quantum number ω and are denoted by (ω). The values of ω contained in a give [M] are

$$\omega = M, M - 2, \ldots, 1 \quad or \quad 0. \tag{8.13}$$

C) The irreps of SO(3) are labelled by the angular momentum quantum number L. The values of L contained in a given (ω) are

$$L = \omega, \omega - 1, \ldots, 1, 0. \tag{8.14}$$

D) The irreps of SO(2) are labelled by the quantum number M, as before.

Thus the complete classification scheme for for chain II is

$$|[M](\omega)LM > . \tag{8.15}$$

8.4 Dynamical symmetries of U(4)

Dynamical symmetries occur when the Hamiltonian can be written in terms of the Casimir operators of the groups appearing in a given group chain. In the vibron model there are obviously two different cases:

8.4.1 Dynamical symmetry I

This dynamical symmetry occurs when the entire Hamiltonian can be written in terms of the Casimir operators of the groups appearing in the group chain

$$U(4) \supset U(3) \supset SO(3) \supset SO(2). \tag{8.16}$$

The most general Hamiltonian in this chain is

$$H^{(I)} = F + \epsilon C_{1U3} + \alpha C_{2U3} + \beta C_{2O3}, \tag{8.17}$$

where

$$F = E_0 + E_1 M + E_2 M^2. \tag{8.18}$$

This Hamiltonian can be diagonalized analytically. In order to do so, one needs to know the eigenvalues of the Casimir invariants, which are found by straightforward techniques. The relevant general formulas can be found in the appendices. The resulting expression is

$$< H^{(I)} >= F + \epsilon n_p + \alpha n_p(n_p + 3) + \beta L(L + 1). \qquad (8.19)$$

8.4.2 Dynamical symmetry II

This dynamical symmetry occurs when the entire Hamiltonian can be written in terms of the Casimir operators of the groups appearing in the group chain

$$U(4) \supset SO(4) \supset SO(3) \supset SO(2). \qquad (8.20)$$

The most general Hamiltonian is

$$H^{(II)} = F + AC_{2SO4} + BC_{2SO3}. \qquad (8.21)$$

Its eigenvalues are found to be

$$< H^{(II)} >= F + A\omega(\omega + 2) + BL(L + 1). \qquad (8.22)$$

In order to see the structure of the corresponding spectrum, it is useful to introduce a quantum number v, defined as

$$v = \frac{M - \omega}{2}, \qquad (8.23)$$

and taking the values

$$v = 0, 1, \ldots, \frac{M}{2} \quad or \quad \frac{M - 1}{2}, \qquad (8.24)$$

for M even or odd. Now the eigenvalue expression can be rewritten as

$$< H^{(II)} >= F' - 4A(M+2)(v+\frac{1}{2}) + 4A(v+\frac{1}{2})^2 + BL(L+1), \qquad (8.25)$$

with

$$F' = F + (M^2 + 4M + 3)A. \qquad (8.26)$$

Typical spectra obtained from the Hamiltonians corresponding to these two limiting symmetries are shown in Fig. 8.2. The O(4) limiting symmetry is shown on the left part of the figure, while the

U(3) symmetry is shown on the right. The common characteristic feature of the two spectra is the presence of low-lying states of negative parity and odd angular momentum. Notice, however, that although states within each band are almost equispaced in the U(3) limit, they look like following an L(L+1) rule in the O(4) case. There seems to be an analogy between the U(3) limit of the U(4) vibron model and the U(5) limit of the U(6) IBM-1. Notice that in both cases the limiting symmetry corresponds to the absence of s bosons.

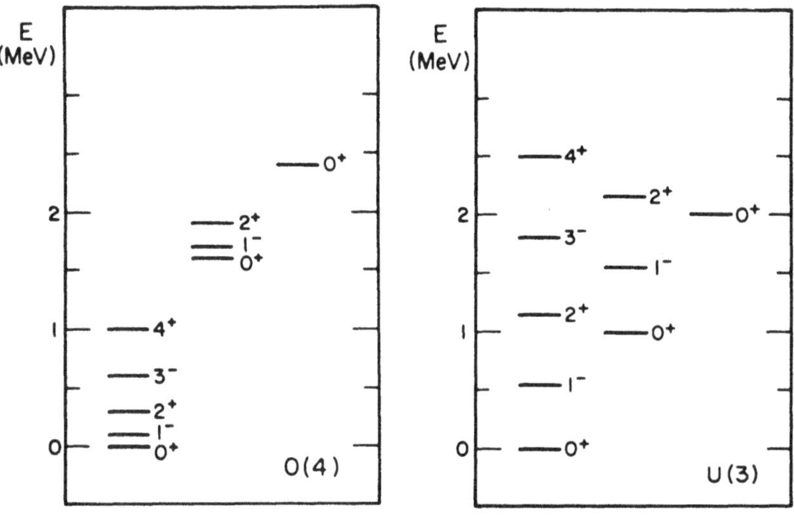

Fig. 8.2 Typical spectra obtained in the O(4) and the U(3) limits of the U(4) model. They both correspond to a system of 4 bosons. (Taken from Iachello (1984b)).

8.5 Physical content of the dynamical symmetries I and II

In chemistry it is known that a simple analysis of molecular spectra with rotational–vibrational character is provided by the Dunham expansion. This is an expansion of the energy levels in terms of vibration–rotation quantum numbers. For diatomic molecules (Iachello and Levine 1982), this yields:

$$E(v, L) = \sum_{ij} y_{ij} (v + \frac{1}{2})^i [L(L+1)]^j, \qquad (8.27)$$

where the coefficients y_{ij} are obtained by a fit to the experimental energy levels. We note that the energy expression obtained in dynamical symmetry II corresponds to the first few terms of the Dunham expansion. This is a result of the limitation that the Hamiltonian contains only one- and two-body terms. Inclusion of higher order terms in the Hamiltonian results in the case of the dynamical symmetry II to an expression for the energy identical to the Dunham expansion.

The Dunham expansion does not contain any information about the wave functions of individual states, so the matrix elements of operators cannot be directly calculated. A more sophisticated analysis is provided by the potential approach, where the energy levels are obtained by solving the Schrödinger equation with an interatomic potential. The potential V is expressed in terms of interatomic variables. One potential used for the description of diatomic molecules is the three-dimensional Morse oscillator with

$$V = V_0[e^{-2a(r-r_0)} - 2e^{-a(r-r_0)}]. \tag{8.28}$$

It can be seen that, in lowest approximation, the energy levels of this potential can be written in the form given by the dynamical symmetry II, with (Iachello and Levine 1982)

$$F' = -V_0, \tag{8.29}$$

$$-4A(M + 2) = \frac{h}{2\pi}a\sqrt{\frac{2V_0}{m}}, \tag{8.30}$$

$$A = -\frac{h^2}{32\pi^2 m}a^2, \tag{8.31}$$

$$B = \frac{h^2}{8m\pi^2 r_0^2}. \tag{8.32}$$

Thus the O(4) dynamical symmetry describes, to some approximation, the energy levels of a three-dimensional Morse oscillator. Its spectrum is typical of a rigid diatomic molecule.

It is interesting to note that several three-dimensional potentials have spectra approximately corresponding to the energy expression obtained in dynamical symmetry I. Among them are the Woods–Saxon potential

$$V(r) = -\frac{V_0}{1 + e^{a(r-r_0)}}, \tag{8.33}$$

and the Pöschl–Teller potential

$$V(r) = -\frac{V_0}{ch^2 ar}. \tag{8.34}$$

The corresponding spectra are typical of a nonrigid (soft) molecule.

8.6 Geometrical analysis of the limits of U(4)

As in the case of IBM-1, it is interesting to study the limiting symmetries of the algebra U(4) in terms of geometrical variables. It turns out that the O(4) symmetry corresponds to two well-separated fragments, and is then appropriate for describing diatomic molecules. The agreement we found between the predictions of this symmetry on one hand and the Dunham expansion and the three-dimensional Morse oscillator on the other reinforces this point. The U(3) symmetry has been found to correspond to two fragments lying very close to each other and thus might be more appropriate for the description of nuclei composed of two clusters (sometimes called dinuclear molecules). Details about methods of studying the classical limit of algebraic Hamiltonians will be given in Chapter 13.

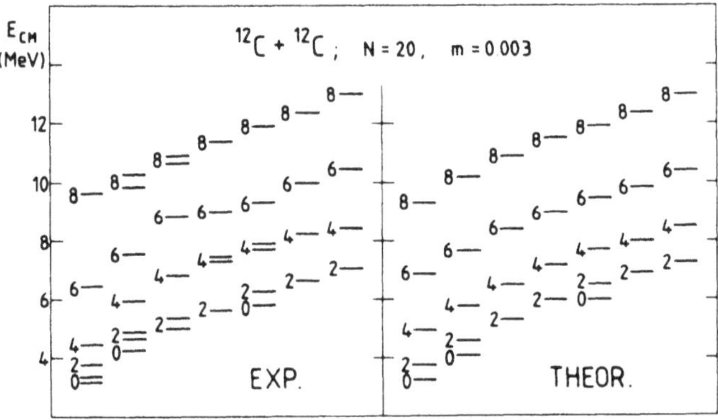

Fig. 8.3 Experimental energies of quasimolecular resonances of the ^{12}C-^{12}C system (left part) compared to the predictions of the vibron model in the U(3) limit (right part). (Taken from Cseh (1985)).

8.7 Quasi-molecular resonances

Light heavy-ions (i.e. heavy ions with $4 \leq A \leq 40$) show characteristic resonances near the Coulomb barrier region. The spectra of these resonances resemble spectra of diatomic molecules, thus they are called quasi-molecular spectra. Since the vibron model has been found to give a good description of molecular spectra, it is expected that it will be able to describe quasi-molecular resonances as well. Both the U(3) and the O(4) limits of the vibron model have been used in this context for the description of the quasi-molecular resonances of the ^{12}C+^{12}C system (Erb and Bromley 1981, Cseh 1985). Results of the U(3) limit of the vibron model are compared to the experimental data in Fig. 8.3, and good agreement is observed.

8.8 Algebraic description of triatomic molecules

The vibron model, used for the description of diatomic molecules, as we have already seen, can be extended in a straightforward manner to describe triatomic molecules (van Roosmalen *et al.* 1983). A triatomic molecule has six internal degrees of freedom that can be considered as the two vectors specifying the interatomic distances. Thus one has twice as many degrees of freedom as in diatomic molecules, where the internal degrees of freedom are characterized by only one vector. Addition of the degrees of freedom described by the two groups G_1 and G_2 is achieved by taking their direct product $G_1 \otimes G_2$. In the case of triatomic molecules the appropriate group is $U_1(4) \otimes U_2(4)$, where the subscripts 1 and 2 are used in order to distinguish the two groups. Possible chains of subgroups are

$$U_1(4) \otimes U_2(4) \supset U_1(3) \otimes U_2(3) \supset O_1(3) \otimes O_2(3) \supset O(3), \quad (8.35)$$

$$U_1(4) \otimes U_2(4) \supset U_1(3) \otimes O_2(4) \supset O_1(3) \otimes O_2(3) \supset O(3), \quad (8.36)$$

$$U_1(4) \otimes U_2(4) \supset O_1(4) \otimes O_2(4) \supset O_1(3) \otimes O_2(3) \supset O(3), \quad (8.37)$$

$$U_1(4) \otimes U_2(4) \supset U_1(3) \otimes U_2(3) \supset U(3) \supset O(3), \quad (8.38)$$

$$U_1(4) \otimes U_2(4) \supset U(4) \supset U(3) \supset O(3), \quad (8.39)$$

$$U_1(4) \otimes U_2(4) \supset O_1(4) \otimes O_2(4) \supset O(4) \supset O(3), \quad (8.40)$$

$$U_1(4) \otimes U_2(4) \supset U(4) \supset O(4) \supset O(3). \quad (8.41)$$

Here we will analyze only the sixth of them, which turns out to be particularly appropriate for the description of triatomic molecules. This is expected since this chain makes maximum use of O(4) subgroups and we have already seen that the O(4) symmetry is the appropriate one for the description of rigid molecules. The quantum numbers necessary to label the irreps of each group are listed below

$$
\begin{array}{cc}
U_1(4) \otimes U_2(4) & [N_1][N_2] \\
O_1(4) \otimes O_2(4) & (\omega_1, 0)(\omega_2, 0) \\
O(4) & (\tau_1, \tau_2) \\
O(3) & L \\
O(2) & M
\end{array}
$$

The Hamiltonian is then written as

$$H = A_1 C_{2O_14} + A_2 C_{2O_24} + AC_{2O4} + BC_{2O3}. \tag{8.42}$$

Its eigenvalues are

$$< H > = A_1\omega_1(\omega_1+2) + A_2\omega_2(\omega_2+2) + A[\tau_1(\tau_1+2)+\tau_2^2] + BL(L+1). \tag{8.43}$$

A typical spectrum obtained from this Hamiltonian is shown in Fig. 8.4. Notice the large number of bands, characteristic of a triatomic molecule.

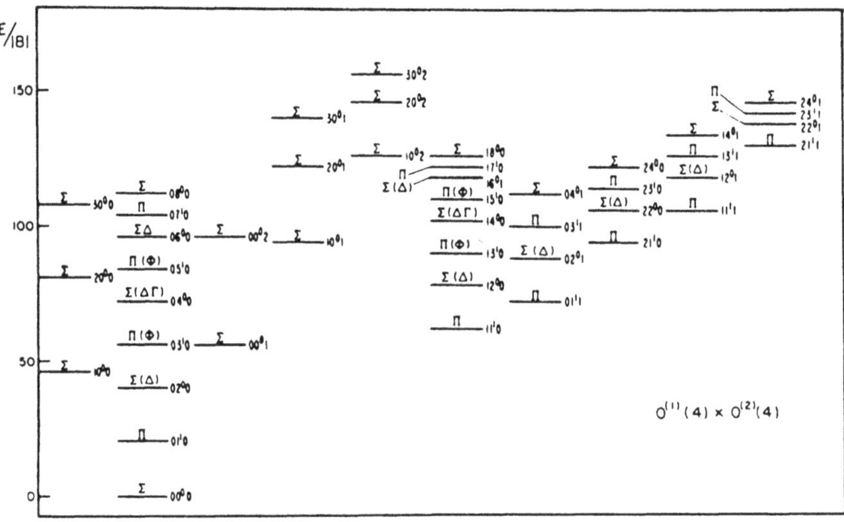

Fig. **8.4** Typical spectrum with $O_1(4) \otimes O_2(4)$ symmetry. The states are labelled by the vibrational species Σ, Π, Δ, ..., and by the vibrational quantum numbers v_1, v_2, v_3 (the notation widely used by chemists). This spectrum was obtained with $N_1=6$, $N_2 = 4$. (Taken from Iachello (1983a)).

8.9 Algebraic approach to clustering

Recently it has been suggested that clustering plays an important role in medium and heavy nuclei. Fig. 8.5 enumerates the possible cluster configurations for the actinide nucleus ^{224}Th. This diagram is called an **Ikeda diagram**. The nucleus ^{224}Th has 8 valence protons and 8 valence neutrons outside the doubly closed-shell core of the nucleus ^{208}Pb. The simplest cluster is an α cluster, composed of two protons and two neutrons. Configurations with more than one cluster are also possible. In the following, we make the simplifying assumption that only one cluster occurs in the nucleus under consideration. Furthermore, we will assume that this single cluster is structureless.

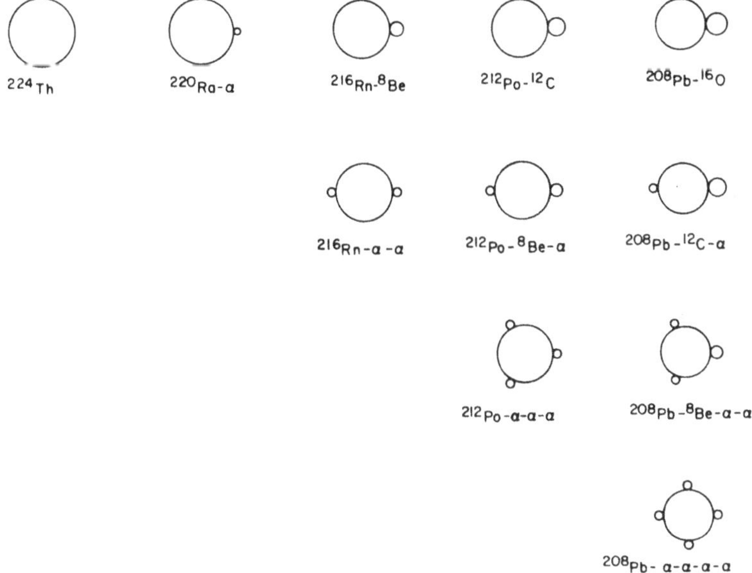

Fig. 8.5 Ikeda diagram showing the possible clustering configurations in the nucleus ^{224}Th. (Taken from Daley and Iachello (1986)).

Since the medium and heavy nuclei where clustering is supposed to be important are often deformed, the study of their structure including clustering degrees of freedom is rather complex. It has been suggested that the use of algebraic techniques may help in attacking this problem. The collective degree of freedom corresponding to the distance separating the two clusters (dipole collectivity) can be described in terms of s' and p bosons forming the group U(4), as in the case of diatomic molecules. In the case of clustering in medium and heavy nuclei it is reasonable to assume that the cluster most likely to be present is an alpha cluster, which can be considered structureless.

The rest of the nucleus forms another cluster, which is too big to be considered structureless, so its internal degrees of freedom have to be taken into account. In the algebraic framework, the internal degrees of freedom of the big cluster can be described by the usual s and d bosons of the IBM-1, which form the group U(6). If several alpha clusters are present, we assume that they join together to form larger structureless clusters (e.g. three alpha clusters are treated as ^{12}C). With these simplifications taken into account, the system of two clusters, one having internal degrees of freedom and the other being structureless, can be described by the group

$$G = U(6) \otimes U(4). \tag{8.44}$$

This is usually referred to as the **Nuclear Vibron Model** (Daley and Iachello 1986). The general Hamiltonian can be written as

$$H = H_a + H_b + H_{ab}, \tag{8.45}$$

where H_a describes the internal degrees of freedom of the big cluster, H_b describes the relative motion of the two clusters, and V_{ab} their interaction. In general, the eigenvalue problem for H has to be solved numerically. A computer program, ALPHA, has been written for this purpose (H. J. Daley, 1982, University of Arizona, USA). However, analytical solutions can be found in certain cases, when dynamic symmetries are present. Dynamic symmetries for coupled systems are much more difficult to study than those for single systems. The technique which one uses is that of finding all possible common subalgebras and combining the appropriate generators for the two (or more) systems.

8.10 The SU(3) limit of the Nuclear Vibron Model

In the present case we will limit ourselves to the study of clustering in deformed nuclei. For deformed nuclei, the appropriate group chain is

$$U_a(6) \supset SU_a(3) \supset O_a(3) \supset O_a(2). \tag{8.46}$$

This chain has several common subalgebras with the chain

$$U_b(4) \supset U_b(3) \supset SU_b(3) \supset O_b(3) \supset O_b(2), \tag{8.47}$$

which is suitable for describing anharmonic motions of a nonrigid dinucleus system. The subscripts a and b have been added to distinguish the members of each chain.

The appropriate group chain of the combined system is

$$U_a(6) \otimes U_b(4) \supset SU_a(3) \otimes U_b(3)$$

$$\supset SU_a(3) \otimes SU_b(3) \supset SU(3) \supset O(3) \supset O(2). \qquad (8.48)$$

For illustrative purposes we give here the explicit expressions for the generators of some of the groups appearing in this chain. The eight generators of $SU_a(3)$ are

$$G_\mu^{(1)a} = \sqrt{10}[d^+ \otimes \tilde{d}]_\mu^1, \qquad (8.49)$$

$$G_\mu^{(2)a} = [d^+ \tilde{s} + S^+ \tilde{d}]_\mu^2 \pm \frac{\sqrt{7}}{2}[d^+ \otimes \tilde{d}]_\mu^2, \qquad (8.50)$$

while the nine generators of $U_b(3)$ are

$$G_0^{(0)b} = [p^+ \otimes \tilde{p}]_0^0, \qquad (8.51)$$

$$G_\mu^{(1)b} = \sqrt{2}[p^+ \otimes \tilde{p}]_\mu^1, \qquad (8.52)$$

$$G_\mu^{(2)b} = \frac{\sqrt{3}}{2}[p^+ \otimes \tilde{p}]_\mu^2. \qquad (8.53)$$

The $\pm$ sign in $G^{(2)a}$ describes oblate $(+)$ or prolate $(-)$ nuclei. The eight generators of the combined $SU(3)$ group are

$$G_\mu^1 = G_\mu^{(1)a}, \qquad (8.54)$$

$$G_\mu^2 = G_\mu^{(2)a} \pm G_\mu^{(2)b}. \qquad (8.55)$$

The $\pm$ sign in G_μ^2 describes parallel $(+)$ or anti-parallel $(-)$ coupling.

To obtain an analytic expression for the energies, we write the Hamiltonian in terms of Casimir operators of the above chain

$$H = \epsilon_p C_{1U_b3} + \alpha_p C_{2SU_b3} + \kappa_d C_{2SU_a3} + \kappa C_{2SU3} + \kappa' C_{2O3}. \qquad (8.56)$$

States can be labelled by the irreps appearing in the chain. The necessary quantum numbers are listed next to each group below

$$
\begin{array}{cc}
U_a(6) \otimes U_b(4) & [N] , [M] \\
SU_a(3) \otimes SU_b(3) & (\lambda, \mu)_a , n_p \\
SU(3) & (\lambda, \mu) \\
O(3) & L \\
O(2) & M .
\end{array}
$$

Because the step from SU(3) to O(3) is not fully decomposable, an additional label χ, introduced by Vergados, is necessary. The values of each of the above quantum numbers have been described earlier. The only additional ingredient needed to construct the coupled states is a rule for the outer products $(\lambda, \mu)_a \otimes (n_p, 0)_b$, which can be obtained using standard techniques. For example,

$$(\lambda, \mu)_a \otimes (1, 0)_b = (\lambda + 1, \mu) \oplus (\lambda - 1, \mu + 1) \oplus (\lambda, \mu - 1), \quad (8.57)$$

with obvious restrictions when either λ or μ is equal to zero.

The eigenvalues of the Hamiltonian in these coupled states are

$$< H >= \epsilon_p n_p + \alpha_p n_p(n_p + 3) + \kappa_d C(\lambda_\alpha, \mu_\alpha) + \kappa C(\lambda, \mu) + \kappa' L(L+1), \tag{8.58}$$

where

$$C(\lambda, \mu) = \lambda^2 + \mu^2 + \lambda\mu + 3(\lambda + \mu). \tag{8.59}$$

A typical spectrum obtained from this Hamiltonian is shown in Fig. 8.6. Notice the rotational structure of each band. Also notice the presence of very low lying negative parity bands, occurring in energy lower than the β_1 and γ_1 positive parity bands.

8.11 Generalizations of the Nuclear Vibron Model

In describing the dinuclear system in terms of the Nuclear Vibron Model we have made the assumption that one of the two fragments is structureless. If we remove this assumption, we have to take into account the internal degrees of freedom of both clusters, as well as the degrees of freedom corresponding to their separation distance. Using the subscripts 1 and 2 to label the two fragments, the group theoretical structure of the problem becomes

$$U_1(6) \otimes U(4) \otimes U_2(6). \tag{8.60}$$

In the case of three clusters (labelled as 1, 2, 3), all of them having internal structure, one has to take into account, in addition to the internal degrees of freedom of each cluster, the degrees of freedom corresponding to their separations. As in the case of the triatomic molecule, studied earlier, two sets of s and p bosons are necessary (here the two sets will be labelled as a and b, respectively). Thus the overall group theoretical structure of the problem becomes

$$U_1(6) \otimes U_a(4) \otimes U_2(6) \otimes U_b(4) \otimes U_3(6). \tag{8.61}$$

Further generalizations following this path are possible. None of
the generalizations mentioned in this section has yet been applied.

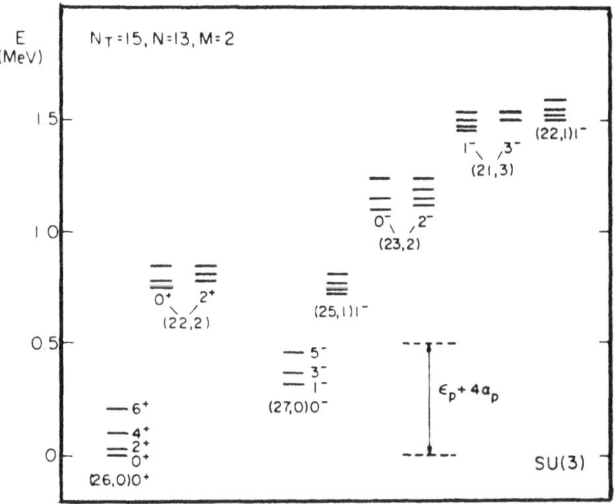

Fig. 8.6 Typical spectrum in the SU(3) limit of the Nuclear Vibron Model.
Bands are labelled by the quantum numbers (λ, μ) and χ. Only a portion
of the spectrum is shown, obtained with boson numbers $N = 13$, $M = 2$.
(Taken from Daley and Iachello (1986).

References

The present account of the vibron model for diatomic molecules has
been based on Iachello (1985a) and Iachello and Levine (1982). Much
information about the application of the vibron model in triatomic
molecules can be found in van Roosmalen *et al.* (1983). The present
account of the nuclear vibron model has been based on Daley and
Iachello (1986). Another detailed description of this model can be
found in Daley and Barrett (1986).

Baktybaev, K B, and Kabulov, A B, 1985. *J. Phys.*, **G11**, L63.
Bonetti, R, 1985. *Varenna 1985*, 477.
Bonin, W, Dahlinger, M, Glienke, S, Kankeleit, E, Krämer, M,
    Habs, D, Schwartz, B, and Backe, H, 1983.
    Z. Phys., **A310**, 249.
Butler, P A, 1986. *Dubrovnik 1986*, **1**, 101.

Chasman, R R, 1986. *Dubrovnik 1986*, **1**, 5.
Cindro, N, 1982. *Poiana Brasov 1982*, 397.
Cindro, N, and Greiner, W, 1983. *J. Phys.*, **G9**, L175.
Cottle, P D, and Bromley, D A, 1986. *Phys. Lett.*, **182B**, 129.
Cseh, J, 1983. *Phys. Rev.*, **C27**, 2991.
Cseh, J, 1984. *Debrecen 1984*, 643.
Cseh, J, 1985. *Phys. Rev.*, **C31**, 692.
Cseh, J, 1987. *Oaxtepec 1987*, 57.
Cseh, J, and Suhonen, J, 1985. *Phys. Rev.*, **C33**, 1553.
Daley, H J, 1984. *Gull Lake 1984*, 377.
Daley, H J, 1985. *Legnaro 1985*, 129.
Daley, H J, 1986. *J. Phys.*, **G12**, L51.
Daley, H J, and Barrett, B R, 1986. *Nucl. Phys.*, **A449**, 256.
Daley, H J, and Gai, M, 1984. *Phys. Lett.*, **149B**, 13.
Daley, H J, and Iachello, F, 1983. *Phys. Lett.*, **131B**, 281.
Daley, H J, and Iachello, F, 1986. *Ann. Phys.*, **167**, 73.
Daley, H J, and Nagarajan, M A, 1986. *Phys. Lett.*, **166B**, 379.
Erb, K A, and Bromley, D A, 1981. *Phys. Rev.*, **C23**, 2781.
Gai, M, 1984a. *Debrecen 1984*, 331.
Gai, M, 1984b. *Gull Lake 1984*, 283.
Gai, M, 1985. *Legnaro 1985*, 13.
Gai, M, 1986. *Dubrovnik 1986*, **1**, 93.
Gilmore, R, and Draayer, J P, 1985. *J. Math. Phys.*, **26**, 3053.
Iachello, F, 1981a. *Chem. Phys. Lett.*, **78**, 581.
Iachello, F, 1981b. *Phys. Rev.* , **C23**, 2778.
Iachello, F, 1983a. *Drexel 1983*, 3.
Iachello, F, 1983b. *Nucl. Phys.*, **A396**, 233c.
Iachello, F, 1984a. *Chester 1984*, 101.
Iachello, F, 1984b. *Debrecen 1984*, 631.
Iachello, F, 1984c. *Nucl. Phys.*, **A421**, 97c.
Iachello, F, 1985a. *La Rábida 1985*, 242.
Iachello, F, 1985b. *Varenna 1985*, 442.
Iachello, F, 1985c. *Phys. Lett.*, **160B**, 1.
Iachello, F, and Jackson, A D, 1982. *Phys. Lett.*, **108B**, 151.
Iachello, F, and Levine, R D, 1982. *J. Chem. Phy.*, **77**, 3046.
Levit, S, and Smilansky, U, 1982. *Nucl. Phys.*, **A389**, 56.
Lomnitz-Adler, J, and van Isacker, P, 1982. *Oaxtepec 1982*, 137.
Satpathy, L, Sarangi, P, and Faessler, A, 1986. *J. Phys.*, **G12**, 201.
Talmi, I, 1981. *Phys. Lett.*, **103B**, 177.
van Roosmalen, O S, and Dieperink, A E L, 1982. *Ann. Phys.*, **139**, 198.

van Roosmalen, O S, Iachello, F, Levine, R D, and Dieperink,
    A E L, 1983. *J. Chem. Phys.*, **79**, 2515.
Yang, B J, and Hwang, Z, 1987. *Phys. Rev.*, **C35**, 786.

9

ALGEBRAIC DESCRIPTION
OF ASYMMETRIC SHAPES IN NUCLEI

9.1 Introduction

Several medium and heavy nuclei are known to show low-lying states
of negative parity. In the actinide region such states occur as low as
250 keV, while in the rare earth region such states have been observed
around 800 keV.

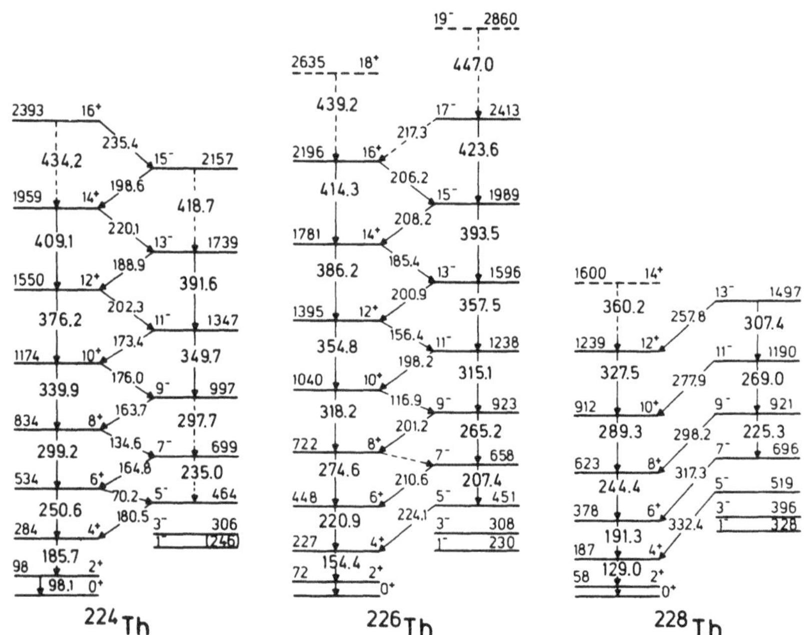

Fig. 9.1 Experimental level schemes of some neutron-deficient thorium
isotopes. (Taken from Schüler *et al.* (1986)).

Three different interpretations of the structure of these states
have been put forward. In one of them, the occurence of low-lying
negative parity states is attributed to the formation of small clusters
at the nuclear surface. In the second, these states are thought to

152

arise from octupole vibrations around a spherical or quadropoly deformed ground state. In the third, the nucleus is thought to acquire a permanent octupole deformation. The first interpretation can be described algebraically in the framework of the nuclear vibron model, discussed in Chapter 8. The algebraic description of the second interpretation is achieved with the introduction of a negative parity f-boson, discussed in Chapter 3. In this chapter we will deal with the algebraic formulation of the third interpretation. Before doing so, some discussion of the experimental data is appropriate. One must bear in mind that the regions of applicability of the different descriptions is still an open problem. Recent work on this problem can be found in Dubrovnik 1986.

9.2 Low-lying bands with octupole deformation

In Fig. 9.1 the level schemes of a few neutron-deficient thorium isotopes are given. In addition to the well-known ground state bands formed by levels of even angular momentum and positive parity, there are other bands formed by levels of odd angular momentum and negative parity. Similar observations can be made in Fig. 9.2, which shows the level schemes of some neutron-rich barium isotopes. It is interesting to study the energy displacement between the positive- and negative-parity bands. Quantitatively this can be achieved by introducing the function (Nazarewicz and Olanders 1985)

$$\delta E_L = E_{L^-} - \frac{(L+1)E_{(L-1)^+} + LE_{(L+1)^+}}{2L+1}, \qquad (9.1)$$

which represents the displacement of the positive and negative parity bands. This function has been plotted versus the angular momentum for some nuclei in Fig. 9.3. It is clear that with increasing angular momentum this function goes to zero. Thus as the spin increases the negative- and positive-parity levels become interlaced with spacings roughly appropriate to a single band. The fact that the levels in these bands are connected by strong electric quadrupole and strong electric dipole transitions further strengthens the argument. Such patterns of interlaced negative- and positive-parity bands having energies appropriate for a single band are known in reflection-asymmetric molecules. It is therefore assumed that nuclei showing such patterns acquire at high enough angular momentum stable octupole deformation, i.e. reflection asymmetric shapes.

In Fig. 9.3 it is clear that the evolution towards octupole deformation occurs at lower angular momentum in the neutron-deficient actinides than in the neutron-rich rare earths. Experimentally these effects were first identified in the actinide region, and subsequently the theoretical predictions for their presence in the rare-earth region were fulfilled.

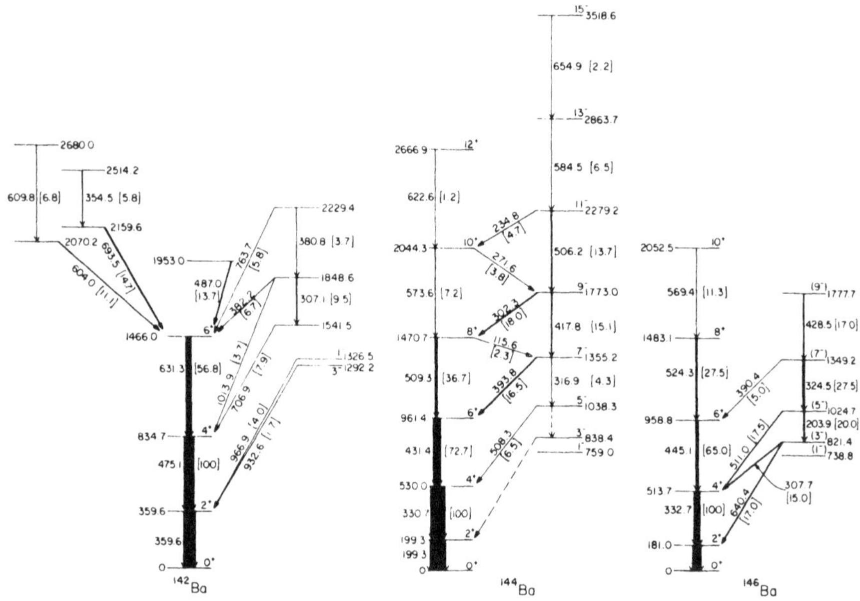

Fig. 9.2 Experimental energy levels of some neutron-rich barium isotopes. (Taken from Phillips *et al.* (1986)).

It is instructive to discuss briefly here the geometric approach (Leander *et al.* 1982) to octupole deformation, which was historically the first to be introduced. It has been suggested that nuclei with reflection-asymmetric shapes can be described by a potential looking like a double well (see Fig. 9.4). Octupole bands occur when the barrier separating the two wells is very high. If the barrier is not very high, the positive- and negative-parity bands are displaced relative to each other. If the barrier disappears entirely, an octupole vibrational spectrum is obtained. A double-well potential which has been used in this context is

$$V(\epsilon_3) = \frac{1}{2}C\epsilon_3^2 + D(e^{-\epsilon_3^2/\alpha^2} - 1), \tag{9.2}$$

illustrated in Fig. 9.5.

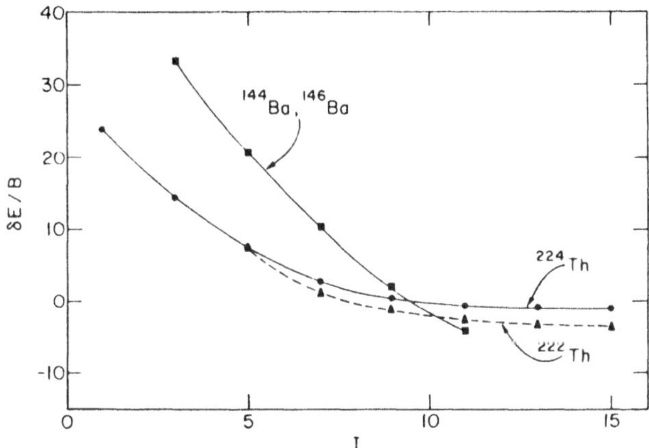

Fig. 9.3 The energy difference function $\delta E/B$ versus angular momentum. B is the rotational constant, deduced from the energy splittings of the 10^+ and 8^+ levels in each nucleus. (Taken from Phillips *et al.* (1986)).

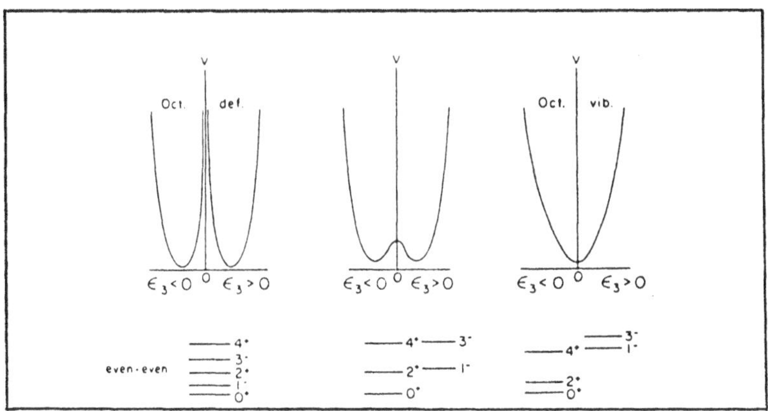

Fig. 9.4 Schematic representation of the dependence of a rotational band on the shape of the nuclear potential for reflection asymmetric deformation. (Taken from Leander *et al.* (1982)).

The breaking of reflection symmetry (Leander and Sheline 1984) is illustrated along with the breaking of rotational symmetry in Fig. 9.6. A spherical nucleus has rotational symmetry. This symmetry is broken if the nucleus is deformed, acquiring a prolate or oblate shape though still remaining reflection-symmetric. The middle figure shows the nucleus in its own frame of reference (called the **intrinsic frame**). The figure on the right is a picture of the same nucleus in the laboratory frame. The nucleus is rotating, giving rise to a rotational spectrum (i.e. to a sequence of energy levels 0^+, 2^+, 4^+, ..., obeying the L(L+1) rule). Breaking of the reflection symmetry occurs when the nucleus is deformed in a way that it is not reflection-symmetric any more, as illustrated in the middle of Fig. 9.6. In the laboratory frame (illustrated at the right part of Fig. 9.6), the kind of oscillation undertaken by this reflection-asymmetric nucleus is shown. This kind of oscillation gives rise to bands of both positive and negative parity. This effect is called **parity doubling**.

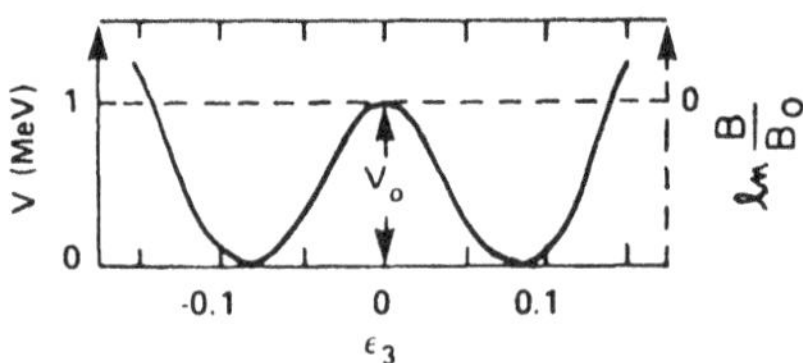

Fig. 9.5 Schematic representation of the double-well potential $V(\epsilon_3)$ used for the description of nuclei with reflection-asymmetric shapes. (Taken from Leander *et al.* (1982)).

9.3 The U(8) model

Quadrupole collective states have been described in IBM-1 in terms of s and d bosons of positive parity. It has been suggested that quadrupole–octupole collective spectra may be similarly described by using, in addition to the positive parity s and d bosons, bosons of negative parity carrying angular momentum 1 (p bosons) and 3 (f bosons).

The simplest model suitable to describe low-lying negative parity states of octupole nature would have been a model including s and f bosons only, spanning the dynamical group U(8) (1+7=8) (Engel and Iachello 1985). However, this model leads to bands with energies increasing as L(L+15) instead of L(L+1). Thus this model is not suitable for the description of nuclei, although it might be of interest in molecular physics.

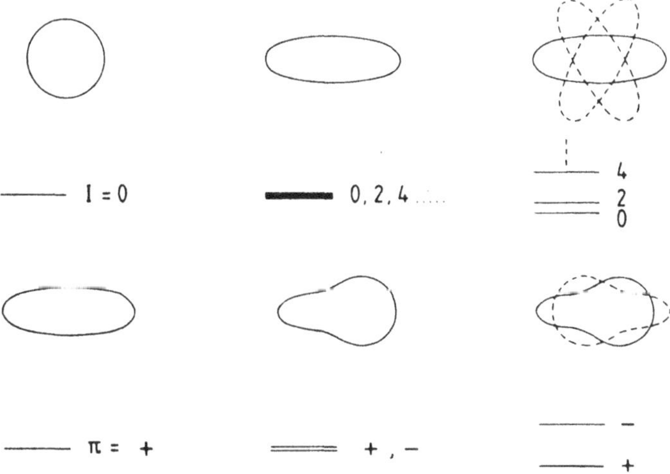

Fig. 9.6 Schematic representation of breaking of rotational symmetry (above) and breaking of reflection symmetry (below). The figures in the middle are in the intrinsic frame of reference, while those on the right are in the laboratory frame. (Taken from Leander and Sheline (1984)).

9.4 The U(11) model

The next simplest model suitable for the description of low-lying negative parity spectra is formed by considering s, p, and f bosons (Engel and Iachello 1985). It turns out that in this case rotational bands obeying the L(L+1) rule can be produced in a straightforward manner. This model has a U(11) dynamical group, spanned by the s, p, and f boson operators (1+3+7=11). States are thus characterized by the totally symmetric representations [N] of U(11), where N is the total number of bosons. Among the various group chains of U(11), the following chain exists

$$U(11) \supset U(10) \supset SU(3) \supset O(3) \supset O(2). \qquad (9.3)$$

The states corresponding to this group chain are fully determined as follows

$$|[N](N_b)\omega_b(\lambda_b, \mu_b)K_b L M >, \qquad (9.4)$$

where:

N is the total number of bosons labelling the irreps of U(11),

N_b is the total number of negative parity bosons (p and f) labelling the irreps of U(10),

(λ_b, μ_b) are the Elliott quantum numbers labelling the irreps of SU(3),

L is the angular momentum quantum number labelling the irreps of O(3),

M is the z-component of the angular momentum labelling the irreps of O(2),

and ω_b (K_b) is the additional quantum number required to fully determine the states, since the step from U(10) to SU(3) (from SU(3) to O(3)) is not fully decomposable.

In order to determine the allowed values of the quantum numbers (λ_b, μ_b) one needs the branching rules for $U(10) \supset SU(3)$, which can be found in Elliott (1958a, 1958b).

Standard techniques allow one to construct the Hamiltonian corresponding to the dynamic symmetry of the above chain. Its eigenvalues are

$$< H >= \alpha + \beta N_b + \gamma N_b^2 + \kappa_b C(\lambda_b, \mu_b) + \kappa' L(L + 1), \qquad (9.5)$$

with

$$C(\lambda, \mu) = \lambda^2 + \mu^2 + \lambda\mu + 3(\lambda + \mu). \qquad (9.6)$$

This spectrum resembles that of a strongly deformed, axially symmetric, reflection-asymmetric shape. Notice that since positive-parity states occur when N_b even, while negative-parity states arise when N_b odd, bands of opposite parity are displaced with respect to one another. However parity doubling, which is expected according to the geometrical arguments of Sec. 9.2, does not automatically occur in this model. In any case, exact parity doubling does not occur in real even–even nuclei. But, if we wish, we can obtain it here by introducing signature-dependent interactions, which act differently on states with odd or even N_b. In the description of nuclear rotational spectra, **signature** is a quantum number defined as $\sigma = (-1)^{L+K}$. Thus within a band of fixed K, the signature quantum number takes alternating signs in successive values of the angular momentum L. As a result, bands with $K \neq 0$ tend to separate into two families, distinguished by σ (see Bohr and Mottelson (1975) for more details). In the present case, signature will depend on N and N_b. Allowing signature-dependent terms, we obtain

$$< H >= \alpha + \beta N_b + \gamma [1 - \frac{(-1)^{N+N_b}}{N_b}] N_b^2$$

$$+\kappa_b[1 - \frac{(-1)^{N+N_b}}{N_b}]C(\lambda_b,\mu_b) + \kappa'L(L+1). \qquad (9.7)$$

A typical spectrum obtained with this Hamiltonian is shown in Fig. 9.7. We note that a rotational band is built on each vibrational band-head. In the case of a reflection-symmetric body, described in terms of s and d bosons only, all the odd parity bands are missing, as well as all the bands with K=odd.

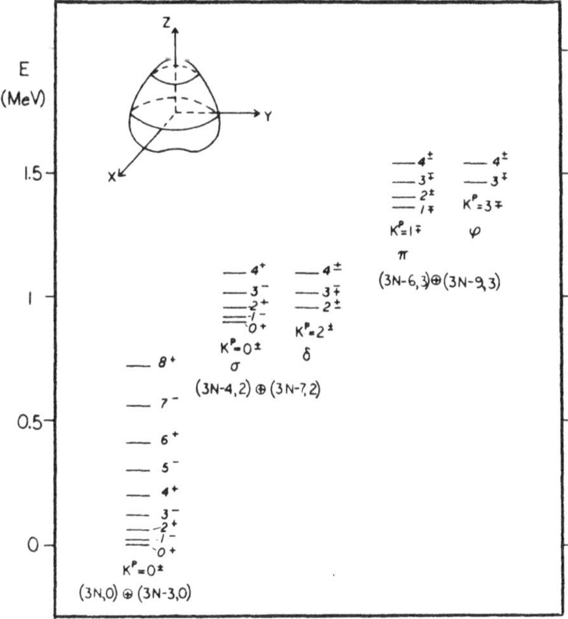

Fig. 9.7 Typical spectrum of a nucleus with octupole deformation, obtained from the U(11) model at the limit of large boson number. (Taken from Iachello (1985)).

9.5 The U(16) model

A more realistic model can be obtained by including d bosons along with the s, p, and f bosons (Engel and Iachello 1985). The dynamical group is now U(16). States are characterized by the totally symmetric representations [N] of that group. The positive-parity bosons, s and d, span a $U_a(6)$ subgroup, while the negative-parity bosons, p and f, span a $U_b(10)$ subgroup. The subscripts a and b are used to distinguish subgroups composed of bosons with positive and negative parity, respectively. It is an interesting feature of the model that both the $U_a(6)$ and $U_b(10)$ subgroups contain the same subgroup chains U(5), SU(3), and O(6). Here we are going to discuss only one of the many possible chains. Since we are mainly interested in describing axially symmetric rotations, we consider the chain

$$U(16) \supset U_a(6) \otimes U_b(10) \supset SU_a(3) \otimes SU_b(3) \supset SU(3) \supset O(3) \supset O(2).$$
$$(9.8)$$

Here SU(3) denotes the group formed by addition of the generators of $SU_a(3)$ and $SU_b(3)$. The states are fully determined as

$$|[N](N_a)(N_b)\omega_b(\lambda_a,\mu_a)(\lambda_b,\mu_b)(\lambda,\mu)KLM>, \qquad (9.9)$$

where N_a is the number of positive-parity bosons labelling the irreps of $U_a(6)$, while N_b is the number of negative-parity bosons labelling the irreps of $U_b(10)$. The rest of the quantum numbers are analogous to the quantum numbers used in labelling the states of U(11). In determining the allowed values of the quantum numbers (λ_a,μ_a) and (λ_b,μ_b), the branching rules for $U_a(6) \supset SU_a(3)$ and $U_b(10) \supset SU_b(3)$ are needed, respectively. In determining the allowed values of (λ,μ) one also needs the rules for decomposing the tensor product $(\lambda_a,\mu_a) \otimes (\lambda_b,\mu_b)$. These can be found by consideration of the Young tableaux.

It is a straightforward task to see that the Hamiltonian corresponding to the dynamical symmetry of the above chain has eigenvalues

$$< H >= \alpha + \beta N_b + \gamma N_b^2 + \kappa C(\lambda_a,\mu_a)$$

$$+\kappa_b C(\lambda_b,\mu_b) + \kappa C(\lambda,\mu) + \kappa' L(L+1). \qquad (9.10)$$

As in the case of U(11), parity doubling does not occur automatically. Again, it can be obtained if we allow for signature-dependent interactions to be present, when the expression of the eigenvalues of the Hamiltonian becomes

$$< H >= \alpha+\beta N_b+\gamma[1-\frac{(-1)^{N+N_b}}{N_b}]N_b^2+\kappa_a[1-\frac{(-1)^{N+N_b}}{N_a}]C(\lambda_a,\mu_a)$$

$$+\kappa_b[1-\frac{(-1)^{N+N_b}}{N_b}]C(\lambda_b,\mu_b)+\kappa[1-\frac{(-1)^{N+N_b}}{3N_b+2N_a}]C(\lambda,\mu)+\kappa' L(L+1).$$
$$(9.11)$$

A typical spectrum obtained with this Hamiltonian is shown in Fig. 9.8. Only the vibrational bandheads are shown. On top of each bandhead a rotational band is built, as in the case of U(11). There are three types of vibrations, arising from the three terms in the Hamiltonian proportional to κ, κ_a, κ_b. The first are labelled by the symbols Σ, Π, Δ, Φ, Γ, ..., the second (which are the usual quadrupole modes) are labelled by β and γ, while the third are labelled by σ, π, δ, ϕ, ...

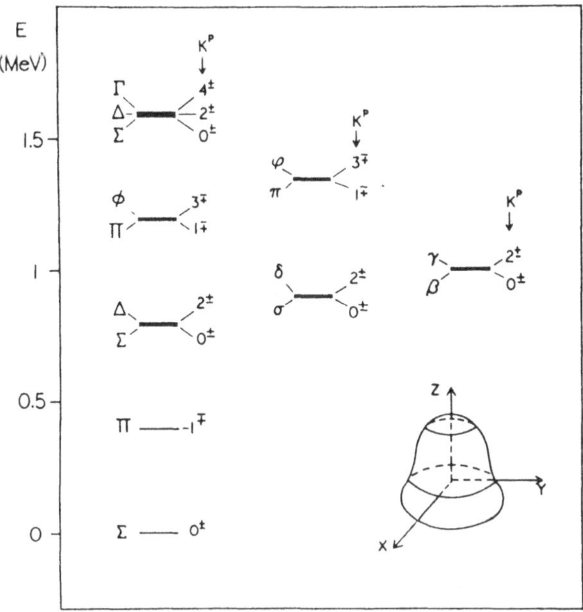

Fig. 9.8 Typical spectrum of a nucleus with both octupole and quadrupole deformation, obtained from the U(16) model at the limit of large boson number. (Taken from Iachello (1985)).

The reduction $U_b(10) \supset SU_b(3)$ was contained in the subgroup chain studied here. Thus this chain corresponds to the rotational limit. Other subgroup chains of U(16) can be obtained. An interesting one occurs by making the reduction $U_b(10) \supset U_b(3) \otimes U_b(7)$. This corresponds to the vibrational limit. It turns out that the spectra provided by the potential of Sec. 9.2 are intermediate between the spectra provided by these two chains. This is a subject of ongoing research.

References

The discussion of the various algebraic models used in this chapter has been based on Engel and Iachello (1985), and Iachello (1985). Much material on permanent octupole deformation in nuclei can be found in Dubrovnik 1986. The U(16) model has been recently extended to the description of odd mass nuclei (Engel, Frank and Pittel 1987).

Baranowski, R, Böning, K, and Sobiczewski, A, 1986. *Dubrovnik 1986*, **1**, 121.

Barrett, B R, 1986. *Dubrovnik 1986*, **2**, 1010.

Bohr, A, and Mottelson, B R, 1975. *Nuclear Structure*, Vol II. Benjamin, Reading.

Briançon, C, and Mikhailov, I N, 1986. *Dubrovnik 1986*, **1**, 131.

Butler, P A, *et al.*, 1986. *Dubrovnik 1986*, **1**, 101.

Castaños, O, Frank, A, Hess, P O, and Ogura, H, 1986. *Phys. Rev. Lett.*, **56**, 400.

Chasman, R R, 1986. *Dubrovnik 1986*, **1**, 5.

Cottle, P D, *et al.*, 1987. *Phys. Rev.*, **C35**, 1939.

Cottle, P D, and Bromley, D A, 1986. *Phys. Lett.*, **182B**, 129.

Eid, S A, Briançon, C, Hamilton, W D, Liang, C F, Walen, R J, and Green, V, 1986. *Dubrovnik 1986*, **1**, 126.

Elliott, J P, 1958a. *Proc. R. Soc. London*, Ser A **245**, 128.

Elliott, J P, 1958b. *Proc. R. Soc. London*, Ser A **245**, 562.

Engel, J, 1986. *Dubrovnik 1986*, **1**, 73.

Engel, J, and Iachello, F, 1985. *Phys. Rev. Lett.*, **54**, 1126.

Engel, J, Frank, A, and Pittel, S, 1987. *Phys. Rev.*, **C35**, 1973.

Gai, M, 1986. *Dubrovnik 1986*, **1**, 93.

Góźdź, A, Szpikowski, S, and Zając, K, 1980. *Nukleonika*, **25**, 1055.

Iachello, F, 1985. *Legnaro 1985*, 27.

Leander, G A, and Sheline, R K, 1984. *Nucl.Phys.*, **A413**, 375.

Leander, G A, Sheline, R K, Möller, P, Olanders, P, Ragnarsson, I, and Sierk, A J, 1982. *Nucl. Phys.*, **A388**, 452.

Nazarewicz, W, and Olanders, P, 1985. *Nucl. Phys.*, **A441**, 420.

Neugart, R, 1986. *Dubrovnik 1986*, **1**, 114.

Phillips, W R, Ahmad, I, Emling, H, Holzmann, R, Janssens, R V F, Khoo, T L, and Drigert, M W, 1986. *Phys. Rev. Lett.*, **57**, 3257.

Schüler, P, *et al.*, 1986. *Phys. Lett.*, **174B**, 241.
Sheline, R K, and Sood, P C, 1986. *Phys. Rev.*, **C34**, 2362.
Zylicz, J, 1986. *Dubrovnik 1986*, **1**, 79.

10

**ALGEBRAIC DESCRIPTION
OF GIANT DIPOLE RESONANCES**

10.1 Introduction

Giant dipole resonances have been known in nuclear physics for a long time. It is observed that the cross-section for the absorption of γ-rays by nuclei has a very big maximum at an energy of 20 to 25 MeV. This maximum has been associated with the existence in this energy region of states with isospin T=1 and high electric dipole moments, connected through very strong electric dipole (E1) transitions to the ground state. It is believed that these high electric dipole moments are created by the proton and the neutron distributions moving against each other. The associated states have then odd parity, being 1^- states. More than one 1^- state is observed, split by an energy difference of the order of 3 MeV. The distribution of the E1 strength among these states has also been studied.

Giant dipole resonances correspond to modes of nuclear excitation lying high in energy. In the IBM framework we have discussed so far only low-lying modes of excitation. In order to be able to describe the high-lying modes, additional bosons will be needed. In general they are denoted by capital letters, S^+, P^+, D^+, ..., in order to distinguish them from the bosons s^+, p^+, d^+, ..., used for the description of low-lying modes. By coupling the high-lying and low-lying modes of excitation one expects to be able to describe various related phenomena like the splitting of the giant resonances observed in triaxially deformed nuclei, as well as the ratio of dipole strengths for each state.

10.2 The P boson

The simplest algebraic description of giant dipole resonances involves the usual positive parity s and d bosons of IBM-1 for the description of low-lying excitation modes, as well as a P boson of odd parity for the description of the high-lying modes. Notice that the parity of the P boson has to be odd, since it is associated with states of odd parity. The s and d bosons span the group $U_a(6)$, while the P bosons

164

span a group $U_b(3)$. Analytical solutions (Rowe and Iachello 1983) can be found if one uses the group chain

$$U_a(6) \otimes U_b(3) \supset SU_a(3) \otimes SU_b(3) \supset SU(3) \supset O(3) \supset O(2). \quad (10.1)$$

Notice the formal similarity with the nuclear vibron model chain encountered in Sec. 8.10.

The P boson belongs to the (1,0) irrep of $SU(3)_b$. We assume that only one P boson is enough in order to describe the giant dipole resonance states. If (λ, μ) is the irrep describing the low-lying states, we get the following selection rules for the irreps of the coupled system

$$(\lambda, \mu) \otimes (1,0) = (\lambda + 1, \mu) \oplus (\lambda - 1, \mu + 1) \oplus (\lambda, \mu - 1). \quad (10.2)$$

This relation holdes if $\mu \neq 0$. In the case of $\mu = 0$, the following simpler rule holds

$$(\lambda, 0) \otimes (1,0) = (\lambda + 1, 0) \oplus (\lambda - 1, 1). \quad (10.3)$$

Remember that axially symmetric nuclei have ground state bands belonging to irreps with $\mu = 0$, while the ground state band belongs to irreps with $\mu \neq 0$ in nuclei with triaxial deformation. In the case of axially symmetric nuclei, one obtains after the coupling two bands, with K^π equal to 0^- and 1^-, respectively. Both of them contain $L^\pi = 1^-$ states. In the case of triaxial nuclei, several bands are obtained. However, only three of them contain 1^- states. To see this, consider as an example the case $\mu = 2$. From the right hand side of the selection rule we see that we obtain irreps with $\mu = 2$ (which contains $K^\pi = 0^-, 2^-$), $\mu = 3$ (which contains $K^\pi = 1^-, 3^-$), and $\mu = 1$ (which contains $K^\pi = 1^-$). Out of these 5 bands, only the bands with $K^\pi = 0^-, 1^-, 1^-$, contain states with $L^\pi = 1^-$.

10.3 Splitting of giant dipole resonances

10.3.1 An exactly soluble Hamiltonian

In what follows, we will try to give the simplest possible description (Rowe and Iachello 1983) of the splitting of giant resonances and the distribution of E1 strength in the algebraic framework defined so far. Although we can write down in the usual way the most general Hamiltonian corresponding to the dynamic symmetry of the chain given above, we follow a slightly different path, which is physically more transparent, with the aim to keep the number of free parameters in the Hamiltonian to an absolute minimum. Two terms are absolutely necessary in the Hamiltonian: a term giving rise to the high-lying vibrations of odd parity, which will obviously contain only the P boson, and a term describing the coupling between the high-lying and the low-lying modes, which will contain all bosons, s, d, and P. The first term should essentially provide the excitation energy of the group of the 1^- states, typically 20–25 MeV. The second term should account for the splitting of the 1^- states (typically of the order of 3 MeV). Terms describing the low-lying modes (i.e. terms involving s and d bosons only) are expected to be much smaller in magnitude in comparison to the two terms already mentioned and so they can be neglected. Remember that the low-lying modes give rise to states with excitation energy typically of a few keV. Taking all this into account, we write the Hamiltonian

$$H = \frac{h}{2\pi}\omega n_P - 2\kappa(Q_a \odot Q_b), \qquad (10.4)$$

where n_P is the number of P-bosons, Q_a is the quadrupole generator of $SU_a(3)$ (corresponding to the s-d degrees of freedom), and Q_b is the quadrupole generator of $SU_b(3)$ (corresponding to the P degree of freedom), according to the standard definitions. Defining

$$Q = Q_a + Q_b, \qquad (10.5)$$

$$L = L_a + L_b, \qquad (10.6)$$

and using the relation

$$C_{2SU3} = 2(Q \odot Q) + \frac{3}{4}(L \odot L), \qquad (10.7)$$

and the corresponding relations for $SU_a(3)$ and $SU_b(3)$, we find that

$$(Q_a \odot Q_b) = \frac{1}{2}(C_{2SU3} - C_{2SU_a3} - C_{2SU_b3}) - \frac{3}{8}(L_a \odot L_b). \quad (10.8)$$

The $(L_a \odot L_b)$ term is expected to be much smaller than the other terms, and so it is omitted. (Notice that this term is not diagonal in the basis provided by the above chain. This is so because L_a and L_b are not quantum numbers used in characterizing the states, but $L = L_a + L_b$ is.) After making this assumption, the Hamiltonian can be diagonalized exactly. Since we assume that only one P boson is present, the eigenvalue of the first term will simply be $\frac{h}{2\pi}\omega$. For convenience, we will denote the eigenvalue of the C_{2SU3} Casimir operator in the (λ,μ) irrep as $e(\lambda,\mu)$, i.e.

$$e(\lambda,\mu) = \lambda^2 + \mu^2 + \lambda\mu + 3(\lambda + \mu). \tag{10.9}$$

The eigenvalues of the Casimir operators of $SU_a(3)$ and $SU_b(3)$ will also be denoted in the same fashion. In this notation the eigenvalues of the Hamiltonian are written as

$$E = \frac{h}{2\pi}\omega + \kappa[e(\lambda_u,\mu_u) + e(\lambda_b,\mu_b) - c(\lambda,\mu)]. \tag{10.10}$$

Since we assumed that only one P boson is present, we always have $(\lambda_b,\mu_b) = (1,0)$. In the following subsections the interesting cases of axially symmetric nuclei and of triaxial nuclei will be discussed.

10.3.2 Axially symmetric nuclei

For axially symmetric nuclei, where $(\lambda_a,\mu_a) = (\Lambda,0)$, the giant dipole resonance is predicted to split into two states with energies

$$E_+ - \frac{h}{2\pi}\omega = \kappa[e(\Lambda,0) + e(1,0) - e(\Lambda - 1,1)] = \kappa(\Lambda + 3), \tag{10.11}$$

$$E_- - \frac{h}{2\pi}\omega = \kappa[e(\Lambda,0) + e(1,0) - e(\Lambda + 1,0)] = -2\kappa\Lambda. \tag{10.12}$$

For $\Lambda >> 3$, one obtains

$$\frac{E_+ - \frac{h}{2\pi}\omega}{E_- - \frac{h}{2\pi}\omega} = -\frac{1}{2}. \tag{10.13}$$

This result has also been obtained theoretically by the geometrical method. The ratio of the dipole strengths going to these two states can be calculated analytically, and the final result is

$$\frac{S_+}{S_-} = \frac{2\Lambda}{\Lambda + 3}. \tag{10.14}$$

At the limit $\Lambda >> 3$ one obtains the value 2 for this ratio, as in the geometrical picture.

10.3.3 Triaxial nuclei

For triaxial nuclei, where $(\lambda_a, \mu_a) = (\Lambda, M)$, i.e. $\mu_a \neq 0$, the giant dipole resonance is predicted to split into three states with energies

$$E_1 - \frac{h}{2\pi}\omega = \kappa[e(\Lambda, M) + e(1,0) - e(\Lambda + 1, M)] = -\kappa(2\Lambda + M),$$
$$(10.15)$$
$$E_2 - \frac{h}{2\pi}\omega = \kappa[e(\Lambda, M) + e(1,0) - e(\Lambda - 1, M + 1)] = -\kappa(\Lambda - M + 3),$$
$$(10.16)$$
$$E_3 - \frac{h}{2\pi}\omega = \kappa[e(\Lambda, M) + e(1,0) - e(\Lambda, M - 1)] = \kappa(\Lambda + 2M + 6).$$
$$(10.17)$$

The ratio of the dipole strengths can again be calculated analytically. For example, in the case of $M = 2$ and $\Lambda >> 6$ one obtains

$$S_1 : S_2 : S_3 = 3 : 4 : 2. \tag{10.18}$$

10.4 The general Hamiltonian

We studied here a special case of the algebraic model which could be solved analytically. In the general case this is not possible , and numerical calculations have to be undertaken. Computer programs, called PBOSON and PBOSONT (P. van Isacker, 1980, University of Ghent, Belgium) exist for this purpose. In the general case, the Hamiltonian can be written as

$$H = H_{sd} + H_P + H_{Psd}, \tag{10.19}$$

where H_{sd} describes the low-lying modes in terms of the usual s and d bosons of IBM-1, H_P describes the high-lying modes in terms of the P boson of odd parity, and H_{Psd} describes the coupling between the two modes. Of them, H_{sd} has the form studied in Chapter 1, H_P takes the simple vibrational form

$$H_P = \epsilon_P n_P, \tag{10.20}$$

where n_P is the number of P-bosons, and the coupling Hamiltonian H_{Psd} can be written as

$$H_{Psd} = b_0((d^+ \otimes \tilde{d})^0 \odot (P^+ \otimes \tilde{P})^0) + b_1((d^+ \otimes \tilde{d})^1 \odot (P^+ \otimes \tilde{P})^1)$$

$$+b_2([(s^+ \otimes \tilde{d} + d^+ \otimes \tilde{s})^2 + \chi(d^+ \otimes \tilde{d})^2] \odot (P^+ \otimes \tilde{P})^2). \qquad (10.21)$$

10.5 E1 transition probabilities

In order to calculate E1 transition probabilities, one needs the expression of the electric dipole operator in terms of the bosons, which is

$$D_\mu^1 = D_0(P^+ \otimes \tilde{s} + s^+ \otimes \tilde{P})_\mu^1 + D_2(P^+ \otimes \tilde{d} + d^+ \otimes \tilde{P})_\mu^1. \qquad (10.22)$$

The dipole strength to the nth 1^- state is then given by the square of the reduced matrix element of this operator between the ground state 0_1^+ and the 1^- state

$$S_n = |<1_n^-||D^1||0_1^+>|^2. \qquad (10.23)$$

After finding the dipole strengths, several properties of the giant dipole resonances can be calculated, such as the photoabsorption cross sections and elastic and inelastic photon scattering cross sections. Here, we will not pursue this problem further, addressing the interested reader to the original references (see, for example, Maino, Ventura, Zuffi, and Iachello 1984, 1985).

10.6 Giant monopole and quadrupole resonances

So far only the giant dipole resonance has been discussed here. A similar algebraic approach to the high-lying giant monopole and quadrupole resonances can be formulated. In that case one will have to introduce S and D bosons of odd parity (we use capital letters in order to distinguish them from the s and d bosons of IBM-1, which have even parity), which belong to the (2,0) irrep of SU(3). We will not treat this problem here, either, referring the interested reader to Ventura *et al.* (1986).

10.7 Excitation mechanisms giving rise to 1^- states

It is instructive at this point to consider at once all possible excitation mechanisms whice give rise to 1^- states in nuclei. Fig. 10.1 gives a schematic representation of them, as well as of the electric dipole (E1) strength connecting the 1^- states to the ground state, for the nucleus ^{150}Nd. Lowest in energy one finds the quadrupole excitations, which do not give rise to odd parity states. Then, one encounters the (reflection-asymmetric) states with octupole defomation. Subsequently, the states due to clustering are encountered. Finally, at higher energies, one meets the giant dipole resonance states.

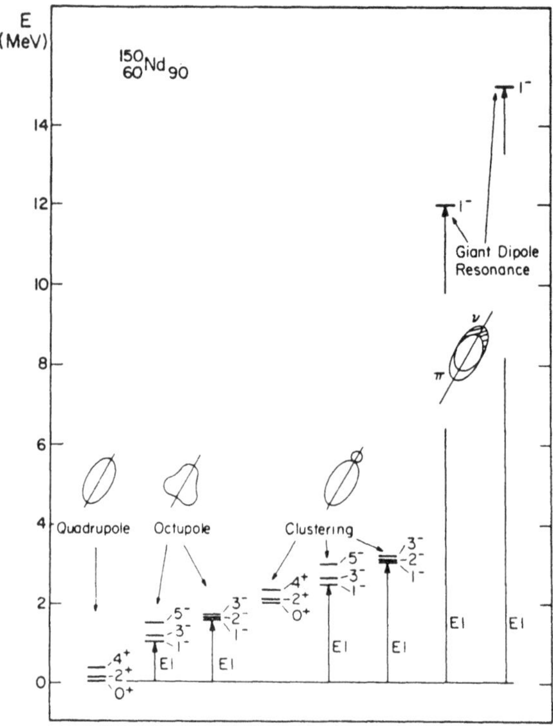

Fig. **10.1** Schematic representation of the distribution of E1 strength expected in ^{150}Nd. (Taken from Iachello (1985)).

It should be stressed here that ^{150}Nd is a nucleus for which there is experimental evidence that all these types of excitation coexist. This is not necessarily true for any nucleus. In fact, the distinction between octupole and cluster configurations is a point of high interest in current research (Dubrovnik 1986). In ^{18}O, for example, it is thought that clustering plays a dominant role, but in ^{218}Ra both mechanisms have been suggested. Finally, the *negative* parity $L = 1$ states considered here must not be confused with the *positive* parity $L = 1$ mixed symmetry states encountered in Chapter 5.

References

This chapter has been based on Rowe and Iachello (1983). The "state of the art" can be found in Dubrovnik 1986.

Chen, H T, 1981. *Phys. Rev. Lett.*, **47**, 1382.
Hahne, F J W, and Scholtz, F G, 1986. *Dubrovnik 1986*, **2**, 963.
Iachello, F, 1985. *Legnaro 1985*, 27.
Maino, G, Ventura, A, Zuffi, L, and Iachello, F, 1984. *Phys. Rev.*, **C30**, 2101.
Maino, G, Ventura, A, Zuffi, L, and Iachello, F, 1985. *Phys. Lett.*, **152B**, 17.
Maino, G, Ventura, A, Zuffi, L, and van Isacker, P, 1985. *Varenna 1985*, 423.
Morrison, I, and Weise, J, 1982. *J. Phys.*, **G8**, 687.
Nathan, A M, 1986. *Dubrovnik 1986*, **1**, 385.
Rowe, D J, and Iachello, F, 1983. *Phys. Lett.*, **130B**, 231.
Scholtz, F G, 1985. *Phys. Lett.*, **151B**, 87.
Scholtz, F G, and Hahne, F J W, 1983. *Phys. Lett.*, **123B**, 147.
Scholtz, F G, and Hahne, F J W, 1986. *Phys. Rev.*, **C34**, 693.
Turrini, S, Maino, G, and Ventura, A, 1986. *Dubrovnik 1986*, **1**, 397.
Ventura, A, Maino, G, van Isacker, P, and Zuffi, L, 1986. *Dubrovnik 1986*, **2**, 957.

11

THE INTERACTING BOSON–FERMION MODEL

11.1 Introduction

The Interacting Boson Model (IBM) and its various extensions discussed so far, are capable of describing even–even nuclei only. In the case of odd nuclei, bosons alone are not adequate in describing their spectra, because at least one fermion (the "odd" one) will always remain uncoupled. An extended algebraic model, including both bosons and fermions, is necessary then. The extension of IBM including a single fermion in addition to the bosons is called the **Interacting Boson–Fermion Model (IBFM)** (Iachello 1982, 1984) and will be discussed here.

In its simplest version, IBFM assumes that low-lying spectra of odd–even nuclei are dominated by excitations due to the valence particles, while the core is assumed to be completely inert. Furthermore, it is assumed that, whenever possible, the valence particles join together into correlated pairs with angular momenta 0 and 2. As in the case of IBM, the pairs are treated as bosons, denoted s and d, respectively. The subscript π (ν), distinguishes protons and neutrons. In odd–even nuclei, at least one particle must be necessarily unpaired. Thus, a description of these nuclei requires the simultaneous introduction of both bosons (to account for correlated pairs) and fermions (to account for the unpaired particles). As in IBM, because of the particle–hole conjugation properties, bosons and fermions are counted from the nearest closed shells.

Example 1 The nucleus ^{115}Cs has 55 protons and 60 neutrons. Thus it has $55 - 50 = 5$ valence protons and $60 - 50 = 10$ valence neutrons. In IBFM this nucleus is treated as composed of $N_\pi = 2$ proton bosons plus $N_\nu = 5$ neutron bosons plus $M_\pi = 1$ proton fermions plus $M_\nu = 0$ neutron fermions.

Example 2 The nucleus ^{125}Xe has 54 protons and 71 neutrons. Thus it has $54 - 50 = 4$ valence protons and $82 - 71 = 11$ valence neutrons. In IBFM this nucleus is treated as composed of $N_\pi = 2$ proton bosons plus $N_\nu = 5$ neutron bosons plus $M_\pi = 0$ proton fermions plus $M_\nu = 1$ neutron fermion.

The version of the model in which distinction is made between protons and neutrons is called **IBFM-2**, or proton–neutron IBFM (pn-IBFM). Sometimes, a simpler version of the model is used, in

172

which no distinction between protons and neutrons is made. This simpler model is called **IBFM-1**. In IBFM-1 the number of bosons is chosen to be $N = N_\pi + N_\nu$, while the number of fermions is chosen to be $M = M_\pi + M_\nu$. In these notes, only IBFM-1 will be considered. It must be emphasized that the IBFM model is equivalent to the Odd Truncated Quadrupole Model (OTQM) of Paar *et al.* (Paar 1984, Paar *et al.* 1982) (otherwise called the Interacting Boson–Fermion Fermion Model (IBFFM)) (Lopac *et al.* 1986).

11.2 The most general Hamiltonian

The most general Hamiltonian in IBFM-1 will contain three parts: a purely bosonic part, a purely fermionic part, and a boson fermion part describing the interaction between bosons and fermions. Thus we can write symbolically

$$H = H_B + H_F + V_{BF}. \tag{11.1}$$

If we make again the assumption that only one- and two-body terms are present in H, then the bosonic part of the Hamiltonian is exactly the Hamiltonian of the corresponding IBM-1 problem. Here we will focus our attention on the last two parts of H. In order to avoid confusion, the superscript B will be attached to groups associated with bosons, while the superscript F will be used in conjunction with groups associated with fermions. It turns out that the specific form of the last two terms in the Hamiltonian explicitly depends on the microscopic levels the odd fermion can occupy. In order to be able to write it down, we need to introduce suitable fermion operators first.

11.3 Definitions of fermion operators

In IBFM-1, besides the boson operators introduced in IBM-1, we need to introduce fermion operators as well. While in the case of bosons the angular momentum values were always 0 and 2, in the case of fermions the values of angular momentum depend on the particular orbitals available to the single particle. Therefore the fermion creation operators will be denoted by $a_{j\mu}^+$ ($\mu = -j, -j + 1, \ldots, -1/2, +1/2, \ldots, j-1, j$) and the annihilation operators by $a_{j\mu}$. Again, spherical tensor operators can be constructed from the annihilation operators by using

$$\tilde{a}_{j,\mu} = (-1)^{j-\mu} a_{j,-\mu}. \tag{11.2}$$

The anti-commutation relations of the fermion operators can be written as

$$\{a_{j\mu}, a^+_{j'\mu'}\} = \delta_{jj'}\delta_{\mu\mu'}. \tag{11.3}$$

As in the case of bosons, we introduce the operators

$$A^k_\kappa(jj') = [a^+_j \otimes \tilde{a}_{j'}]^k_\kappa. \tag{11.4}$$

It so happens that the commutators of these operators among themselves satisfy the Lie algebra U(m), the group of unitary transformations in m dimensions, where $m = \sum_j(2j+1)$. Using the notation introduced above, we will denote it by $U^F(m)$. The summation in the definition of m indicates that all the orbitals available to the single particle have to be taken into account.

Example In Fig. 1.1, the valence neutrons occupy levels with $j = \frac{5}{2}, \frac{7}{2}, \frac{11}{2}, \frac{1}{2}, \frac{3}{2}$.

11.4 The fermion algebra $SU^F(4)$

Here we will consider the simple case (Iachello 1984) in which the single particle can occupy only a $j = 3/2$ level, implying $m = 4$. Thus in this special case the relevant fermion algebra is $U^F(4)$. This algebra contains $16 = 4^2$ operators $A^k_\kappa(jj')$, which, written down explicitly , read

$$A^0_0(\frac{3}{2}, \frac{3}{2}) = [a^+ \otimes \tilde{a}]^0_0, \tag{11.5}$$

$$A^1_\mu(\frac{3}{2}, \frac{3}{2}) = [a^+ \otimes \tilde{a}]^1_\mu, \tag{11.6}$$

$$A^2_\mu(\frac{3}{2}, \frac{3}{2}) = [a^+ \otimes \tilde{a}]^2_\mu, \tag{11.7}$$

$$A^3_\mu(\frac{3}{2}, \frac{3}{2}) = [a^+ \otimes \tilde{a}]^3_\mu. \tag{11.8}$$

11.5 The most general Hamiltonian for $U^B(6) \otimes U^F(4)$

When the boson operators form the algebra $U^B(6)$ and the fermion operators form the algebra $U^F(4)$, the boson–fermion system has the symmetry $U^B(6) \otimes U^F(4)$. The purely fermionic part of the Hamiltonian, including only one-body and two-body terms, by analogy to the bosonic part, will be

$$H_F = E'_0 + \eta \sum_\mu a^+_\mu a_\mu + c_1((a^+ \otimes a^+)^1 \otimes (\tilde{a} \otimes \tilde{a})^1)^0$$

$$+c_3((a^+ \otimes a^+)^3 \otimes (\tilde{a} \otimes \tilde{a})^3)^0. \tag{11.9}$$

As we have already remarked, in the IBFM approach an odd nucleus is always described by a system of bosons plus one fermion. Since only one fermion is present, the two-body interactions indicated by the last two terms can be neglected. Then the only new parameter introduced by going from an even–even nucleus (bosons only) to an odd–even nucleus is the single particle energy of the sublevel where the odd proton is lying. Similarly, in the case of an even–odd nucleus, the only new parameter is the single particle energy of the sublevel where the odd neutron is lying. This parameter has been called η in the above Hamiltonian. If one has a situation where fermions occupy more than one sublevel, the single particle energies of all these sublevels will be needed. These can be taken from shell-model calculations. For example, the single particle energies of the proton sublevels in the 50–82 major shell are listed below in MeV (counted from the $3s_{1/2}$ level)

$$
\begin{array}{ll}
3s_{1/2} & 0 \\
2d_{3/2} & -0.35 \\
1h_{11/2} & -1.34 \\
2d_{5/2} & -1.67 \\
1g_{7/2} & -3.48
\end{array}
$$

The part of the Hamiltonian describing the boson–fermion interaction will be

$$V_{BF} = w_0'((s^+ \otimes \tilde{s})^0 \otimes (a^+ \otimes \tilde{a})^0)^0 + w_2'((d^+ \otimes \tilde{s} + s^+ \otimes \tilde{d})^2 \otimes (a^+ \otimes \tilde{a})^2)^0$$

$$+ w_0((d^+ \otimes \tilde{d})^0 \otimes (a^+ \otimes \tilde{a})^0)^0 + w_1((d^+ \otimes \tilde{d})^1 \otimes (a^+ \otimes \tilde{a})^1)^0$$

$$+ w_2((d^+ \otimes \tilde{d})^2 \otimes (a^+ \otimes \tilde{a})^2)^0 + w_3((d^+ \otimes \tilde{d})^3 \otimes (a^+ \otimes \tilde{a})^3)^0. \tag{11.10}$$

Remark that even in the simplest case of only one fermion in a single sublevel, which we are studying, 6 new parameters appear in this term of the Hamiltonian. It is clear that with greater number of sublevels available to the odd fermion the number of parameters will become too large for making any phenomenological calculations. Thus, in what follows we will concentrate our efforts in these cases in which the symmetry of the problem allows for the exact diagonalization of the Hamiltonian to be carried out.

11.6 Chains of subalgebras of $U^F(4)$

Once the algebraic structure of the participating fermions has been identified, one needs to find all possible chains of subalgebras, as in the case of bosons. This problem can be very difficult, especially for large values of m. Here we will consider again as an example the special case $j = 3/2$ (i.e. m=4). In this case, only one possible chain of subalgebras exists.

A) From the 16 operators given above, delete $A_0^0(3/2, 3/2)$. The remaining 15 operators form the algebra SU(4), the group of special unitary transformations in 4 dimensions.

B) From the remaining 15 operators, delete $A_\mu^2(3/2, 3/2)$. The remaining 10 operators are the generators of the algebra Sp(4), the group of symplectic transformations in 4 dimensions.

C) From the remaining 10 operators, delete $A_\mu^3(3/2, 3/2)$. The remaining three operators, $A_\mu^1(3/2, 3/2)$, form the algebra SU(2).

D) From the remaining three operators keep only $A_0^1(3/2, 3/2)$. This operator is the generator of the algebra SO(2).

Thus the fermionic chain in the case m=4 is

$$U^F(4) \supset SU^F(4) \supset Sp^F(4) \supset SU^F(2) \supset SO^F(2). \qquad (11.11)$$

11.7 Labelling the basis of $U^F(4)$

This group chain can now be used to classify states and to provide a basis in which numerical diagonalization of the Hamiltonian can be undertaken. The quantum numbers needed in the complete classification scheme are given below. Again, the fact that we are dealing with a system of fermions only, which thus has to be totally antisymmetric, results in certain simplifications, as will be indicated below.

A) The irreps of $U^F(4)$ and $SU^F(4)$ are labelled by the total number of fermions M, and are denoted by $\{M\}$. Notice that in general 4 quantum numbers are needed for the full characterization of the irreps of U(4), according to the Appendices. However, because here we deal with a fermionic system, the possible irreps have to be totally antisymmetric. It is easy to see that only the irreps represented by a Young tableau having only one column are totally antisymmetric. But these irreps can be fully characterized by only one quantum number, the number of boxes in this single column. We see that the pure fermionic nature of the system under study results in a major simplification. Similar simplifications occur for the irreps of $Sp^F(4)$, which follow below.

B) The irreps of $Sp^F(4)$ are labelled by the quantum number Θ, denoted by $(\Theta, 0)$.

C) The irreps of $SU^F(2)$ are labelled by the angular momentum J.

D) The irreps of $SO^F(2)$ are labelled by the z-component of the angular momentum, denoted by M_J.

The classification scheme for the fermion group chain is given below.

U(4)	Sp(4)	SU(2)
M	$(\Theta,0)$	J
0	(0,0)	0
1	(1,0)	3/2
2	(2,0)	1 , 3
$3\equiv\bar{1}$	(1,0)	3/2
$4\equiv0$	(0,0	0

11.8 An example of dynamical symmetry

Our next task will be to solve the eigenvalue problem for the boson–fermion Hamiltonian. This can be solved analytically only in special cases. One such special case occurs when the boson and fermion group chains are identical. We say then that a dynamical symmetry of the **Class BF-1** occurs. This is in fact the case in the problem under study. (Another case, referred to as dynamical symmetries of **Class BF-2** will be treated later on). Before attacking the real problem, we will try to clarify our method through an example familiar from simple angular momentum algebra.

Example Consider the Hamiltonian (Iachello 1984)

$$H = a_L \vec{L}^2 + a_S \vec{S}^2 + a_{LS} \vec{L} \cdot \vec{S}. \tag{11.12}$$

Coupling the angular momenta

$$\vec{J} = \vec{L} + \vec{S}, \tag{11.13}$$

one can obtain a complete classification scheme for the states. The states are then labelled as $|LSJ>$. Furthermore, one can rewrite H in terms of the Casimir invariants $\vec{L}^2$, $\vec{S}^2$, $\vec{J}^2$, as follows

$$H = (a_L - \frac{1}{2}a_{LS})\vec{L}^2 + (a_S - \frac{1}{2}a_{LS})\vec{S}^2 + \frac{1}{2}a_{LS}\vec{J}^2. \tag{11.14}$$

By now it is apparent that H is diagonal in the basis $|LSJ>$, yielding for the eigenvalues the result

$$E(L,S,J) = (a_L - \frac{1}{2}a_{LS})L(L+1)$$

$$+(a_S - \frac{1}{2}a_{LS})S(S+1) + \frac{1}{2}a_{LS}J(J+1). \tag{11.15}$$

The values of J are given by the well known rule

$$|L - S| \leq J \leq |L + S|. \tag{11.16}$$

11.9 A dynamical symmetry of the Class BF-1

11.9.1 The boson–fermion group chain

In group theoretical language, the addition of two angular momenta is the combination of operators having the same Lie algebra. Thus, this method can be applied if the boson and fermion chains are identical (i.e., in group theoretical language if the groups forming them are isomorphic). In the special fermionic case studied here, this condition is satisfied, if we consider that the boson Hamiltonian has the $O^B(6)$ dynamical symmetry. In fact, the algebras $O^B(6)$ and $SU^F(4)$ are isomorphic , and so are $O^B(5)$ and $Sp^F(4)$, $O^B(3)$ and $SU^F(2)$, $O^B(2)$ and $O^F(2)$. These isomorphisms are denoted in symbolic form as follows

$$O^B(6) \approx SU^F(4), \tag{11.17}$$

$$O^B(5) \approx Sp^F(4), \tag{11.18}$$

$$O^B(3) \approx SU^F(2), \tag{11.19}$$

$$O^B(2) \approx O^F(2). \tag{11.20}$$

A consequence of these isomorphisms is that one can combine boson and fermion operators in a way resembling addition of angular momenta. In particular, one can form the following sums of generators of $O^B(6)$ and generators of $SU^F(4)$ (using appropriate numerical coefficients)

$$D_\mu^1 = G_\mu^1(dd) - \frac{1}{\sqrt{2}}A_\mu^1(\frac{3}{2},\frac{3}{2}), \tag{11.21}$$

$$D_\mu^2 = G_\mu^2(ds) + G_\mu^2(sd) + A_\mu^2(\frac{3}{2},\frac{3}{2}), \tag{11.22}$$

$$D_\mu^3 = G_\mu^3(dd) + \frac{1}{\sqrt{2}}A_\mu^3(\frac{3}{2},\frac{3}{2}). \tag{11.23}$$

These 15 operators happen to be the generators of the spinor group Spin(6). Deleting the 5 operators D_μ^2 one obtains the 10 generators of the spinor group Spin(5). Deleting from the remaining 10 operators the seven operators D^3_μ one is left with the 3 generators of Spin(3). Finally, deleting the 2 operators $D^1_{\pm 1}$ one is left with D^1_0, the generator of Spin(2). These considerations lead to the following group chain (Iachello and Kuyucak 1981)

$$U^B(6) \otimes U^F(4) \supset O^B(6) \otimes SU^F(4) \supset Spin(6)$$

$$\supset Spin(5) \supset Spin(3) \supset Spin(2). \tag{11.24}$$

11.9.2 Labelling of the basis

Having found a common boson–fermion chain, we have now to construct the basis. This is done by taking products of boson and fermion representations (in a way similar to the derivation of the values of J given in the example). First, one has to convert the representations of $O^B(6)$ and $SU^F(4)$ into representations of Spin(6).

Example The representation [N,0,0] of $O^B(6)$ yields the tensor representation [N,0,0] of Spin(6), while the representation [1,0,0,0] of $SU^F(4)$ yields the fundamental spinor representation [1/2,1/2,1/2] of Spin(6).

Taking products of boson and fermion representations one can then construct a classification scheme for the combined chain. The necessary quantum numbers are given next to each group in the list below

$$
\begin{array}{ll}
U^B(6) & [N] \\
U^F(4) & \{M\} \\
O^B(6) & \Sigma \\
Spin(6) & (\sigma_1, \sigma_2, \sigma_3) \\
Spin(5) & (\tau_1, \tau_2) \\
Spin(3) & J \\
Spin(2) & M_J
\end{array}
$$

The values of the quantum numbers Σ, $(\sigma_1, \sigma_2, \sigma_3)$, (τ_1, τ_2), J, M_J contained in each representation $[N] \otimes \{M\}$ of $O^B(6) \otimes U^F(4)$ can be found by standard group theoretical methods. Due to the fact that the step from Spin(5) to Spin(3) is not fully decomposable, an additional quantum number, ν_Δ, is needed to fully classify the states, as in the case of the $O^B(6)$ chain. The complete classification scheme for the states is then

$$|N, M, \Sigma, (\sigma_1, \sigma_2, \sigma_3), (\tau_1, \tau_2), \nu_\Delta, J, M_J > . \tag{11.25}$$

11.9.3 The Hamiltonian and its eigenvalues

Analytic solutions to the eigenvalue problem of the Hamiltonian can then be obtained by writing H in terms of Casimir operators of the groups occuring in the above chain. Assuming again that only one- and two-body terms are present in H, only linear and quadratic Casimir operators have to be employed. One begins with the linear and quadratic Casimir operators of $U^B(6)$ and $U^F(4)$. Within the representations $[N]$ and $\{M\}$, these operators have eigenvalues which depend only on the number of bosons N and the number of fermions M. Since M and N are constant in a given nucleus, these terms only contribute an overall constant in the calculated energy levels. Thus they can be omitted if one is interested only in excitation energies.

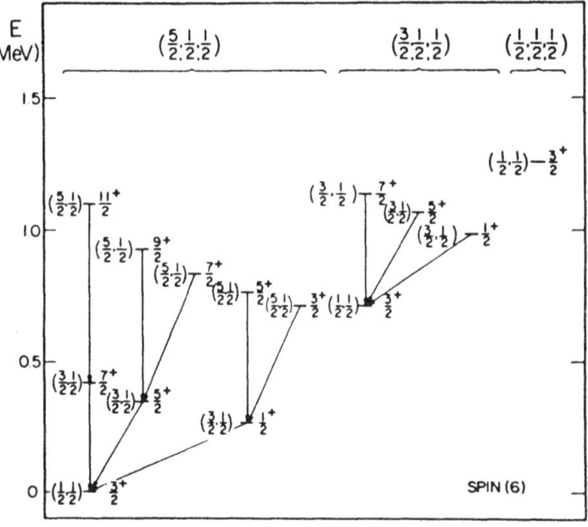

Fig. 11.1 A typical spectrum of an odd–even nucleus with Spin(6) symmetry, obtained with N=2, M=1. The numbers on top of the figure denote the Spin(6) quantum numbers $(\sigma_1, \sigma_2, \sigma_3)$. The numbers in parentheses next to each level denote the Spin(5) quantum numbers (τ_1, τ_2). The ground state is taken as zero of the energy. (Taken from Iachello (1984)).

Remembering that only the groups $U(n)$ possess linear Casimir operators, we conclude that for the present problem the Hamiltonian can be written as

$$H = e_0 + aC_{2\,O^B6} + a'C_{2\,Spin6} + bC_{2\,Spin5} + cC_{2\,Spin3}. \qquad (11.26)$$

Its eigenvalues in the basis given above are

$$E(N, M, \Sigma, (\sigma_1, \sigma_2, \sigma_3), (\tau_1, \tau_2), \nu_\Delta, J, M_J) = e_0 + a\Sigma(\Sigma + 4)$$

$$+a'[\sigma_1(\sigma_1+4)+\sigma_2(\sigma_2+2)+\sigma_3^2]+b[\tau_1(\tau_1+3)+\tau_2(\tau_2+1)]+cJ(J+1). \tag{11.27}$$

In the above expression the term e_o contributes only to the binding energy of the nucleus. It has the form

$$e_0 = E_0 + E_1 N + E_2 N^2 + E_3 M + E_4 M^2 + E_5 MN. \tag{11.28}$$

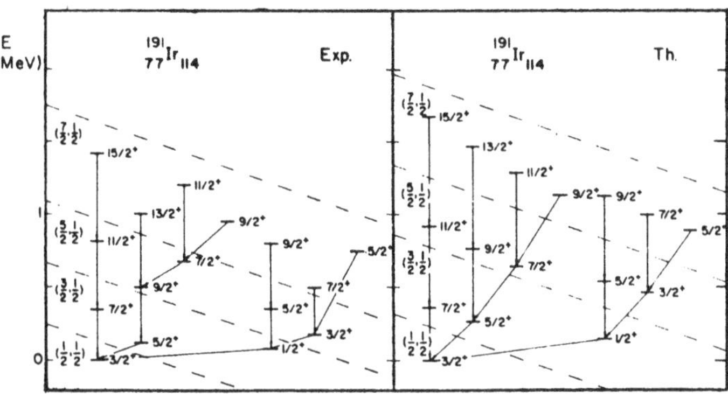

Fig. 11.2 Comparison between the predictions of the Spin(6) limit of IBFM and experiment for the nucleus $^{191}_{77}\mathrm{Ir}_{114}$, which has N=8, M=1. (Taken from Iachello (1984)).

A typical spectrum obtained with this Hamiltonian is shown in Fig. 11.1. Remember that this Hamiltonian was obtained by using the O(6) symmetry for the bosonic system, and in addition, assuming that the odd nucleon occupies a j=3/2 sublevel. It is interesting to look for real nuclei with spectra resembling those of the Hamiltonian derived above. Possible candidates should be looked for in areas of the periodic table where even–even nuclei with axial deformation (best described in the O(6) limit of IBM-1) occur, since we want the bosonic system to be as close to the O(6) symmetry as possible. In addition, the odd nucleon has to lie in a j=3/2 sublevel. The first condition is fulfilled in the $_{76}$Os–$_{78}$Pt region. From the list of single particle levels given in Sec. 11.5, notice that the 6 valence proton holes of $_{76}$Os occupy the $3s_{1/2}$ and $2d_{3/2}$ sublevels. Adding one more proton (i.e. considering the $_{77}$Ir isotopes), means that only five valence proton holes are left, the odd proton hole being in the $2d_{3/2}$ level, which is a j=3/2 level. Thus $_{77}$Ir isotopes are particularly good candidates for being adequately described by the above Spin(6) Hamiltonian. In Figs. 11.2 and 11.3 the experimental spectra of ^{191}Ir and ^{193}Ir are compared to the predictions of the Spin(6) Hamiltonian. Quite good agreement between theory and experiment is observed.

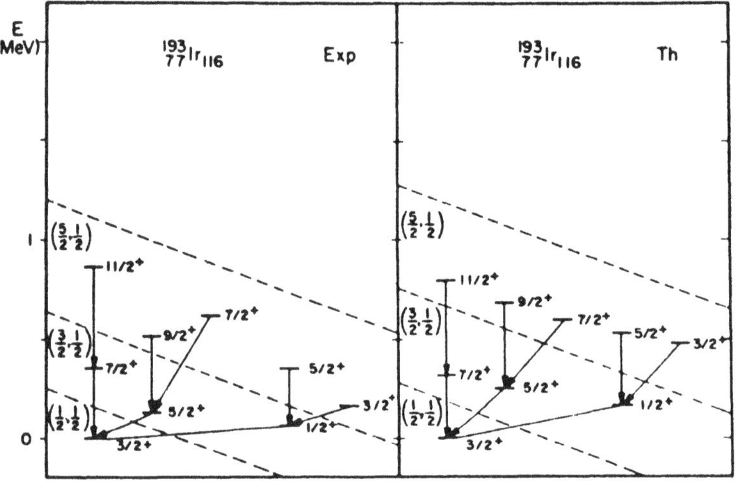

Fig. **11.3** Comparison between the predictions of the Spin(6) limit of IBFM and experiment for the nucleus $^{193}_{77}$Ir$_{116}$, which has N=7, M=1. (Taken from Iachello (1984)).

11.10 Other dynamical symmetries of the Class BF-1

As it has been mentioned above, the Class BF-1 of dynamical symmetries takes place when the angular momentum of the fermions is such that the fermion group chain is isomorphic to a boson group chain. Once the method has been understood, it can be applied to numerous cases. Two examples which have been worked out will be given here. In both of these examples the bosonic part of the Hamiltonian shows the $U^B(5)$ dynamical symmetry, while the single fermion can occupy a $j = 3/2$ or a $j = 1/2$ level. More examples can be found in the references listed at the end of this chapter.

11.10.1 Example 1

As in the case already studied, assume that the single fermion can occupy only a $j = 3/2$ level, so that the relevant fermion algebra is again $U^F(4)$ and the fermionic chain is

$$U^F(4) \supset SU^F(4) \supset Sp^F(4) \supset SU^F(2) \supset SO^F(2). \qquad (11.29)$$

Furthermore, assume that the bosonic part of the Hamiltonian shows the $U^B(5)$ dynamical symmetry, corresponding to the bosonic chain

$$U^B(6) \supset U^B(5) \supset O^B(5) \supset O^B(3) \supset O^B(2). \qquad (11.30)$$

Taking advantage of the isomorphisms

$$O^B(5) \approx Sp^F(4), \qquad (11.31)$$

$$O^B(3) \approx SU^F(2), \qquad (11.32)$$

$$O^B(2) \approx O^F(2), \qquad (11.33)$$

we can produce the boson–fermion chain (Bijker and Kota 1984)

$$U^B(6) \otimes U^F(4) \supset U^B(5) \otimes SU^F(4) \supset O^B(5) \otimes Sp^F(4)$$

$$\supset Spin(5) \supset Spin(3) \supset Spin(2). \qquad (11.34)$$

The quantum numbers needed to specify the irreps of each group are listed below

$$
\begin{array}{ll}
U^B(6) \otimes U^F(4) & [N]\ \{M\} \\
U^B(5) \otimes SU^F(4) & n_d \\
O^B(5) \otimes Sp^F(4) & v\ (t_1, t_2) \\
Spin(5) & (v_1, v_2) \\
Spin(3) & J \\
Spin(2) & M_J.
\end{array}
$$

Since the step from Spin(5) to Spin(3) is not fully decomposable, an additional quantum number n_Δ is required at this point. Thus the full classification scheme for the chain is

$$|[N], \{M\}, n_d, v, (t_1, t_2), (v_1, v_2), n_\Delta, J, M_J > . \qquad (11.35)$$

The eigenvalues of the Hamiltonian are then given by

$$E(N, M, v, (t_1, t_2), (v_1, v_2), n_\Delta, J, M_J) = \epsilon_0(N, M) + \epsilon n_d$$

$$+\alpha n_d(n_d + 4) + \beta v(v + 3) + \beta'[t_1(t_1 + 4) + t_2(t_2 + 2)]$$

$$+\beta''[v_1(v_1 + 3) + v_2(v_2 + 1)] + \gamma J(J + 1). \tag{11.36}$$

Experimental examples with this symmetry have been found in the $_{29}$Cu region (Bijker and Kota 1984).

11.10.2 Example 2

As in the previous case, assume that the bosonic part of the Hamiltonian shows the $U^B(5)$ dynamical symmetry. For the single fermion assume that it can occupy a level with $j = 1/2$, so that the corresponding group is $U^F(2)$. Using the isomorphisms

$$O^B(3) \approx SU^F(2), \tag{11.37}$$

$$O^B(2) \approx O^F(2), \tag{11.38}$$

we obtain the following group chain (Bijker and Kota 1984)

$$U^B(6) \otimes U^F(2) \supset U^B(5) \otimes U^F(2) \supset O^B(5) \otimes U^F(2)$$

$$\supset O^B(3) \otimes SU^F(2) \supset Spin(3) \supset Spin(2). \tag{11.39}$$

The quantum numbers characterizing the representations of each group are shown below

$$
\begin{array}{cc}
U^B(6) \otimes U^F(2) & [N] \, , \, \{M\} \\
U^B(5) \otimes U^F(2) & n_d \\
O^B(5) \otimes U^F(2) & v \\
O^B(3) \otimes SU^F(2) & L \, , \, S \\
Spin(3) & J \\
Spin(2) & M_J
\end{array}
$$

Since the step from $O^B(5)$ to $O^B(3)$ is not fully decomposable, an additional quantum number n_Δ is needed at this point. Finally, the eigenvalues of the Hamiltonian are

$$E(N, M, n_d, v, n_\Delta, L, S, J, M_J) = \epsilon_0(N, M) + \epsilon n_d + \alpha n_d(n_d + 4)$$

$$+\beta v(v + 3) + \gamma L(L + 1) + \gamma' S(S + 1) + \gamma'' J(J + 1). \tag{11.40}$$

Experimental examples of this chain have been found in the $_{45}$Rh region (Bijker and Kota 1984).

11.11 Class BF-2 of dynamical symmetries

This class of dynamical symmetries occurs whenever the angular momenta, j, of the orbitals which can be occupied by the unpaired fermions can be split into a pseudo-orbital part, k, and a spin part, $s = 1/2$, in a way that k gives a fermionic chain isomorphic to the chain corresponding to some bosonic dynamical symmetry (Iachello 1982). Two examples of dynamical symmetries of this class will be considered here.

11.11.1 Example 1

Suppose that the bosons have the $O^B(6)$ dynamical symmetry, while the fermions can occupy the levels with $j = 1/2, 3/2, 5/2$ (van Isacker, Frank, and Sun 1984). Then $m = 12$ and the corresponding fermion group is $U^F(12)$. However, in this case the fermion angular momentum can be split into a pseudo angular momentum part $k = 0$, 2 and a spin part $s = 1/2$. Then the boson group chain is

$$U^B(6) \supset O^B(6) \supset O^B(5) \supset O^B(3) \supset O^B(2), \qquad (11.41)$$

while the fermion chain is

$$U^F(12) \supset U^F(6) \otimes SU^F(2) \supset O^F(6) \otimes SU^F(2) \supset O^F(5) \otimes SU^F(2)$$

$$\supset O^F(3) \otimes SU^F(2) \supset SU^F(2) \supset O^F(2). \qquad (11.42)$$

These two chains can be combined into a chain whose members are listed below, followed by the quantum numbers labelling their irreps

$$
\begin{array}{ll}
U^B(6) \otimes U^F(12) & \{N\}\ \{M\} \\
U^B(6) \otimes U^F(6) \otimes SU^F(2) & \\
O^B(6) \otimes O^F(6) \otimes SU^F(2) & \Sigma\ (\Sigma_1', \Sigma_2', \Sigma_3') \\
O^{B+F}(6) \otimes SU^F(2) & (\sigma_1, \sigma_2, \sigma_3) \\
O^{B+F}(5) \otimes SU^F(2) & (\tau_1, \tau_2)\ \nu_\Delta \\
O^{B+F}(3) \otimes SU^F(2) & L\ S \\
SU^{B+F}(2) & J \\
O^{B+F}(2) & M_J
\end{array}
$$

The Hamiltonian is written as

$$H = AC_{2O^B6} + A'C_{2O^F6} + A''C_{2O^{B+F}6} + BC_{2O^{B+F}5}$$

$$+CC_{2O^{B+F}3} + C'C_{2SU^F2} + C''C_{2SU^{B+F}2}. \qquad (11.43)$$

Its eigenvalues can be written as

$$E = E_0(N, M) + A\Sigma(\Sigma + 4) + A'[\Sigma_1'(\Sigma_1' + 4) + \Sigma_2'(\Sigma_2' + 2) + (\Sigma_3')^2]$$

$$+ A''[\sigma_1(\sigma_1 + 4) + \sigma_2(\sigma_2 + 2) + \sigma_3^2] + B[\tau_1(\tau_1 + 3) + \tau_2(\tau_2 + 1)]$$

$$+ CL(L + 1) + C'S(S + 1) + C''J(J + 1). \tag{11.44}$$

Experimental examples of this symmetry have been found in the $_{78}$Pt region (Balantekin *et al.* 1983). One of them is shown in Fig. 11.4.

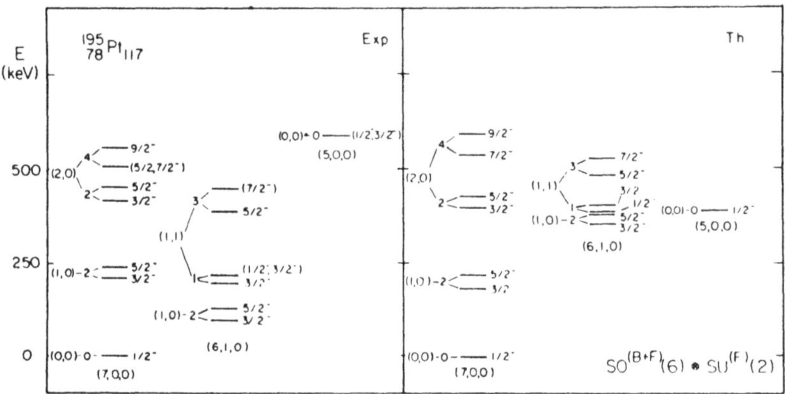

Fig. 11.4 Comparison between the predictions of the $O^{B+F}(6)$ limit of IBFM and experiment for the nucleus $^{195}_{78}$Pt$_{117}$, which has N=6, M=1. All states belong to the irrep $\Sigma = 6$. (Taken from Iachello (1982)).

11.11.2 Example 2

Suppose that the bosons have the $SU^B(3)$ dynamical symmetry, while the fermions can occupy the levels with $j = 1/2, 3/2$. Then $m = 6$ and the corresponding fermion group is $U^F(6)$. However, one can split the fermion angular momentum into a pseudo angular momentum part $k = 1$ and a spin part $s = 1/2$. Then the boson chain is

$$U^B(6) \supset SU^B(3) \supset O^B(3) \supset O^B(2), \tag{11.45}$$

while the fermion chain is

$$U^F(6) \supset SU^F(3) \otimes SU^F_S(2) \supset O^F(3) \otimes SU^F_S(2) \supset SU^F_J(2) \supset O^F(2).$$
$$(11.46)$$

The two chains can be combined into one chain (Iachello 1982), whose members are listed below, accompanied by the quantum numbers needed to specify the irreps

$$
\begin{array}{cc}
U^B(6) \otimes U^F(6) & \{N\}\ \{M\} \\
SU^B(3) \otimes SU^F(3) \otimes SU^F(2) & (\lambda_B, \mu_B)\ (\lambda_F, \mu_F) \\
SU^{B+F}(3) \otimes SU^F(2) & (\lambda, \mu)\ \mathrm{K} \\
O^{B+F}(3) \otimes SU^F(2) & \mathrm{L\ S} \\
SU^{B+F}(2) & \mathrm{J} \\
O^{B+F}(2) & M_J
\end{array}
$$

The Hamiltonian can be put in the form

$$H = \beta C_{2SU^B3} + \beta' C_{2SU^F3} + \beta'' C_{2SU^{B+F}3}$$

$$+\alpha C_{2O^{B+F}3} + \alpha' C_{2SU^F2} + \alpha'' C_{2SU^{B+F}2}. \qquad (11.47)$$

Its eigenvalues are

$$E = E_0(N, M) + \beta(\lambda_b^2 + \mu_B^2 + \lambda_B\mu_B + 3\lambda_B + 3\mu_B)$$

$$+\beta'(\lambda_F^2 + \mu_F^2 + \lambda_F\mu_F + 3\lambda_F + 3\mu_F) + \beta''(\lambda^2 + \mu^2 + \lambda\mu + 3\lambda + 3\mu)$$

$$+\alpha L(L+1) + \alpha' S(S+1) + \alpha'' J(J+1). \qquad (11.48)$$

11.12 Electromagnetic transition rates

As in the case of IBM, electromagnetic transition rates can be calculated in IBFM as well. The necessary transition operators will be composed by two parts, one bosonic and one fermionic

$$T^L = T^L_B + T^L_F. \qquad (11.49)$$

The explicit form of the bosonic part of the various transition operators is the same as in IBM-1. In lowest order, only one-body operators are included. The fermionic part, as in the case of the Hamiltonian, depends on the specific sublevels occupied by the single fermion. In the case of a fermion occupying a $j = 3/2$ sublevel, which was studied in detail earlier, the fermionic part of the electric quadrupole operator in lowest order reads

$$T^{E2}_{F,\mu} = \gamma_2 (a^+ \otimes \tilde{a})^2_\mu. \tag{11.50}$$

The corresponding bosonic operator, already encountered in IBM-1, is

$$T^{E2}_{B,\mu} = \alpha_2 (d^+ \otimes \tilde{s} + s^+ \otimes \tilde{d})^2_\mu + \beta_2 (d^+ \otimes \tilde{d})^2_\mu. \tag{11.51}$$

Once the transition operators have been written down, one needs to calculate their matrix elements between appropriate states. If the combined boson–fermion operators can be written as generators of subgroups appearing in the boson–fermion chain, the matrix elements can be calculated analytically. For example, in the case of the T^{E2} operators of the specific example mentioned here, analytic calculations are possible if the total operator can be written in the form

$$T^{E2}_\mu = (d^+ \otimes \tilde{s} + s^+ \otimes \tilde{d})^2_\mu + (a^+ \otimes \tilde{a})^2_\mu, \tag{11.52}$$

which is a generator of the group Spin(6), as we have already seen.

11.13 Numerical calculations

As in the other models studied so far, in IBFM, too, dynamic symmetries agree with experiment only in a few cases. For the vast majority of nuclei, numerical calculations have to be undertaken. A computer program, called ODDA (O. Scholten, 1980, University of Groningen, The Netherlands) has been written for this purpose. However, an additional difficulty has to be overcome in the case of IBFM, that of the very large number of free parameters in the boson–fermion interaction term of the Hamiltonian, already encountered in Sec. 11.5. Since IBFM is supposed to be a truncation of the shell-model, one way to get some guidance on which terms are more important, is to look for a microscopic determination of the parameters appearing in the Hamiltonian. Assuming that odd fermions can occupy various levels of different j, it has been found that the more important terms in the boson–fermion interaction part of the Hamiltonian are

$$V'_{BF} = \sum_j A_j ((d^+ \otimes \tilde{d})^0 \otimes (a^+_j \otimes \tilde{a}_j)^0)^0$$

$$+ \sum_{jj'} B_{jj'} (\{ s^+ \otimes \tilde{d} + d^+ \otimes \tilde{s})^2 + \chi (d^+ \otimes \tilde{d})^2 \} \otimes (a^+_j \otimes \tilde{a}_{j'})^2)^0$$

$$+ \sum_{jj'j''} C_{jj'j''} : ((a_j^+ \otimes \tilde{d})^{j''} \otimes (d^+ \otimes \tilde{a}_{j'})^{j''})^0 : . \qquad (11.53)$$

In the last term the symbol : indicates normal ordering, i.e. that when calculating matrix elements all the creation operators must be on the left of all annihilation operators. The terms in V'_{BF} are called the monopole, the quadrupole, and the exchange term, respectively.

References

This chapter has been based on Iachello (1982, 1984). Similar discussions of IBFM can also be found in Iachello (1980b, 1980c). Many details on the various chains considered here, as well as for other chains, can be found in Iachello and Kuyucak (1981), Bijker and Kota (1984), van Isacker, Frank, and Sun (1984), Bijker and Iachello (1985). A recent list of boson–fermion symmetries studied so far can be found in Vervier (1986). Accounts of early experimental tests of IBFM can be found in Casten (1980a, 1980c). The "state of the art" can be found in Dubrovnik 1986. Much more work on a variety of theoretical considerations as well as numerous experimental tests of IBFM symmetries can be found in the extensive list of references given here.

Alonso, C E, Arias, J M, Bijker, R, and Iachello, F, 1984.
  Phys. Lett., **144B**, 141.
Arias, J M, and Alonso, C E, 1985. *La Rábida 1985*, 612.
Arias, J M, Alonso, C E, and Bijker, R, 1985. *Nucl. Phys.*,
  A445, 333.
Arias, J M, Alonso, C E, and Lozano, M, 1986. *Phys. Rev.*,
  C33, 1482.
Arias, J M, Alonso, C E, and Lozano, M, 1987. *Nucl. Phys.*,
  A466, 295.
Arima, A, and Iachello, F, 1976. *Phys. Rev.*, **C14**, 761.
Arima, A, Gelberg, A, and Scholten, O, 1987. *Phys. Lett.*,
  185B, 259.
Árvay, Z, Alikov, B, Kvasil, J, Nazmitdinov, R, Sharonov, J,
  and Wawryszczuk, J, 1984. *Debrecen 1984*, 755.
Balantekin, A B, and Paar, V, 1986a. *Phys. Lett.*, **169B**, 9.
Balantekin, A B, and Paar, V, 1986b. *Phys. Rev.*, **C34**, 1917.
Balantekin, A B, Bars, I, Bijker, R, and Iachello, F, 1983.
  Phys. Rev., **C27**, 1761.
Barci, V, Gizon, J, Gizon, A, Crawford, J, Genevey,
  J, Płochocki, A, and Cunningham, M A, 1982. *Nucl. Phys.*,

A383, 309.

Bijker, R, 1984. *Drexel 1984*, 627.

Bijker, R, 1986. *Dubrovnik 1986*, **1**, 193.

Bijker, R, and Dieperink, A E L, 1982. *Nucl. Phys.*, **A379**, 221.

Bijker, R, and Iachello, F, 1985. *Ann. Phys.*, **161**, 360.

Bijker, R, and Kota, V K B, 1984. *Ann. Phys.*, **156**, 110.

Blasi, N, and van der Werf, S Y, 1986. *Nucl. Phys.*, **A456**, 397.

Blasi, N, and Lo Bianco, G, 1987. *Phys. Lett.*, **185B**, 254.

Brant, S, Paar, V, and Vretenar, D, 1984. *Z. Phys.*, **A319**, 355.

Brant, S, Paar, V, Vretenar, D, and Meyer, R A, 1986. *Phys. Rev.*, **C34**, 341.

Bruce, A M, Hicks, D, and Warner, D D, 1987. *Nucl. Phys.*, **A465**, 221.

Bruce, A M, Gelletly, W, Lukasiak, J, Phillips, W R, and Warner, D D, 1985. *Phys. Lett.*, **165B**, 43.

Bucurescu, D, Cata, G, Cutoiu, D, Constantinescu, G, Ivaşcu, M, and Zamfir, N V, 1983. *Nucl. Phys.*, **A401**, 22.

Bucurescu, D, Cata, G, Cutoiu, D, Constantinascu, G, Ivaşcu, M, and Zamfir, N V, 1985. *Nucl. Phys.* , **A443**, 217.

Casten, R F, 1980a. *Drexel 1980*, 369.

Casten, R F, 1980b. *Erice 1980*, 317.

Casten, R F, 1980c. *Nucl. Phys.*, **A347**, 173.

Casten, R F, and Smith, G J, 1979. *Phys. Rev. Lett.*, **43**, 337.

Cizewski, J A, 1980. *Erice 1980*, 389.

Cizewski, J A, 1983. *Drexel 1983*, 175.

Cunningham, M A, 1981. *Phys. Lett.*, **106B**, 11.

Cunningham, M A, 1982a. *Nucl. Phys.*, **A385**, 204.

Cunningham, M A, 1982b. *Nucl. Phys.*, **A385**, 221.

de Gelder, P, *et al.*, 1983. *Nucl. Phys.*, **A401**, 397.

Dukelsky, J, and Lima, C, 1986. *Phys. Lett.*, **182B**, 116.

du Marchie van Voorthuysen, E H, de Voigt, M J A, Blasi, N, and Jansen, J F W, 1981. *Nucl. Phys.*, **A355**, 93.

Elliott, J P, 1985. *Rep. Prog. Phys.*, **48**, 171.

Faessler, A, Kuyucak, S, and Wakai, M, 1986. *Nucl. Phys.*, **A458**, 381.

Feng, D H, Sun, H Z, and Vallières, M, 1986. *Phys. Rev.*, **C33**, 1471.

Fogelberg, B, and Hoff, P, 1982. *Nucl. Phys.*, **A391**, 445.

Frank, A, Pittel, S, Warner, D D, and Engel, J, 1986. *Phys. Lett.*, **182B**, 233.

Fransson, K, Oms, J, Abreu, M C, and the ISOCELE Collaboration, 1987. *Nucl. Phys.*, **A469**, 323.

Gelberg, A, 1983. *Z. Phys.*, **A310**, 117.

Gelberg, A, 1984. *Z. Phys.*, **A315**, 119.

Ghaleb, H H, and Krane, K S, 1984. *Nucl. Phys.*, **A426**, 20.

Heyde, K, and Paar, V, 1986. *Phys. Lett.*, **179B**, 1.

Hippe, D, Schuh, H W, Kaup, U, Zell, K O, von Brentano, P, Fossan, D B, 1983. *Z. Phys.*, **A311**, 329.

Hübsch, T, and Paar, V, 1984. *Z. Phys.* , **319**, 111.

Hübsch, T, and Paar, V, 1985. *Phys. Lett.*, **151B**, 1.

Hübsch, T, Paar, V, and Vretenar, D, 1985. *Phys. Lett.*, **151B**, 320.

Iachello, F, 1980a. *Erice 1980*, 273.

Iachello, F, 1980b. *Erice 1980*, 365.

Iachello, F, 1980c. *Nucl. Phys.*, **A347**, 51c.

Iachello, F, 1982. *Erice 1982*, 5.

Iachello, F, 1983. *Florence 1983*, 145.

Iachello, F, 1984. *Trieste 1984*, **2**, 875.

Iachello, F, and Kuyucak, S, 1981. *Ann. Phys.*, **136**, 19.

Iachello, F, and Scholten, O, 1979. *Phys. Rev. Lett*, **43**, 679.

Iachello, F, and Scholten, O, 1980. *Phys. Lett.*, **91B**, 189.

Jolie, J, van Isacker, P, Heyde, K, Moreau, J, van Landeghem, G, Waroquier, M, and Scholten, O, 1985. *Nucl. Phys.* , **A438**, 15.

Kaup, U, Gelberg, A, Gast, W, and von Brentano, P, 1979. *Rhodes 1979*, 236.

Kaup, U, Gelberg, A, von Brentano, P, and Scholten, O, 1980. *Phys. Rev.*, **C22**, 1738.

Kaup, U, Vorwerk, R, Hippe, D, Schuh, H W, von Brentano, P, and Scholten, O, 1981. *Phys. Lett.*, **106B**, 439.

Kota, V K B, 1986a. *Dubrovnik 1986*, **1**, 259.

Kota, V K B, 1986b. *Phys. Rev.*, **C33**, 2218.

Kuyucak, S, Faessler, A, and Wakai, M, 1984. *Nucl. Phys.*, **A420**, 83.

Lo Bianco, G, 1980. *Erice 1980*, 309.

Lo Bianco, G, 1981. *Trieste 1981*, 71.

Lo Bianco, G, Molho, N, Moroni, A, Bracco, A, and Blasi, N, 1981. *J. Phys.*, **G7**, 219.

Lo Bianco, G, Molho, N, Moroni, A, Angius, S, Blasi, N, Ferrero, A, 1979. *J. Phys.*, **G5**, 697.

Lopac, V, Brant, S, Paar, V, Schult, O W B, Seyfarth, H, and Balantekin, A B, 1986. *Z. Phys.*, **A323**, 491.

McGowan, F K, Johnson, N R, Lee, I Y, Milner, W T, Roulet, C, Hattula, J, Fewell, M P, Ellis-Akovali, Y A, Diamond, R M, Stephens, F S, and Guidry, M W, 1986. *Phys. Rev.* , **C33**, 855.

Michailova, M M, 1987. *J. Phys.*, **G13**, L149.
Mundy, S J, Lukasiak, J, and Phillips, W R, 1984. *Nucl. Phys.*, **A426**, 144.
Paar, V, 1984. *Debrecen 1984*, 675.
Paar, V, and Brant, S, 1984. *Phys. Lett.*, **143B**, 1.
Paar, V, Brant, S, Canto, L F, Leander, G, and Vonk, M, 1982. *Nucl. Phys.*, **A378**, 41.
Panqueva, J, Hellmeister, H P, Bergmeister, F J, and Lieb, K P, 1981. *Phys. Lett.*, **98B**, 248.
Panqueva, J, Hellmeister, H P, Lühmann, L, Bergmeister, F J, Lieb, K P, and Otsuka, T, 1982. *Nucl. Phys.*, **A389**, 424.
Pfeiffer, B, Brant, S, Kratz, K L, Meyer, R A, and Paar, V, 1986. *Z. Phys.*, **A325**, 487.
Pinkston, W T, 1984. *Gull Lake 1984*, 204.
Pinkston, W T, and Feng, D H, 1984. *Phys. Rev.* , **C30**, 1431.
Pittel, S, 1987. *Oaxtepec 1987*, 257.
Scholten, O, 1980a. *Drexel 1980*, 503.
Scholten, O, 1980b. *Drexel 1980*, 285.
Scholten, O, 1982. *Phys. Lett.*, **108B**, 155.
Scholten, O, 1984. *Drexel 1984*, 133.
Scholten, O, 1985. In *Progress in Particle and Nuclear Physics*, (ed. A. Faessler) Vol 14, p 189. Pergamon, Oxford.
Scholten, O, and Blasi, N, 1982. *Nucl. Phys.*, **A380**, 509.
Scholten, O, and Ozzelo, T, 1983. *Phys. Lett.*, **125B**, 106.
Scholten, O, and Ozzello, T, 1984. *Nucl. Phys.*, **A424**, 221.
Scholten, O, and Warner, D D, 1984. *Phys. Lett.*, **142B**, 315.
Scholten, O, Harakeh, M N, van der Plicht, J, Put, L W, Siemssen, R H, van der Werf, S Y, and Sekiguchi, M, 1980. *Nucl. Phys.*, **A348**, 301.
Sistemich, K, Kawade, K, Lawin, H, Lhersonneau, G, Ohm, H, Paffrath, U, Lopac, V, Brant, S, and Paar, V, 1986. *Z. Phys.*, **A325**, 139.
Sunko, D K, Brant, S, Vretenar, D, and Paar, V, 1986. *Dubrovnik 1986*, **1**, 240.
Szpikowski, S, Kłosowski, P, and Próchniak, L, 1986. *Dubrovnik 1986*, **1**, 265.
Van der Jeugt, J, 1985. *J. Phys.*, **A18**, L745.
Vanhorenbeeck, J, Duhamel, P, Del Marmol, P, Fettweis, P, and Heyde, K, 1983. *Nucl. Phys.*, **A408**, 265.
van Isacker, P, 1987. *J. Math. Phys.*, **28**, 957.
van Isacker, P, Frank, A, and Sun, H Z, 1984. *Ann. Phys.*, **157**, 183.
Vervier, J, 1983. *Phys. Lett.*, **133B**, 135.

Vervier, J, 1984. *Phys. Lett.*, **149B**, 267.

Vervier, J, 1986. *Dubrovnik 1986* **1**, 175.

Vervier, J, van Isacker, P, Jolie, J, Kota, V K B, and Bijker, R, 1985. *Phys. Rev.*, **C32**, 1406.

von Brentano, P, Gelberg, A, and Kaup, U, 1980. *Erice 1980*, 303.

Warner, D D, 1984. *Phys. Rev. Lett.*, **52**, 259.

Warner, D D, 1986. *Dubrovnik 1986*, **1**, 148.

Warner, D D, and Bruce, A M, 1984. *Phys. Rev.*, **C30**, 1066.

Warner, D D, van Isacker, P, Jolie, J, and Bruce, A M, 1984. *Drexel 1984*, 156.

Warner, D D, van Isacker, P, Jolie, J, and Bruce, A M, 1985. *Phys. Rev. Lett.*, **54**, 1365.

Warner, D D, Casten, R F, Stelts, M L, Börner, H G, and Barreau, G, 1982. *Phys. Rev.*, **C26**, 1921.

Wood, J L, 1980. *Drexel 1980*, 451.

Wood, J L, 1980. *Erice 1980*, 321.

Wood, J L, 1980. *Erice 1980*, 381.

Zell, K O, Harter, H, Hippe, D, Schuh, H W, and von Brentano, P, 1984. *Z. Phys.*, **A316**, 351.

Zhou, S H, Frank, A, and van Isacker, P, 1984. *Phys. Rev.*, **C27**, 2430.

12

NUCLEAR SUPERSYMMETRIES (SUSYs !)

12.1 Introduction

In all models discussed so far, each nucleus has to be treated separately. However, one may try to consider symmetries which include several nuclei at the same time. For example, these can be even–even and odd–even nuclei. The simultaneous description of even–even and odd–even (or even–odd) nuclei has been achieved in the framework of nuclear supersymmetries, which will be described in this chapter.

Consider the neighboring nuclei $^{190}_{76}\text{Os}_{114}$ and $^{191}_{77}\text{Ir}_{114}$. The first is an even–even nucleus, which has $82 - 76 = 6$ valence proton holes and $126 - 114 = 12$ valence neutron holes. Thus it can be described as a system of 3 proton bosons and 6 neutron bosons in IBM-2, or as a system of 9 bosons in IBM-1. The second nucleus is an odd–even nucleus, having $82 - 77 = 5$ valence proton holes and 12 valence neutron holes. This nucleus can be described in IBFM-1 as a system of $2 + 6 = 8$ bosons and a single fermion (proton in this case), which happens to occupy a sublevel with $j = 3/2$, as we have already seen in Chapter 11.

Suppose now we want to describe these two nuclei simultaneously, i.e. in the framework of the same model. Starting from ^{190}Os we need in order to go to ^{191}Ir an operator annihilating a (proton) boson and creating a single fermion (proton) in its place. Clearly, such an operator cannot be accommodated in the framework of Lie algebras studied so far. Thus the introduction of "graded" Lie algebras becomes necessary, which will be done now.

12.2 The graded Lie algebra U(6/m)

In IBM-1 one has 6 boson operators (s and d_μ, denoted briefly as b_i, i=1,2,...,6), so that one can form 36 bilinear products of the form $b_i^+ b_j$, $i, j = 1, 2, \ldots, 6$, which generate the Lie algebra U(6). In the case of fermions occupying various j_i sublevels, one has

$$m = \sum_i (2j_i + 1) \tag{12.1}$$

194

operators a_n^+, n=1,2,...,m. Thus one can form m^2 bilinear operators of the form $a_{n_1}^+ a_{n_2}$, $n_1, n_2 = 1, 2, \ldots, m$, which generate the Lie algebra U(m). Now consider at once all possible bilinear products of the b_i and a_n operators, i.e. in addition to the above mentioned boson-boson and fermion–fermion products consider the 6m fermion–boson products $a_n^+ b_i$ and the 6m boson–fermion products $b_i^+ a_n$. Taken together all these $36 + m^2 + 12m = (6 + m)^2$ operators generate the **"graded" Lie algebra** U(6/m) (Iachello 1982a). The 36 boson-boson operators, as well as the m^2 fermion–fermion operators satisfy boson-like commutation relations, thus they are called bosonic generators. (It may sound strange to the reader that fermion–fermion operators satisfy boson commutation relations, but this is in fact the case !). The 6m boson–fermion operators as well as the 6m fermion–boson operators are called fermionic generators. By anti-commuting them one obtains bosonic generators, thus the algebra closes.

Once the algebra U(6/m) has been identified, we need to specify its irreps. The fundamental representation of U(6/m) will have 6+m dimensions, and is denoted as □. If the system under study contains N particles, the Young supertableau of the most symmetric U(6/m) irrep will have N □'s in a row. The N particles can be all bosons, or one can have $N - 1$ bosons and one single fermion, or $N - 2$ bosons and 2 uncoupled fermions, etc. The last term of this sequence will be N fermions, if the value of m allows them (i.e. if $m \geq N$). Otherwise the last term will be $(N - m, m)$. Notice that the conserved quantity is the total number of bosons+uncoupled fermions, not the total number of nucleons or nucleon pairs, as it used to be in models studied previously.

In order to fully specify the states, one has to consider chains of subalgebras of U(6/m). An obvious subalgebra is $U^B(6) \otimes U^F(m)$. The irreps of $U^B(6) \otimes U^F(m)$ contained in the above mentioned Young supertableau of N particles are $(N, 0)$, $(N - 1, 1)$, $\ldots$, $(0, N)$, if $m \geq N$, otherwise the sequence stops at $(N - m, m)$. The first number in the parentheses labels the $U^B(6)$ irreps, the second number labels the $U^F(m)$ irreps. Notice that only one number is needed in order to specify the $U^B(6)$, the total number of bosons. This is because here only the most symmetric irreps of $U^B(6)$ occur, since we are dealing with a pure bosonic system. These $U^B(6)$ irreps are represented by one-line Young tableaux, the number of boxes in the Young tableau being equal to the number of bosons. Similarly, the irreps of $U^F(m)$ are characterized by one label only, the total number of fermions. This is because the $U^F(m)$ irreps have to be fully anti-symmetric, since they describe a fully-fermionic system. Thus they are represented by one-column Young tableaux, having as many boxes as the number of uncoupled fermions present.

12.3 The graded Lie algebra U(6/4)

We try to clarify these concepts using the above mentioned example (Bars 1983), where the uncoupled fermions can occupy a $j = 3/2$ level ($m = 4$). Notice that a $j = 3/2$ level can only accomodate up to $2j + 1 = 4$ fermions. In this case the superalgebra is U(6/4). With 9 particles present, the possible $U^B(6) \otimes U^F(4)$ irreps are (9,0), (8,1), (7,2), (6,3), (5,4). The first corresponds to a system of 9 bosons. In the example studied above, this represents the nucleus ^{190}Os. The next irrep corresponds to a system of 8 bosons and 1 uncoupled fermion. In the example mentioned above this stands for the nucleus ^{191}Ir, where the uncoupled fermion is a proton. The next irrep stands for a system of 7 bosons and 2 uncoupled fermions (protons in this case). By counting the number of protons and fermions, we find that this configuration corresponds to the nucleus $^{192}_{78}$Pt$_{114}$. However, we expect the ground state of this nucleus to be described in IBM-1 by a system of $2 + 6 = 8$ bosons. Thus, the 7 boson plus two uncoupled fermion configuration is expected to describe two quasiparticle states in this nucleus. Thus we use the symbol ^{192}Pt* for this member of the sequence. Similarly, the irrep (6,3) corresponds to a system of 6 bosons and 3 uncoupled fermions (protons), which turns out to be ^{193}Au* (the star indicates a 2+1 quasiparticle state in this case), while the (5,4) irrep corresponds to a system of 5 bosons and 4 uncoupled fermions, which turns out to be ^{194}Hg** (notice that two stars are used to indicate a 4-quasiparticle state). All these nuclei can be connected through supersymmetric transformations within the $N = 9$ irrep of U(6/4).

How about the ground states of ^{192}Pt, ^{193}Au, ^{194}Hg ? They simply belong to other U(6/4) irreps. The ground state of $^{192}_{78}$Pt$_{114}$ requires $2 + 6 = 8$ bosons, thus it will belong to the $(8, 0)$ $U^B(6) \otimes U^F(4)$ irrep contained in the $N = 8$ irrep of U(6/4). The $(7, 1)$ irrep of this sequence will represent ^{193}Au, while the $(6, 2)$ irrep will represent ^{194}Hg*. The ground state of $^{194}_{80}$Hg$_{114}$ requires $1 + 6 = 7$ bosons, thus it is represented by the $(7, 0)$ $U^B(6) \otimes U^F(4)$ irrep of the $N = 7$ irrep of U(6/4).

12.4 An example of dynamical supersymmetry

Once we have clarified which nuclei are described by each U(6/4) irrep, we proceed in producing an energy formula describing their excitation spectra. Dynamic supersymmetries, which are formally similar to dynamic symmetries, will be exploited for this purpose. A possible chain of subgroups of U(6/4) is shown below, along with the quantum numbers needed to specify the irreps at each stage

$$
\begin{array}{ll}
U(6/4) & N \\
U^B(6) \otimes U^F(4) & N' \\
SO^B(6) \otimes SU^F(4) & \Sigma \\
Spin(6)^{B+F} \approx SU(4)^{B+F} & \sigma_1, \sigma_2, \sigma_3 \\
Spin(5)^{B+F} \approx Sp(4)^{B+F} & \tau_1, \tau_2 \; \nu_\Delta \\
Spin(3)^{B+F} \approx SU(2) & J
\end{array}
$$

By N' we indicate the number of bosons. The eigenvalues of the Hamiltonian, in which we assume that only first- and second-order Casimir invariants appear, will be

$$
E = E_0 + E_1 N + E_2 N^2 + E_3 N' + E_4 (N')^2 + E_5 N N' + A_1 \Sigma(\Sigma + 4)
$$

$$
+ A_2(\sigma_1(\sigma_1 + 4) + \sigma_2(\sigma_2 + 2) + \sigma_3^2) + B(\tau_1(\tau_1 + 3) + \tau_2(\tau_2 + 1)) + CJ(J + 1).
\tag{12.2}
$$

A few comments are appropriate at this point:

i) The first five terms determine the position of the ground state of each nucleus participating in a supersymmetric multiplet. Thus only four free parameters (coming from the last four terms) are left for describing the excitation spectra of all members in a multiplet.

ii) The same parameters apply to all nuclei in a supersymmetric multiplet.

iii) The specific chain studied above contains the $O^B(6)$ subgroup of $U^B(6)$, thus it is expected to work best for nuclei near this symmetry (γ-unstable nuclei). This condition is in fact fulfilled in the Os–Pt region under consideration.

iv) Notice the great similarity between the chain studied here and the chain studied in Sec. 11.9, in the IBFM framework. Except for the fact that in the present case $U^B(6) \otimes U^F(4)$ comes from U(6/4), the rest of the chain is exactly the same.

v) As a consequence of the similarity of the chains, the Hamiltonians obtained in the two cases are formally very similar. The terms describing excitation energies are exactly the same.

vi) If a supersymmetry is present, the quantity

$$
N = N' + M,
\tag{12.3}
$$

where N' is the total number of bosons and M is the total number of uncoupled fermions, is conserved within an irrep of U(6/m).

vii) The difference between the supersymmetric chain and the IBFM chain is that the latter is capable of describing only odd-A nuclei, while the former can describe a series of nuclei, both even–even and odd-A (odd–even or even–odd).

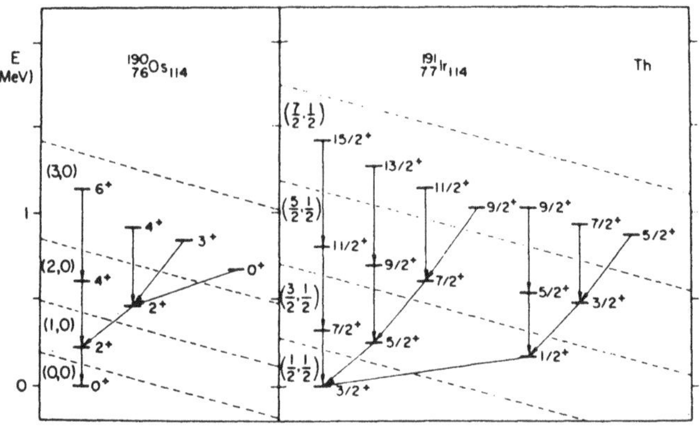

Fig. 12.1 Theoretical spectra of the nuclei $^{190}_{76}$Os$_{114}$ and $^{191}_{77}$Ir$_{114}$, predicted by the supersymmetry U(6/4). (Taken from Iachello (1982a)).

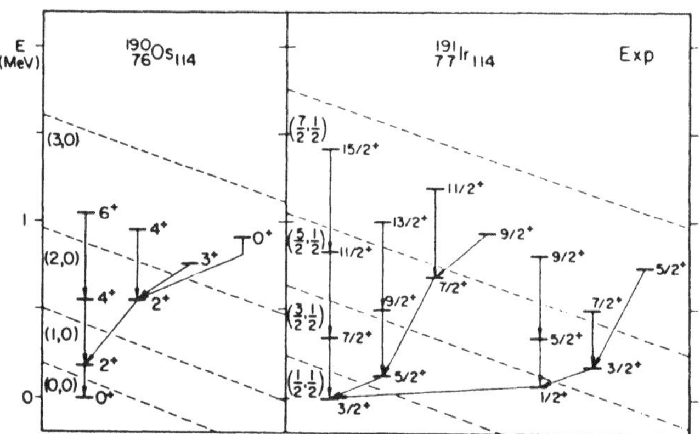

Fig. 12.2 Experimental spectra of the nuclei $^{190}_{76}$Os$_{114}$ and $^{191}_{77}$Ir$_{114}$, to be compared with the theoretical spectra of Fig. 12.1. (Taken from Iachello (1982a)).

The theoretical predictions obtained with this Hamiltonian for the above mentioned nuclei ^{190}Os and ^{191}Ir are shown in Fig. 12.1,

while the experimental spectra of these two nuclei are shown in Fig.
12.2.Some members of the $N = 9$ and $N = 8$ irreps of SU(6/4) are
shown in Fig. 12.3.

12.5 Additional examples of dynamical supersymmetry

The U(6/4) supersymmetry studied so far had group theoretical
structure formally similar to a dynamical symmetry of the class BF-
1. One can easily obtain supersymmetric chains formally similar to
dynamical symmetries of the class BF-2. Two examples will be given
here.

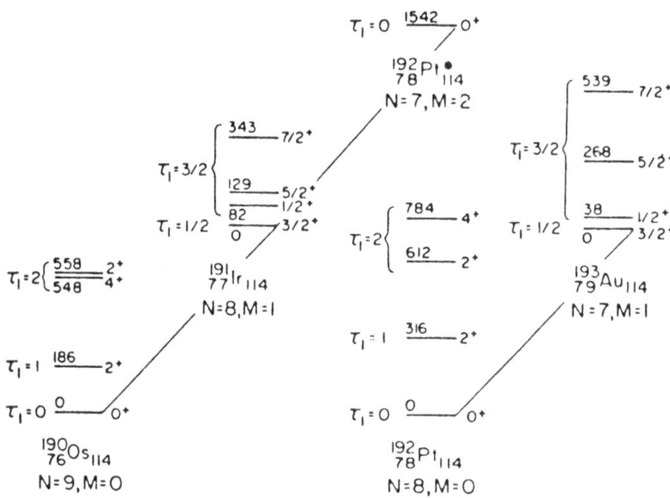

Fig. 12.3 Supersymmetry U(6/4) in heavy nuclei. A part of the $N = 9$
multiplet ($^{190}_{76}$Os$_{114}$ ($M = 0$), $^{191}_{77}$Ir$_{114}$ ($M = 1$), $^{192}_{78}$Pt$^{*}_{114}$ ($M = 2$)) and a
part of the $N = 8$ multiplet ($^{192}_{78}$Pt$_{114}$ ($M = 0$), $^{193}_{79}$Au$_{114}$ ($M = 1$)) are
shown. (Taken from Bars (1983)).

Example 1 Consider the nucleus $^{194}_{78}$Pt$_{116}$. In terms of IBM-
1, this is described as a system of $2 + 5 = 7$ bosons. Its neighbor,
^{195}Pt, has an uncoupled fermion (neutron hole), which can occupy
sub-levels with $j = 1/2$, $j = 3/2$, $j = 5/2$. Thus $m = 12$ and the
supersymmetry group is U(6/12). Using first the decomposition

$$U(6/12) \supset U^{B}(6) \otimes U^{F}(12) \qquad (12.4)$$

one can then proceed as in subsec. 11.11.1, by splitting the fermion angular momentum into a pseudo angular momentum part and a spin part. The Hamiltonian obtained has the same excitation terms as the one in subsec. 11.11.1, while its terms determining the ground state energies of nuclei in the same supermultiplet are the same as the corresponding terms of the Hamiltonian of the previous subsection. Thus it will not be repeated here.

Example 2 Suppose that the uncoupled fermion(s) can occupy sub-levels with $j = 1/2$, $j = 3/2$. Then the supersymmetric group is $U(6/6)$. Starting with the decomposition

$$U(6/6) \supset U^B(6) \otimes U^F(6), \tag{12.5}$$

one can proceed as in subsec. 11.11.2, by splitting the fermion angular momentum into a pseudo angular momentum part and a spin part. The excitation spectrum is the same as in subsec. 11.11.2, the terms determining the ground state energies of the various members of each supermultiplet are the same as in the previous subsection.

So far we have considered only chains starting with the decomposition

$$U(6/m) \supset U^B(6) \otimes U^F(m), \tag{12.6}$$

which lead to chains very similar to the ones obtained in IBFM, since we immediately break the "graded" Lie algebra into its ordinary Lie subalgebras. Other chains are also possible, starting with sub-superalgebras, i.e. subalgebras of $U(6/m)$ which are "graded" Lie algebras themselves. As an example, we mention here the chain

$$SU(6/4) \supset OSp(6/4) \supset O(6) \otimes Sp(4)$$

$$\supset O(5) \otimes Sp(4) \supset Sp(4) \supset SU(2). \tag{12.7}$$

We are not going to pursue further this problem here. A list of many possible chains is give by Bars (1983).

12.6 Odd–odd nuclei

So far we have applied supersymmetries to series of even–even plus odd–even nuclei (e.g. ^{196}Pt, ^{197}Au, ^{198}Hg*, ...) or series of even–even plus even–odd nuclei (e.g. ^{196}Pt, ^{197}Pt, ^{198}Pt*, ...). In the first series the uncoupled fermions (proton holes) occupy a $j = 3/2$ sublevel, thus the appropriate supersymmetry is $U(6/4)$. In the second series the uncoupled fermions (neutron holes) occupy sublevels with $j = 1/2$, $j = 3/2$, $j = 5/2$, thus the appropriate supersymmetry is U(6/12). Notice that the nucleus ^{196}Pt is described by both supersymmetries. This suggests the possibility of introducing a larger supersymmetry, U(6/16), being able to describe both of the above mentioned series at once. This larger supersymmetry can describe odd–odd nuclei, too (Bars 1983). For example, the odd–odd nucleus ^{198}Au can be obtained from ^{196}Pt by a combination of a U(6/4) transformation (introducing the uncoupled proton) and a U(6/12) transformation (introducing the uncoupled neutron). Starting from ^{196}Pt one has to possibilities for obtaining ^{198}Au:

i) First construct ^{197}Au through a U(6/4) transformation. Then go from ^{197}Au to ^{198}Au through a U(6/12) transformation.

ii) First construct ^{197}Pt through a U(6/12) transformation. Subsequently go from ^{197}Pt to ^{198}Au through a U(6/4) transformation.

U(6/16) is a larger group of transformations, including U(6/4) and U(6/12) as subgroups. Remember that U(6/4) contains a U(4) fermion subgroup, which describes uncoupled protons in a $j = 3/2$ shell. For this subgroup we will use the symbol $U_p^F(4)$. Similarly, U(6/12) contains a U(12) fermion subgroup, which describes uncoupled neutrons in $j = 1/2$, $j = 3/2$, $j = 5/2$ sublevels. For this subgroup we will use the symbol $U_n^F(12)$. One possible subgroup chain for U(6/16) is

$$U(6/16) \supset U_{p+n}^B(6) \otimes U_p^F(4) \otimes U_n^F(12)$$

$$\supset U_{p+n}^B(6) \otimes U_p^F(4) \otimes SU_n^F(6) \otimes SU_n^F(2)$$

$$\supset SO_{p+n}^B(6) \otimes U_p^F(4) \otimes SO_n^F(6) \otimes SU_n^F(2)$$

$$\supset SO_{p+n}^B(6) \otimes SU_{p+n}^F(4) \otimes SU_n^F(2)$$

$$\supset Spin(6)_{p+n}^{B+F} \otimes SU_n^F(2) \supset Spin(5)_{p+n}^{B+F} \otimes SU_n^F(2)$$

$$\supset Spin(3)_{p+n}^{B+F} \otimes SU_n^F(2) \supset SU(2). \qquad (12.8)$$

The last SU(2) subgroup is the angular momentum group. The most symmetric U(6/16) irrep is represented by a one-line supertableau, the number of boxes in the tableau being the conserved quantity. Irreps of $U_{p+n}^B(6)$ are characterized by one-line Young tableaux (where the number of boxes equals the total number of bosons), while irreps of $U_p^F(4)$ or $U_n^F(12)$ are characterized by one-column Young tableaux (where the number of boxes equals the number of uncoupled protons or neutrons, respectively), as we have already seen. Thus, the lowest sets of subgroup irreps contained in the $\{N\}$ irrep of U(6/16) will be $(N,0,0)$, $(N-1,1,0)$, $(N-1,0,1)$, $(N-2,1,1)$, where the three numbers in the parentheses characterize the irreps of $U^B_{p+n}(6)$, $U^F_p(4)$, $U^F_n(12)$, respectively. In the particular example considered above, we have $N = 6$ and $(6,0,0)$ represents the nucleus ^{196}Pt, while $(5,1,0)$ represents ^{197}Au, $(5,0,1)$ represents ^{197}Pt, and $(4,1,1)$ represents ^{198}Au. Higher subgroup irreps, like $(4,2,0)$, $(4,0,2)$, $(3,2,1)$, $(3,1,2)$, ..., are also possible, but they correspond to nuclei with at least two uncoupled protons or two uncoupled neutrons. Thus they are not describing the ground states of these nuclei, but excited states (two-quasiparticle states in the present examples).

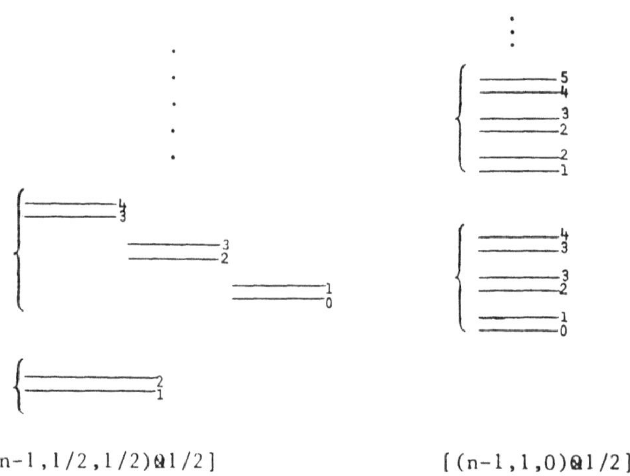

Fig. 12.4 Theoretical prediction (off scale) for the spectrum of an odd–odd nucleus, provided by the U(6/16) supersymmetry. (Taken from Bars (1983)).

A schematic representation (off scale) of the spectrum predicted by the chain studied above for odd–odd nuclei (like ^{198}Au in the studied example) is shown in Fig. 12.4. The doublet structure comes from the coupling of the neutron pseudo-spin $(SU^F_n(2))$ in the last step. We see that there are low lying states with angular momentum 0, 1, or 2. Depending on the values of the parameters in the Hamiltonian, any of these can be made the ground state (i.e. can be made to lie lowest). Experimental data for odd–odd nuclei in this region are scarce. From existing data, however, one can see that the prediction of a ground state having angular momentum 0, 1, or 2 is fulfilled.

12.7 U(12) systematics

We saw in this chapter that the idea of supersymmetry allows for the
simultaneous description of neighboring nuclei, by considering the
U(6/m) dynamical supergroup, which is subsequently decomposed
into the group $U^B(6) \otimes U^F(m)$, which is the same as the group
appearing in IBFM.

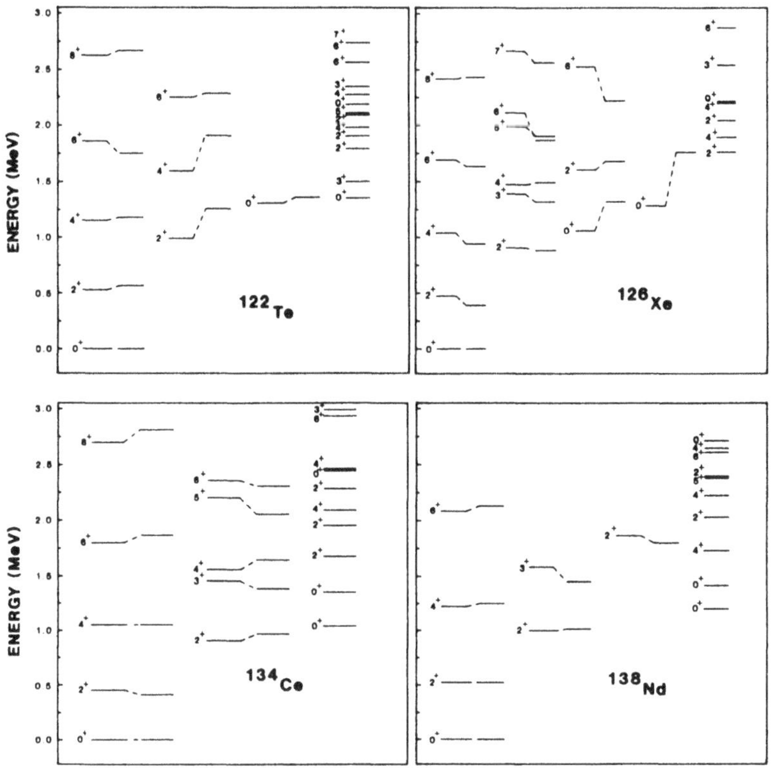

Fig. 12.5 Comparison between experimental (right) and theoretical (left)
spectra of four members of a $N = 7$ multiplet in the rare earth region.
The theoretical calculations have been performed with an IBM-2 Hamiltonian with constant parameters. At the far right of each panel are shown
predicted levels which have no experimental counterparts. (Taken from
Solari *et al.* (1987)).

The role of the supergroup U(6/m) is to provide a Hilbert space corresponding to more than one nucleus. The simultaneously described nuclei are characterized by the same total number of bosons plus uncoupled fermions.

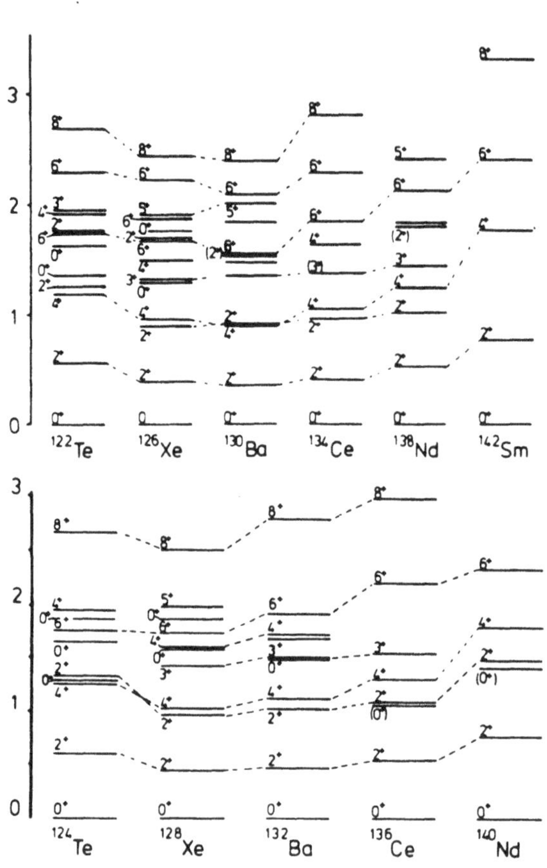

Fig. 12.6 Experimental spectra of a $N = 7$ (upper part) and a $N = 6$ (lower part) F-spin multiplet in the rare earth region. All known levels of positive parity are shown up to 2 MeV. The ground state band and the γ_1 band are shown up to the energy of the 8_1^+ level. (Taken from Harter *et al.* (1985)).

The idea of supersymmetry can be used in a different direction in the framework of IBM-2, by considering the group U(12), which is subsequently decomposed into $U_p(6) \otimes U_n(6)$. The role of the group

U(12) is to allow for the simultaneous description of even–even nuclei with the same total boson number $N = N_p + N_n$. We will call such nuclei members of a **U(12) multiplet** (Solari *et al.* 1987). In quarter shells where both valence protons and valence neutrons are particle-like or hole-like, the U(12) multiplet is transverse to the valley of stability, so that usually no more than two of its members are known experimentally well enough in order to allow for a numerical test of the model. In this case the members of a U(12) multiplet are isobars, characterized by (A, Z), $(A, Z+2)$, $(A, Z+4)$, ... However, in quarter shells where valence protons are particle-like and valence neutrons hole-like, or vice versa, the members of the U(12) multiplet lie along the valley of stability, and several of them can be known experimentally quite well in order to allow for a meaningful test of the theory. In this case the members of a U(12) multiplet differ from their neighbors in the multiplet by a quartet (two protons plus two neutrons). They are characterized by (A, Z), $(A + 4, Z + 2)$, $(A + 8, Z + 4)$, ...

As an example, consider the quarter shell with $50 \leq Z \leq 66$, $66 \leq N \leq 82$. In this quarter shell the $N = 7$ U(12) multiplet contains the members ^{122}Te, ^{126}Xe, ^{130}Ba, ^{134}Ce, ^{138}Nd, and ^{142}Sm. The spectra of these nuclei have been fitted by an IBM-2 Hamiltonian similar to this of Sec. 5.6, using constant parameters for all of them. Some of the results of the fit are shown in Fig. 12.5. The calculated levels appear on the left, the experimental levels on the right. Levels predicted by the model which cannot be identified with experimental levels are shown at the far right in each panel. Very good agreement between theory and experiment is observed.

It is interesting to remark that nuclei in the same U(12) multiplet have the same number of valence bosons, the same value of F-spin, and varying values of the z-component of F-spin, F_0. Thus the U(12) multiplets are identical to the **F-spin multiplets**, which had been considered earlier by Harter *et al.* (1985). It has been argued that members of an F-spin multiplet have similar structure. This can be seen in Fig. 12.6, where the experimental spectra for the above mentioned $N = 7$ multiplet, as well as for the $N = 6$ multiplet are shown.

References

The discussion of supersymmetries in this chapter has been based on Bars (1983) and Iachello (1982a). The "state of the art" can be found in Dubrovnik 1986. Much work on theoretical developments as well as on experimental tests of nuclear supersymmetries can be found in the extensive list of references given here.

Árvay, Z, 1986. *Dubrovnik 1986*, **1**, 246.
Baake, M, and Reinicke, P, 1986. *J. Math. Phys.*, **27**, 1430.
Baake, M, Reinicke, P, and Gelberg, A, 1986. *Phys. Lett.*, **166B**, 10.
Balantekin, A B, 1986a. *Dubrovnik 1986*, **1**, 223.
Balantekin, A B, 1986b. *Oaxtepec 1986*, 25.
Balantekin, A B, and Bars, I, 1982. *J. Math. Phys.*, **23**, 1239.
Balantekin, A B, Bars, I, and Iachello, F, 1981a. *Phys. Rev. Lett.*, **47**, 19.
Balantekin, A B, Bars, I, and Iachello, F, 1981b. *Nucl. Phys.*, **A370**, 284.
Balantekin, A B, Bars, I, Bijker, R, and Iachello, F, 1983. *Phys. Rev.* , **C27**, 1761.
Bars, I, 1983. *Drexel 1983*, 155.
Bijker, R, and Scholten, O, 1985. *Phys. Rev.*, **C32**, 591.
Blasi, N, Bijker, R, Harakeh, M N, Iwasaki, Y, Sterrenburg, W A, van der Werf, S Y, and Vergnes, M, 1982. *Nucl. Phys.*, **A388**, 77.
Bruce, A M, and Warner, D D, 1984. *Gull Lake 1984*, 155.
Casten, R F, 1983. *Oaxtepec 1983*, 1.
Casten, R F, 1984. *Nucl. Phys.*, **A421**, 27c.
Casten, R F, Colvin, G G, and Schreckenbach, K, 1987. *J. Phys.*, **G13**, 221.
Casten, R F, Warner, D D, Gowdy, G M, Rofail, N, and Lieb, K P, 1983. *Phys. Rev.* , **C27**, 1310.
Cizewski, J A, 1984a. *Debrecen 1984*, 299.
Cizewski, J A, 1984b. *Gull Lake 1984*, 103.
Cizewski, J A, 1984c. *Oaxtepec 1984*, 23.
Cizewski, J A, 1986. *Dubrovnik 1986*, **1**, 181.
Cizewski, J A, Burke, D G, Brown, R E, and Sunier, J W, 1987. *Phys. Rev.*, **C36**, 822.
Cizewski, J A, Burke, D G, Flynn, E R, Brown, R E, and Sunier, J W, 1981. *Phys. Rev. Lett.*, **46**, 1264.
Cizewski, J A, Burke, D G, Flynn, E R, Brown, R E, and Sunier, J W, 1983. *Phys. Rev.* , **C27**, 1040.
Cizewski, J A, Colvin, G G, Börner, H G, Hoyler, F, Kerr, S A, and Schreckenbach, K, 1987. *Phys. Rev. Lett.*, **58**, 10.
Colvin, G G, Börner, H G, Geltenbort, P, Hoyler, F, Kerr, S A, Schreckenbach, K, and Cizewski, J A, 1987. *Nucl. Phys.*, **A465**, 240.

Corwin, L, Ne'eman, Y, and Sternberg, S, 1975. *Rev. Mod. Phys.*, **47**, 573.

Elliott, J P, 1985. *Rep. Prog. Phys.* , **48**, 171.

Feng, D H, Sun, H Z, Vallières, M, Gilmore, R, Frank, A, and van Isacker, P, 1984. *Nucl. Phys.*, **A421**, 167c.

Frank, A, 1985. *Oaxtepec 1985*, 137.

Frank, A, 1987. *Oaxtepec 1987*, 129.

Harakeh, M N, Goldhoorn, P, Iwasaki, Y, Lukasiak, J, Put, L W, van der Werf, S Y, and Zwarts, F, 1980. *Phys. Lett.*, **97B**, 21.

Harter, H, von Brentano, P, Gelberg, A, and Casten, R F, 1985. *Phys. Rev.*, **C32**, 631.

Hübsch, T, 1986. *Dubrovnik 1986*, **1**, 253.

Iachello, F, 1980. *Phys. Rev. Lett.*, **44**, 772.

Iachello, F, 1982a. *Erice 1982*, 5.

Iachello, F, 1982b. *Nucl. Phys.*, **A374**, 635c.

Iwasaki, Y, Aarts, E H L, Harakeh, M N, Siemssen, R H, and van der Werf, S Y, 1981. *Phys. Rev.*, **C23**, 1477.

Jolie, J, 1985. *La Rábida 1985*, 623.

Jolie, J, 1986. *Dubrovnik 1986*, **1**, 205.

Jolie, J, Heyde, K, van Isacker, P, and Frank, A, 1987. *Nucl. Phys.*, **A466**, 1.

Jolie, J, van Isacker, P, Heyde, K, and Frank, A, 1985a. *Phys. Rev. Lett.*, **55**, 1457.

Jolie, J, van Isacker, P, Heyde, K, and Frank, A, 1985b. *Phys. Rev. Lett.*, **55**, 2739.

Kitipova, V, 1986. *Z. Phys.*, **A323**, 247.

Ling, Y S, Zhang, M, Xu, J M, Vallières, M, Gilmore, R, Feng, D H, and Sun, H Z, 1984. *Phys. Lett.*, **148B**, 13.

Mauthofer, A, Stelzer, K, Gerl, J, Elze, T W, Happ, T, Eckert, G, Faestermann, T, Frank, A, and van Isacker, P, 1986. *Phys. Rev.*, **C34**, 1958.

Morrison, I, and Jarvis, P D, 1985. *Nucl. Phys.*, **A435**, 461.

Mundy, S J, Lukasiak, J, and Phillips, W R, 1984. *Nucl. Phys.*, **A426**, 144.

Paar, V, 1984. *Debrecen 1984*, 675.

Paar, V, Brant, S, and Kraljević, H, 1982. *Phys. Lett.*, **110B**, 181.

Solari, H G, Gilmore, R, and Vallières, M, 1987. *Phys. Rev.*, **C35**, 320.

Sun, H Z, Vallières, M, Feng, D H, Gilmore, R, and Casten, R F, 1984. *Phys. Rev.* , **C29**, 352.

Sun, H Z, Feng, D H, Vallières, M, Gilmore, R, van Isacker,

P, and Frank, A, 1985. *Phys. Rev.*, **C31**, 1899.
Sunko, D K, and Paar, V, 1984. *Phys. Lett.*, **146B**, 279.
Vallières, M, 1984. *Gull Lake 1984*, 120.
Vallières, M, Sun, H Z, Feng, D H, Gilmore, R, and Casten, R F, 1984. *Phys. Lett.*, **135B**, 339.
van Isacker, P, 1986. *Dubrovnik 1986*, **1**, 231.
van Isacker, P, Frank, A, and Sun, H Z, 1984. *Ann. Phys.*, **157**, 183.
van Isacker, P, Jolie, J, Heyde, K, and Frank, A, 1985. *Phys. Rev. Lett.*, **54**, 653.
van Isacker, P, Jolie, J, Heyde, K, Waroquier, M, Moreau, J, and Scholten, O, 1984. *Phys. Lett.*, **149B**, 26.
Vergnes, M, 1984. *Gull Lake 1984*, 91.
Vergnes, M, 1986. *Dubrovnik 1986*, **1**, 187.
Vergnes, M, Berrier-Ronsin, G, and Bijker, R, 1983. *Phys. Rev.*, **C28**, 360.
Vergnes, M, Berrier-Ronsin, G, Rotbard, G, Vernotte, J, Maison, J M, and Bijker, R, 1984. *Phys. Rev.*, **C30**, 517.
Vergnes, M, Rotbard, G, Kalifa, J, Berrier-Ronsin, G, Vernotte, J, Seltz, R, and Burke, D G, 1981. *Phys. Rev. Lett*, **46**, 584.
Vergnes, M, Berrier-Ronsin, G, Rotbard, G, Vernotte, J, Langevin-Joliot, H, Gerlic, E, Van de Wiele, J, Guillot, J, and van der Werf, S Y, 1981. *Phys. Lett.*, **107B**, 349.
Vergnes, M, Grafeuille, S, Rotbard, G, Berrier-Ronsin, G, Vernotte, J, Maison, J M, Fortier, S, Tamisier, R, van Isacker, P, and Jolie, J, 1985. *Phys. Rev.* , **C31**, 2071.
Vervier, J, 1981. *Phys. Lett.*, **100B**, 383.
Vervier, J, 1984. *Debrecen 1984*, 663.
Vervier, J, 1986. *Dubrovnik 1986*, **1**, 175.
Vervier, J, and Janssens, R V F, 1982. *Phys. Lett.*, **108B**, 1.
Vervier, J, Holzmann, R, Janssens, R V F, Loiselet, M, and van Hove, M A, 1981. *Phys. Lett.*, **105B**, 343.
Vervier, J, van Isacker, P, Jolie, J, Kota V K B, and Bijker, R, 1985. *Phys. Rev.* , **C32**, 1406.
Warner, D D, 1984. *Gull Lake 1984*, 139.
Warner, D D, 1985. *Oaxtepec 1985*, 377.
Warner, D D, 1986. *Dubrovnik 1986*, **1**, 148.
Warner, D D, Casten, R F, and Frank, A, 1986. *Phys. Lett.*, **180B**, 207.
Wood, J L, 1981. *Phys. Rev.*, **C24**, 1788.
Wu, H, 1985. *Phys. Rev.*, **C32**, 2087.
Yates, S W, and Kleppinger, E W, 1984. *Debrecen 1984*, 381.

Zhou, S H, Frank, A, and van Isacker, P, 1983a. *Oaxtepec 1983*, 386.

Zhou, S H, Frank, A, and van Isacker, P, 1983b. *Phys. Lett.*, **124B**, 275.

13

GEOMETRICAL LIMIT OF ALGEBRAIC MODELS

13.1 Introduction

In the previous chapters several times we referred to the geometrical limit of various algebraic Hamiltonians. Here we will give a brief description of the methods which are usually used for deriving a geometrical Hamiltonian from an algebraic archetype. In the case of IBM-1 these methods demonstrate that its classical limit is equivalent (up to a homomorphism) to the geometrical collective model of Bohr and Mottelson (Bohr and Mottelson 1975, Preston and Bhaduri 1975). The method of Ginocchio and Kirson, as well as the method of Hatch and Levit, make use of coherent states, while the method of Klein, Li and Vallières avoids them.

13.2 The method of Ginocchio and Kirson

This approach to the geometrical limit was introduced by Ginocchio and Kirson (1980a, 1980b), and simultaneously by Dieperink, Scholten, and Iachello (1980).

Finding the classical limit of quantum mechanical operators is an old and difficult problem. This problem is, however, simplified if the quantum mechanical operators of which one wishes to determine the classical limit happen to form a Lie algebra $U(n)$. In this case it is known that the classical limit of the operators can be described in terms of $n - 1$ classical complex variables (the remaining variable is eliminated by insisting that one remains within the totally symmetric irreducible representations of U(n)). In the case of IBM-1, where the relevant Lie algebra is U(6), one thus needs 5 classical complex variables.

In order to introduce geometric (classical) variables in IBM-1, one can consider the **coherent state**

$$|N, \alpha_\mu> = (s^+ + \sum_\mu \alpha_\mu d_\mu^+)^N |0>, \qquad (13.1)$$

where N is the total number of bosons and α_μ ($\mu = -2, -1, 0, 1, 2$) are 5 complex variables. For static problems it turns out that these 5 variables can be chosen to be real. In addition, one can perform a suitable transformation, in order to relate the 5 real variables α_μ to the classical variables β and γ (describing the shape of the nucleus) and the three Euler angles (describing the orientation of the nucleus in space), for which we use the collective symbol Ω. Since

$$\alpha_0 = \beta cos\gamma, \tag{13.2}$$

$$\alpha_{\pm 2} = \frac{1}{\sqrt{2}}\beta sin\gamma, \tag{13.3}$$

$$\alpha_{\pm 1} = 0, \tag{13.4}$$

the coherent state takes the form

$$|N,\beta,\gamma> = [s^+ + \beta cos\gamma \quad d_0^+ + \frac{1}{\sqrt{2}}\beta sin\gamma \quad (d_{+2}^+ + d_{-2}^+)]^N|0>.\tag{13.5}$$

The coherent state (13.5) has the advantage that it contains all possible states of IBM-1, i.e. any state of IBM-1 can be obtained from (13.5) with suitable choice of the values of the variables β and γ.

The classical limit of any operator O is defined as its coherent state expectation value (up to a normalization constant) as follows

$$O_{cl} = \frac{<N,\beta,\gamma|O|N,\beta,\gamma>}{<N,\beta,\gamma|N,\beta,\gamma>}.\tag{13.6}$$

Thus the classical limit of the Hamiltonian H is the energy functional

$$E(N,\beta,\gamma) = \frac{<N,\beta,\gamma|H|N,\beta,\gamma>}{<N,\beta,\gamma|N,\beta,\gamma>}.\tag{13.7}$$

By minimizing the energy functional with respect to the classical shape variables β and γ one determines the equilibrium shape of the nucleus. For the most general IBM-1 Hamiltonian of Sec. 1.4 one finds the energy functional (van Isacker and Chen 1981)

$$E(N,\beta,\gamma) = \frac{N}{1+\beta^2}(\epsilon_s + \epsilon_d\beta^2)$$

$$+\frac{N(N-1)}{1+\beta^2}(f_1\beta^4 + f_2\beta^3 cos3\gamma + f_3\beta^2 + \frac{u_0}{2}),\tag{13.8}$$

where

$$f_1 = \frac{c_0}{10} + \frac{c_2}{7} + \frac{9c_4}{35},\tag{13.9}$$

$$f_2 = -2\sqrt{\frac{1}{35}}\tilde{v}_2, \tag{13.10}$$

$$f_3 = \sqrt{\frac{1}{5}}(\tilde{v}_0 + u_2). \tag{13.11}$$

13.2.1 The U(5) limit of IBM-1

The Hamiltonian of the U(5) limit of IBM-1 can be put in the form

$$H^I = \epsilon_d n_d + \sum_{L=0,2,4} \frac{1}{2}\sqrt{2L+1}\,c_L[(d^+ d^+)^L(\tilde{d}\tilde{d})^L]^0. \tag{13.12}$$

From this one obtains the energy functional

$$E^I = \epsilon_d \frac{N\beta^2}{1+\beta^2} + f_1 N(N-1)\frac{\beta^4}{(1+\beta^2)^2}, \tag{13.13}$$

which has a minimum at $\beta = 0$. This energy functional corresponds to a spherical shape, as we have already seen in Sec. 1.8.

13.2.2 The SU(3) limit of IBM-1

The Hamiltonian in the SU(3) limit of IBM-1 can be written in the form

$$H^{II} = -\kappa(Q \cdot Q) - \kappa'(L \cdot L). \tag{13.14}$$

From this one obtains the energy functional

$$E^{II} = -\kappa[\frac{N}{1+\beta^2}(5 + \frac{11}{4}\beta^2)$$

$$+ \frac{N(N-1)}{(1+\beta^2)^2}(\frac{\beta^4}{2} + 2\sqrt{2}\beta^3 cos3\gamma + 4\beta^2)] - \kappa'\frac{6N\beta^2}{1+\beta^2}, \tag{13.15}$$

which has a minimum at $\gamma = 0$ and $\beta \neq 0$. This energy functional corresponds to a deformed body with axial symmetry and prolate shape, as discussed already in Sec. 1.8. For large values of N ($N \to \infty$) the minimum is located at $\beta = \sqrt{2}$.

13.2.3 The O(6) limit of IBM-1

The Hamiltonian in the O(6) limit of IBM-1 can be put in the form

$$H^{III} = AP_6 + BC_{2O5} + CC_{2O3}, \qquad (13.16)$$

where

$$C_{2O3} = Z_3\sqrt{3}[(d^+\tilde{d})^1(d^+\tilde{d})^1]^0, \qquad (13.17)$$

$$C_{2O5} = Z_5(\sqrt{7}[(d^+\tilde{d})^3(d^+\tilde{d})^3]^0 + \sqrt{3}[(d^+\tilde{d})^1(d^+\tilde{d})^1]^0), \qquad (13.18)$$

$$P_6 = \frac{5}{4}[(d^+d^+)^0(\tilde{d}\tilde{d})^0]^0 - \frac{\sqrt{5}}{4}[(d^+d^+)^0(ss)^0 + (s^+s^+)^0(\tilde{d}\tilde{d})^0]^0$$

$$+ \frac{1}{4}[(s^+s^+)^0(ss)^0]^0, \qquad (13.19)$$

where Z_3 and Z_5 are suitable normalization constants. The operator P_6 is simply related to the second order Casimir invariant of O(6) through the relation

$$P_6 = \frac{1}{4}N(N+4) - \frac{1}{8}C_{2O6}. \qquad (13.20)$$

Using the above Hamiltonian one obtains the energy functional

$$E^{III} = (2B + 6C)\frac{N\beta^2}{1+\beta^2} + \frac{A}{4}N(N-1)\left(\frac{1-\beta^2}{1+\beta^2}\right)^2, \qquad (13.21)$$

which is independent of γ and has a minimum at $\beta \neq 0$, corresponding to a γ-unstable deformed shape. For $N \to \infty$ the minimum occurs at $\beta = 1$.

13.2.4 Triaxial shapes

From the results of the last three subsections it is clear that no triaxial shape (i.e. a shape with $\beta \neq 0$, $\gamma \neq 0$, $\pi/3$) occurs in IBM-1. Triaxial shapes occur by adding three-body ("cubic") interactions in the IBM-1 Hamiltonian, which were studied in Sec. 2.2. The classical limit of the "cubic" term mentioned there has been obtained by the present method (van Isacker and Chen 1981).

13.3 The method of Hatch and Levit

The method introduced by Hatch and Levit (1982) also makes use of coherent states.

Consider the most general IBM-1 Hamiltonian H given in Sec. 1.4. Its classical limit can be obtained from the Heisenberg equations of motion by replacing the boson operators s, s^+, d_μ, d_μ^+ by classical commuting variables α_s, α_s^*, α_μ, α_μ^*, respectively. The resulting equations are (considering $h/(2\pi) = 1$)

$$i\dot\alpha_j = \frac{\partial H_c}{\partial \alpha_j^*}, \tag{13.22}$$

$$i\dot\alpha_j^* = -\frac{\partial H_c}{\partial \alpha_j}, \tag{13.23}$$

where $j = s, -2, -1, 0, 1, 2$. These equations have the form of Hamilton equations. The Hamiltonian function H_c in eqs. (13.22), (13.23) is simply obtained from H by replacing each boson operator by its corresponding classical variable, as described above.

The same equations can be obtained by varying the quantity

$$\int dt < \vec\alpha(t)|i\frac{\partial}{\partial t} - H|\vec\alpha(t) > \tag{13.24}$$

with the coherent states

$$|\vec\alpha> = exp(\frac{-|\vec\alpha|^2}{2})exp(\alpha_s s^+ + \sum_\mu \alpha_\mu d_\mu^+)|0> . \tag{13.25}$$

In this approach

$$H_c = < \vec\alpha|H|\vec\alpha > . \tag{13.26}$$

This coincides with the previous definition, since the coherent state is an eigenstate of the boson annihilation operators s and d_μ, i.e.

$$s|\vec\alpha> = a_s|\vec\alpha>, \tag{13.27}$$

$$d_\mu|\vec\alpha> = \alpha_\mu|\vec\alpha> . \tag{13.28}$$

In the previous section the conservation of the boson number has been taken into account already in the boson space. Here we introduce it at the classical level. The boson number operator is

$$N = s^+s + \sum_\mu d_\mu^+ d_\mu. \tag{13.29}$$

Then its classical limit (obtained by the above mentioned replacement) is

$$N(\vec{\alpha}) = \alpha_s^* \alpha_s + \sum_\mu \alpha_\mu^* \alpha_\mu. \qquad (13.30)$$

The variable conjugate to $N(\vec{\alpha})$ is easily found to be

$$Q = i \ \ ln\sqrt{\frac{\alpha_s}{\alpha_s^*}}. \qquad (13.31)$$

Since $N(\vec{\alpha})$ is conserved, Q is a cyclic variable, thus the Hamiltonian does not explicitly depend on it.

The next step is to consider a canonical transformation from the variables α_s, α_s^*, α_μ, α_μ^* to variables N, Q, q_μ and p_μ, choosing q_μ and p_μ to be suitable combinations of α_j and α_j^* resembling nuclear deformation parameters. Since q_μ must satisfy the hermiticity condition

$$q_\mu^* = (-1)^\mu q_{-\mu}, \qquad (13.32)$$

a choice for q_μ is

$$q_\mu = \frac{1}{\sqrt{2}} \left[\alpha_\mu^* \sqrt{\frac{\alpha_s}{\alpha_s^*}} + (-1)^\mu \alpha_{-\mu} \sqrt{\frac{\alpha_s^*}{\alpha_s}} \right]. \qquad (13.33)$$

Then the canonically conjugate momenta are

$$p_\mu = -\frac{i}{\sqrt{2}} \left[\alpha_\mu \sqrt{\frac{\alpha_s^*}{\alpha_s}} - (-1)^\mu \alpha_{-\mu}^* \sqrt{\frac{\alpha_s}{\alpha_s^*}} \right]. \qquad (13.34)$$

Finally one takes advantage of the rotational invariance of the problem by performing a transformation from the space-fixed (laboratory) frame to the body-fixed (intrinsic) frame, using the standard relations

$$q_\mu = \sum_{\nu=-2}^{2} D_{\mu\nu}^{(2)}(\Omega) a_\nu, \qquad (13.35)$$

$$p_\mu = \sum_{\nu=-2}^{2} D_{\mu\nu}^{(2)*}(\Omega) b_\nu, \qquad (13.36)$$

where by Ω we denote the three Euler angles (ψ, θ, ϕ). The intrinsic coordinates a_ν can be parametrized as

$$a_0 = \beta cos\gamma, \qquad (13.37)$$

$$a_{\pm 1} = 0, \tag{13.38}$$

$$a_{\pm 2} = \frac{\beta}{\sqrt{2}} sin\gamma, \tag{13.39}$$

while the five parameters b_ν can be expressed in terms of the momenta as follows

$$b_0 = p_\beta cos\gamma - \frac{p_\gamma}{\beta} sin\gamma, \tag{13.40}$$

$$b_{\pm 1} = -\frac{1}{2\sqrt{2}\beta}\left[\frac{iL_1}{sin(\gamma - 2\pi/3)} \pm \frac{L_2}{sin(\gamma - 4\pi/3)}\right], \tag{13.41}$$

$$b_{\pm 2} = \frac{1}{\sqrt{2}}\left[\frac{p_\gamma}{\beta}cos\gamma + p_\beta sin\gamma \pm \frac{iL_3}{2\beta sin\gamma}\right]. \tag{13.42}$$

In these expressions p_β and p_γ are the momenta conjugate to the β and γ variables, while L_1, L_2, L_3 are the angular momenta in the intrinsic frame, written in terms of the Euler angles (ψ, θ, ϕ) and their conjugate momenta $(p_\psi, p_\theta, p_\phi)$ as follows

$$L_1 = (p_\phi - p_\psi cos\theta)\frac{cos\psi}{sin\theta} - p_\theta sin\psi, \tag{13.43}$$

$$L_2 = (p_\phi - p_\psi cos\theta)\frac{sin\psi}{sin\theta} + p_\theta cos\psi, \tag{13.44}$$

$$L_3 = p_\psi. \tag{13.45}$$

In what follows the results of these transformations for the Hamiltonians of the three limiting symmetries of IBM-1 will be given, and contact with the results of the method of Ginocchio and Kirson will be made.

13.3.1 The U(5) limit of IBM-1

The simplest U(5) Hamiltonian one can use is proportional to the linear Casimir invariant of U(5)

$$H = \epsilon n_d, \tag{13.46}$$

where n_d is the number operator of d-bosons. The classical Hamiltonian corresponding to this boson Hamiltonian is obtained, as already mentioned, by replacing the boson operators d_μ^+, d_μ by the classical variables α_μ^*, α_μ. Performing the above mentioned transformation one then obtains

$$H_c = \frac{\epsilon}{2}(\beta^2 + T), \tag{13.47}$$

where

$$T = p_\beta^2 + \frac{T_\gamma}{\beta^2} \tag{13.48}$$

and

$$T_\gamma = p_\gamma^2 + \frac{1}{4}\sum_{m=1}^{3}\frac{L_m^2}{\sin^2(\gamma - 2\pi m/3)}. \tag{13.49}$$

The general U(5) Hamiltonian has been given in the previous section. For its classical limit we find

$$H_c = \epsilon_d\frac{1}{2}(\beta^2 + T) + A\frac{1}{4}(\beta^2 + T)^2 + BT_\gamma + CL^2 \tag{13.50}$$

where T and T_γ are the same as above and

$$L^2 = L_1^2 + L_2^2 + L_3^2, \tag{13.51}$$

$$A = \frac{9}{35}C_4 + \frac{1}{7}C_2 + \frac{1}{10}C_0, \tag{13.52}$$

$$B = -\frac{3}{70}C_4 + \frac{1}{7}C_2 - \frac{1}{10}C_0, \tag{13.53}$$

$$C = \frac{1}{14}(C_4 - C_2). \tag{13.54}$$

One can obtain a potential function from this classical Hamiltonian by setting all the momenta p_β, p_γ, and L equal to zero. Then one finds

$$V_c(\beta) = \frac{\epsilon_d}{2}\beta^2 + \frac{A}{4}\beta^4. \tag{13.55}$$

This potential function has a minimum at $\beta = 0$, in agreement with the findings of the previous section.

One can actually rewrite this potential function by rescaling the β variable as

$$\beta = \frac{\sqrt{2N}\beta'}{\sqrt{1 + (\beta')^2}}. \tag{13.56}$$

The resulting expression is

$$V_c(\beta') = \epsilon_d \frac{N(\beta')^2}{1 + (\beta')^2} + AN^2 \frac{(\beta')^4}{(1 + (\beta')^2)^2}, \qquad (13.57)$$

which is identical to (13.13) if one makes the usual classical limit replacement of N^2 by $N(N-1)$. (Notice that $A \equiv f_1$).

13.3.2 The SU(3) limit of IBM-1

Using the SU(3) boson Hamiltonian of the previous section we obtain the classical Hamiltonian

$$H_c = -\frac{\kappa}{8}(\beta^2 + T)^2 + \frac{\kappa}{2}T_\gamma - 2\kappa\beta^2[N - \frac{1}{2}(\beta^2 + T)]$$

$$+\kappa[(\frac{p_\gamma^2}{\beta} - \beta p_\beta^2 - \beta^3)cos3\gamma + 2p_\beta p_\gamma sin3\gamma$$

$$+\frac{1}{4\beta}\sum_{m=1}^{3}\frac{L_m^2 cos(\gamma - 2\pi m/3)}{sin^2(\gamma - 2\pi m/3)}]\sqrt{N - \frac{1}{2}(\beta^2 + T)} - (\frac{\kappa}{8} + \kappa')L^2,$$

$$(13.58)$$

where T and T_γ are the same as in the previous subsection.

Letting all the momenta vanish one obtains the potential function

$$V_c(\beta) = -2\kappa N\beta^2 + \frac{7\kappa}{8}\beta^4 - \kappa\beta^3 cos3\gamma\sqrt{N - \frac{1}{2}\beta^2}. \qquad (13.59)$$

This function has a minimum at $\beta = 2\sqrt{N/3}$, $\gamma = 0$, which corresponds to an axially symmetric prolate shape, in agreement with the previous findings. Rescaling the β variable one obtains

$$V_c(\beta') = -\frac{\kappa N^2}{2[1 + (\beta')^2]}[(\beta')^4 + 4\sqrt{2}(\beta')^3 cos3\gamma + 8(\beta')^2]. \qquad (13.60)$$

This function has a minimum at $\beta' = \sqrt{2}$, $\gamma = 0$. For large N this result coincides with the findings of subesc. 13.2.2.

13.3.3 The O(6) limit of IBM-1

Using the O(6) Hamiltonian of the previous section we find the corresponding classical Hamiltonian

$$H_c = \frac{A}{4}[(\beta^2 - N)^2 + \beta^2 p_\beta^2] + \frac{B}{6}T_\gamma + CL^2. \tag{13.61}$$

Letting all momenta to vanish one obtains the potential function

$$V_c(\beta) = \frac{A}{4}(\beta^2 - N)^2, \tag{13.62}$$

which is independent of γ and has a minimum at $\beta = \sqrt{N}$, indicating a γ-unstable shape, in agreement with the previous findings.

13.3.4 The vibron model

The present method has been applied in the study of the classical limit of the vibron model. The results of this study have been discussed in Chapter 8. We do not give any details here, referring the reader to the original paper (Levit and Smilansky 1982).

13.4 The method of Klein, Li, and Vallières

This method (Klein, Li, and Vallières 1982) is different from the previous two methods, since it avoids the use of coherent states. We will illustrate the method by applying it to a specific IBM-1 Hamiltonian of the form

$$H = \epsilon n_d - \kappa(Q \cdot Q)^0, \tag{13.63}$$

where

$$Q = s^+ \tilde{d}_\mu + d_\mu^+ s + \chi(d^+ d)_\mu^2. \tag{13.64}$$

First one makes a transformation from the 6 bosons s, d_μ to 5 bosons b_μ as follows

$$d_\mu^+ d_\nu = b_\mu^+ b_\nu, \tag{13.65}$$

$$d_\mu^+ s = (s^+ d_\mu)^+ = b_\mu^+ \sqrt{N - n_b}, \tag{13.66}$$

$$s^+ s = N - n_d = N - n_b, \tag{13.67}$$

where n_d (n_b) is the number of d (b) bosons. (This transformation is possible because the Lie algebra U(6) admits both a Schwinger boson realization in terms of the 6 bosons s, d_μ and a Holstein-Primakoff boson realization in terms of the 6-1 bosons b_μ. Boson realizations will be further discussed in Chapter 15.)

Subsequently, one makes a transformation into canonical coordinates x_μ and conjugate momenta p_μ

$$b_\mu^+ = \frac{1}{\sqrt{2}}(x_\mu - ip_\mu^+), \tag{13.68}$$

$$\tilde{b}_\mu = \frac{1}{\sqrt{2}}(x_\mu + ip_\mu^+), \tag{13.69}$$

with

$$\tilde{x}_\mu = x_\mu^+, \tag{13.70}$$

$$\tilde{p}_\mu = p_\mu^+, \tag{13.71}$$

$$[x_\mu, p_\nu] = [x_\mu^+, p_\nu^+] = i\delta_{\mu\nu}. \tag{13.72}$$

In what follows, we will confine ourselves to the rotational limit for low-lying energy states. In this case the largest matrix elements of x_μ are of order $\sqrt{N}$, while those of p_μ are of order $1/\sqrt{N}$. Thus in this limit one can use the **adiabatic approximation**, i.e. one can expand in powers of $(p_\mu\tilde{p}_\mu/N)$, keeping terms up to second order in the momenta. Since we are going to use the adiabatic approximation here, it is convenient to rescale coordinates and momenta according to the equations

$$x_\mu \to \sqrt{N}x_\mu, \tag{13.73}$$

$$p_\mu \to \frac{1}{\sqrt{N}}p_\mu, \tag{13.74}$$

and to introduce the geometrical variables β, γ defined in terms of the new variables as follows

$$x_\mu\tilde{x}_\mu = \beta^2, \tag{13.75}$$

$$[(x \otimes x)^2 \otimes x]^0 = -\sqrt{\frac{2}{7}}\beta^3 \, sos3\gamma. \tag{13.76}$$

The rescaling had the purpose to make clearer the relative magnitude of the various terms which will occur in the expansion.

Substituting in the above boson Hamiltonian and keeping for the kinetic energy only the leading term in an expansion in $1/N$, while keeping for the potential energy the first two terms in an expansion in 1/N, we find its classical limit in the form

$$H_c = N\epsilon(\frac{t_0}{N^2} + v_0 + \frac{v_1}{N}),$$ (13.77)

where

$$v_0 = (\frac{1}{2} - 2F)\beta^2 + F\frac{2}{\sqrt{7}}\chi\sqrt{1 - \frac{1}{2}\beta^2}\beta^3 cos3\gamma + F[1 - \frac{\chi^2}{14}]\beta^4,$$ (13.78)

$$v_1 = -6F\beta^2 + \frac{3}{\sqrt{7}}F\chi\frac{1}{\sqrt{1 - \frac{1}{2}\beta^2}}\beta^3 cos\gamma,$$ (13.79)

$$t_0 = \frac{1}{2}p^2 + \frac{1}{2}F\{p^2,\beta^2\} - F\chi\frac{1}{8\sqrt{7}}\{\beta^2 cos3\gamma, \{\sqrt{\frac{1}{1 - \frac{1}{2}\beta^2}}, p^2\}\}$$

$$- \frac{1}{2\sqrt{2}}F\chi\{\sqrt{1 - \frac{1}{2}\beta^2}, \{x_\mu, (p\otimes p)^2_\mu\}\} - \frac{1}{4}F\chi^2\{(x\otimes x)^2_\mu, (p\otimes p)^2_\mu\}.$$ (13.80)

In the above

$$p^2 \equiv p_\mu\tilde{p}_\mu,$$ (13.81)

while F is the dimensionless variable

$$F = \frac{\kappa N}{\epsilon}.$$ (13.82)

In the limit $N \to \infty$ only the v_0 term remains in the classical Hamiltonian, providing the potential energy functional.

In order to make contact with the results of the previous methods we make the transformation

$$\beta = \frac{\sqrt{2}\beta'}{\sqrt{1 + (\beta')^2}}.$$ (13.83)

Then the potential energy functional becomes (dropping the prime from β')

$$E = N\epsilon[(\frac{1}{2} - 2F)\frac{2\beta^2}{1 + \beta^2} + \frac{2F\chi\sqrt{\frac{2}{7}}\beta^3 cos3\gamma}{(1 + \beta^2)^2}$$

$$+F(1 - \frac{\chi^2}{14})\frac{4\beta^4}{(1 + \beta^2)^2}], \qquad (13.84)$$

which contains terms similar to those encountered in the previous methods.

References

The present account of the method of Ginocchio and Kirson has been based on Arima and Iachello (1984) and Iachello (1981). More details can be found in Kirson (1982) and Ginocchio and Kirson (1980b). The present account of the other two methods has been based on Hatch and Levit (1982) and Klein, Li, and Vallières (1982). Extended discussions of the geometrical limits of IBM-1 and IBM-2 can be found in Dieperink (1982a, 1982b). For recent overviews of the relation between algebraic and geometric models see Kirson (1986) and Lipas (1986).

Arima, A, and Iachello, F, 1984. In *Advances in Nuclear Physics*, (ed. J W Negele and E Vogt) Vol 13, p 139. Plenum, New York.
Assenbaum, H J, and Weiguny, A, 1982. *Poiana Brasov 1982*, 192.
Assenbaum, H J, and Weiguny, A, 1983a. *Phys. Lett.*, **120B**, 257.
Assenbaum, H J, and Weiguny, A, 1983b. *Z. Phys.*, **A310**, 75.
Balantekin, A B, and Barrett, B R, 1985. *Phys. Rev.*, **C32**, 288.
Balantekin, A B, and Barrett, B R, 1986. *Phys. Rev.*, **C33**, 1842.
Balantekin, A B, Barrett, B R, and Levit, S, 1983. *Phys. Lett.*, **129B**, 153.
Bijker, R, 1985. *Phys. Rev.*, **C32**, 1442.
Bohr, A, and Mottelson, B R, 1975. *Nuclear Structure*, Vol II. Benjamin, Reading.
Canto, L F, and Paar, V, 1981. *Phys. Lett.*, **102B**, 217.
Castaños, O, Frank, A, and van Isacker, P, 1984. *Phys. Rev. Lett.*, **52**, 263.
Castaños, O, Frank, A, Hess, P, and Moshinsky, M, 1981. *Phys. Rev.*, **C24**, 1367.
Chen, J Q, and van Isacker, P, 1980. *Erice 1980*, 193.
Cohen, T D, 1983. *Phys. Lett.*, **125B**, 433.
Cohen, T D, 1985. *Nucl. Phys.*, **A436**, 165.
Dieperink, A E L, 1982a. *Erice 1982*, 121.
Dieperink, A E L, 1982b. *Poiana Brasov 1982*, 149.
Dieperink, A E L, and Scholten, O, 1980a. *Erice 1980*, 167.

Dieperink, A E L, and Scholten, O, 1980b. *Nucl. Phys.*, **A346**, 125.

Dieperink, A E L, and Scholten, O, 1981. *Oaxtepec 1981*, 4.

Dieperink, A E L, Scholten, O, and Iachello, F, 1980. *Phys. Rev. Lett.*, **44**, 1747.

Elliott, J P, 1985. *Rep. Prog. Phys.*, **48**, 171.

Elliott, J P, Evans, J A, and van Isacker, P, 1986. *Phys. Rev. Lett.*, **57**, 1124.

Elliott, J P, Park, P, and Evans, J A, 1986. *Phys. Lett.*, **171B**, 145.

Faessler, A, 1983. *Nucl. Phys.*, **A396**, 291c.

Fewell, M P, 1987. *Phys. Lett.*, **192B**, 9.

Gilmore, R, and Feng, D H, 1980. *Erice 1980*, 149.

Gilmore, R, and Feng, D H, 1982. *Erice 1982*, 479.

Gilmore, R, Vallières, M, Feng, D H, and Sun, H Z, 1984. *Nucl. Phys.*, **A421**, 141c.

Ginocchio, J N, 1980. *Erice 1980*, 179.

Ginocchio, J N, 1981. *Oaxtepec 1981*, 71.

Ginocchio, J N, 1982. *Nucl. Phys.*, **A376**, 438.

Ginocchio, J N, and Kirson, M W, 1980a. *Phys. Rev. Lett.*, **44**, 1744.

Ginocchio, J N, and Kirson, M W, 1980b. *Nucl. Phys.*, **A350**, 31.

Hatch, R L, and Levit, S, 1982. *Phys. Rev.*, **C25**, 614.

Hess, P O, Moshinsky, M, Castaños, O, and Frank, A, 1982. *Oaxtepec 1982*, 95.

Iachello, F, 1981. *Granada 1981*, 1.

Kirson, M W, 1982. *Ann. Phys.*, **143**, 448.

Kirson, M W, 1986. *Dubrovnik 1986*, **2**, 901.

Kirson, M W, and Leviatan, A, 1985. *Phys. Rev. Lett.*, **55**, 2846.

Kirson, M W, and Leviatan, A, 1986. *Phys. Rev. Lett.*, **56**, 788.

Klein, A, and Vallières, M, 1980. *Drexel 1980*, 487.

Klein, A, and Vallières, M, 1981. *Phys. Rev. Lett.*, **46**, 586.

Klein, A, Li, C T, and Vallières, M, 1982. *Phys. Rev.*, **C25**, 2733.

Klein, A, Rafelski, H, and Rafelski, J, 1981. *Nucl. Phys.*, **A355**, 189.

Klein, A, Li, C T, Cohen, T D, and Vallières, M, 1982. *Erice 1982*, 183.

Leviatan, A, 1986. *Dubrovnik 1986*, **2**, 1030.

Leviatan, A, 1987. *Oaxtepec 1987*, 201.

Levit, S, and Smilansky, U, 1982. *Nucl. Phys.*, **A389**, 56.

Lipas, P O, 1986. *Dubrovnik 1986*, **2**, 921.

Meyer-ter-Vehn, J, 1978. *Erice 1978*, 157.
Meyer-ter-Vehn, J, 1979. *Phys. Lett.*, **84B**, 10.
Moshinsky, M, 1980. *Nucl. Phys.*, **A338**, 156.
Moshinsky, M, 1981. *Nucl. Phys.*, **A354**, 257c.
Moshinsky, M, 1982. *Poiana Brasov 1982*, 97.
Pittel, S, and Dukelsky, J, 1985. *Phys. Rev.*, **C32**, 335.
Preston, M A, and Bhaduri, R K, 1975. *Structure of the Nucleus*, Chapter 9. Addison-Wesley, Reading.
Talmi, I, 1981a. *Oaxtepec 1981*, 397.
Talmi, I, 1981b. *Phys. Lett.*, **103B**, 177.
van Isacker, P, and Chen, J Q, 1981. *Phys. Rev.*, **C24**, 684.
van Roosmalen, O S, and Dieperink, A E L, 1981. *Phys. Lett.*, **100B**, 299.
Walet, N R, Brussaard, P J, and Dieperink, A E L, 1985. *Phys. Lett.*, **163B**, 1.
Weiguny, A, 1981. *Z. Phys.*, **A301**, 335.
Yang, Y T, and Irvine, J M, 1983. *J. Phys.*, **G9**, 185.

14

<h1 style="text-align:center;">GENERALIZED
VARIABLE MOMENT OF INERTIA MODELS</h1>

In this chapter we will discuss some generalizations of the Variable Moment of Inertia (VMI) model, which were inspired by the success of IBM. First, a brief account of the original version of the VMI model will be given. The contents of this chapter are not directly related to the algebraic approach used so far, but they provide useful physical insights.

14.1 The Variable Moment of Inertia model

As we have seen in Chapter 0, ground state bands of rotational nuclei can be described by the simple formula

$$E(J) = \frac{J(J+1)}{2\Theta},\tag{14.1}$$

where Θ is the moment of inertia of the nucleus, which in first approximation is assumed to be constant. In order to improve agreement with experiment, we saw that higher terms need to be added to this formula. Another approach is to let the moment of inertia to vary as a function of the nuclear spin J. One thus obtains the Variable Moment of Inertia (VMI) model (Mariscotti, Scharff-Goldhaber, and Buck 1969), in which only one additional term is required in the expression for the energy. In this model the levels of the ground state band are given by

$$E(J) = \frac{J(J+1)}{2\theta(J)} + \frac{1}{2}c(\theta(J) - \theta_0)^2.\tag{14.2}$$

In this formula $\theta(J)$ is the moment of inertia of the nucleus at the state with spin J, while c and θ_0 are two constant parameters, fitted to the data. The moment of inertia at spin J is determined through the variational condition

$$\frac{\partial E(J)}{\partial \theta(J)}\Big|_J = 0,\tag{14.3}$$

225

which provides a cubic equation for $\theta(J)$, having only one real solution. The constant parameter θ_0 is identified as the ground state moment of inertia of the nucleus. Instead of the constant parameter c, it was found that it is meaningful to use the parameter combination

$$\sigma_{VMI} = \frac{1}{2c\theta_0^3},\qquad (14.4)$$

which is related to the softness of the nucleus.

It is easy to see that the region of validity of the model is

$$2.23 < R_4 < 3.33.\qquad (14.5)$$

Very accurate results have been obtained with the VMI formula for many nuclei, not only in the rotational region, where the model is certainly expected to work, but also in the transitional and even in the vibrational region. Rotational nuclei were found to have high ground state moment of inertia and small softness. Nuclei near the vibrational region are characterized by small ground state moments of inertia and high values of the softness parameter. The moment of inertia is in all cases increasing with J, but this increase is much slower in rotational nuclei than in vibrational ones.

14.2 Generalized VMI models

Although a $J(J+1)$ term is a reasonable starting point for the description of deformed nuclei, this is not the case when one considers vibrational or transitional nuclei. In the case of vibrational nuclei, in particular, as we saw in Chapter 0 a term proportional to J is the reasonable starting point, while the next approximation is given by the **Anharmonic Vibrator Model** (AVM) (Das, Dreizler, and Klein 1970)

$$E(J) \doteq aJ + bJ(J-2).\qquad (14.6)$$

It is interesting to consider this expression from the IBM point of view. The SU(3) (rotational) limit of IBM requires a $J(J+1)$ dependence of $E(J)$, thus it is obtained for $b/a = 1/3$. The SU(5) (vibrational) limit of IBM requires a J dependence of $E(J)$, thus it is obtained for $b \ll a$. Finally, the O(6) (γ-unstable) limit of IBM requires a form

$$E(J) = AJ(J+6) + BJ(J+1),\qquad (14.7)$$

which has the rotational form for $A = 0$, while it is obtained for $b/a = 1/8$ in the case of $B = 0$. The AVM formula thus incorporates all the three limiting symmetries of IBM-1, being a suitable starting point for a unified description of all kinds of nuclei, vibrational, transitional, or rotational.

The most general formula one can obtain from AVM by using the variable moment of inertia concept is (Bonatsos and Klein 1984a)

$$E(J) = \frac{J}{\Phi_1(J)} + \frac{J(J-2)}{\Phi_2(J)} + \frac{1}{2} \sum_{i,j=1}^{2} K_{ij}(\Phi_i(J) - \Phi_{i0})(\Phi_j(J) - \Phi_{j0}),$$

$$(14.8)$$

i.e. one allows for a separate scaling function for each of the J and $J(J-2)$ terms. This formula contains five independent parameters $(\Phi_{10}, \Phi_{20}, K_{11}, K_{22}, K_{12} = K_{21})$. The value of the functions $\Phi_1(J)$ and $\Phi_2(J)$ is determined for each value of J from the variational conditions

$$\frac{\partial E(J)}{\partial \Phi_i(J)}\Big|_J = 0, \quad i = 1, 2. \tag{14.9}$$

Since five parameters are quite many, we choose to reduce their number. Two possible simplified models will be considered here.

14.2.1 The Generalized VMI model

By fixing the ratio of the two scaling functions, one obtains the so-called **Generalized VMI** (GVMI) model (Bonatsos and Klein 1984a)

$$E(J) = \frac{J + XJ(J-2)}{\Phi(J)} + \frac{1}{2}K(\Phi(J) - \Phi_0)^2, \tag{14.10}$$

where X, K, Φ_0 are three constant parameters to be fitted to the data. Notice that for $X = 1/3$ one obtains the original VMI. Thus one expects the parameter X to obtain values close to $1/3$ for rotational nuclei, but gradually smaller values in the transitional and vibrational regions. Φ_0 is the ground state moment of inertia of the nucleus. The variable moment of inertia $\Phi(J)$ is determined for each J by a condition identical to (14.3). Instead of the parameter K, it is again meaningful to use the parameter combination

$$\sigma_{GVMI} = \frac{1 - 2X}{K\Phi_0^3}. \tag{14.11}$$

The region of validity of the model is found to be

$$1.59 < R_4 < 3.33, \qquad (14.12)$$

although in practice the value of R_4 seldom dips below 2.

14.2.2 The Variable Anharmonic Vibrator Model

By setting the scaling function of J equal to a constant, while allowing the scaling function of $J(J-2)$ to vary, we obtain the **Variable Anharmonic Vibrator Model** (VAVM) (Bonatsos and Klein 1984a)

$$E(J) = aJ + \frac{J(J-2)}{\Theta(J)} + \frac{1}{2}C(\Theta(J) - \Theta_0)^2, \qquad (14.13)$$

where a, C, Θ_0 are three constant parameters to be fitted to the data. Notice that in this case $a = E(2)/2$. Θ_0 is the ground state moment of inertia of the nucleus. The variable moment of inertia $\Theta(J)$ is determined for each value of J by a condition identical to (14.3). Instead of the parameter K, it is again meaningful to use the parameter combination

$$\sigma_{VAVM} = -\frac{2}{C\Theta_0^3}. \qquad (14.14)$$

The region of validity of the model is easily found to be

$$2 < R_4 < 3.33. \qquad (14.15)$$

14.2.3 Comparison of the models

A large number of nuclei has been analyzed using the **VMI**, **GVMI**, and **VAVM** models. In the rotational region all three models give equally good results, as expected. In the transitional and the vibrational regions GVMI and VAVM significantly improve the agreement with the data already shown by VMI, indicating that one really has to break the $J(J+1)$ term into two J and $J(J-2)$ terms as one is going away from the rotational limit. Between the two three-parameter models, VAVM gives the best results in these regions, indicating that two separate scaling factors are required in the J and $J(J-2)$ terms. The quality of the fits can be seen in Fig. 14.1, where the $R_{10} = E(10)/E(2)$ values predicted by VAVM and VMI are shown for a series of Yb isotopes. The two VMI parameters were fitted to the E(2) and E(4) levels, while for fitting the three VAVM parameters the E(6) level was used in addition. The results of the two models are identical in the rotational region ($R_4 > 3$), while in the transitional ($2.4 < R_4 < 3$) and vibrational ($R_4 < 2.4$) region VAVM is clearly better.

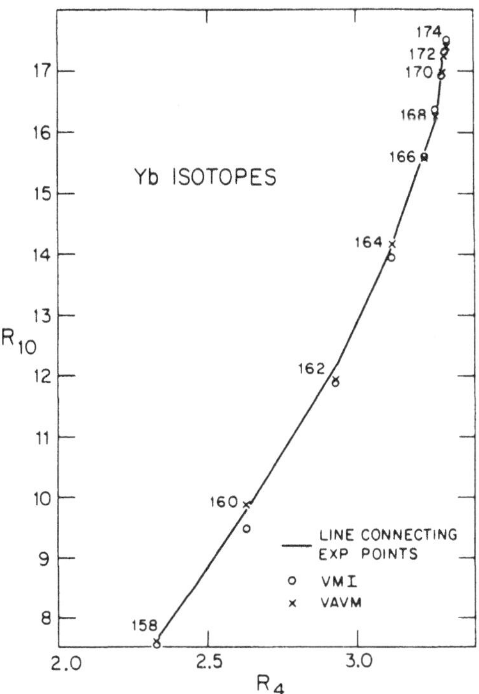

Fig. 14.1 R_{10} vs. R_4 for a series of Yb isotopes. Comparison of experiment with VMI and VAVM model predictions. (Taken from Bonatsos and Klein (1984a)).

14.3 Description of backbending in the VMI framework

The high spin effect of backbending has been described in Sec. 5.10. Here we will show how this effect can be simply described in the VMI framework. As already mentioned in Sec. 5.10, it is generally

thought that backbending arises in the Yrast band when the ground
state band is crossed by a superband. Experimentally it is seen that
the superband has, to a very good approximation, constant moment
of inertia. Thus one is inclined to use the usual VMI (or VAVM
or GVMI) formula for the ground state band, while using for the
superband the simple rotational formula (Bonatsos 1985)

$$E_s(J) = E_0 + A[J(J+1) - K^2], \qquad (14.16)$$

where E_0 is the (constant) bandhead energy, A is the constant moment of inertia, and K is the projection of the angular momentum of
the nucleus along its symmetry axis. K is known to be zero in the
case of ground state or β bands, while it is equal to 2 for γ bands. In
the case of the superband K is not a good quantum number, so that
both of the possible values $K = 0$ and $K = 1$ are used in calculations.

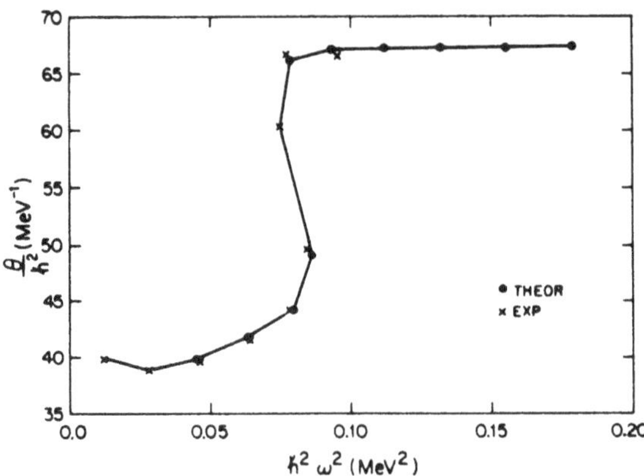

Fig. **14.2** Standard backbending plot for the Yrast band of ^{162}Er. The
solid lines connect the model predictions (for superband with $K = 0$).
Experimental points are represented by x's. (Taken from Bonatsos (1985)).

In addition to the energy formulas for the ground state band and
the superband, one needs to consider the mixing between these two
bands. The exact form of the interband interaction is not known.
Thus we choose to use the standard band coupling formalism, using
an angular momentum dependent interaction

$$V = X\sqrt{J(J+1)} \quad for \quad K = 0 \rightarrow K = 1, \qquad (14.17)$$

$$V = XJ(J+1) \quad for \quad K = 0 \to K = 0. \tag{14.18}$$

In the above X is a parameter measuring the strength of the interband interaction strength. It is known that V has to be small (i.e.
below 50 keV) in all cases that backbending is observed. Large values
of the interband interaction strength smooth out the effect.

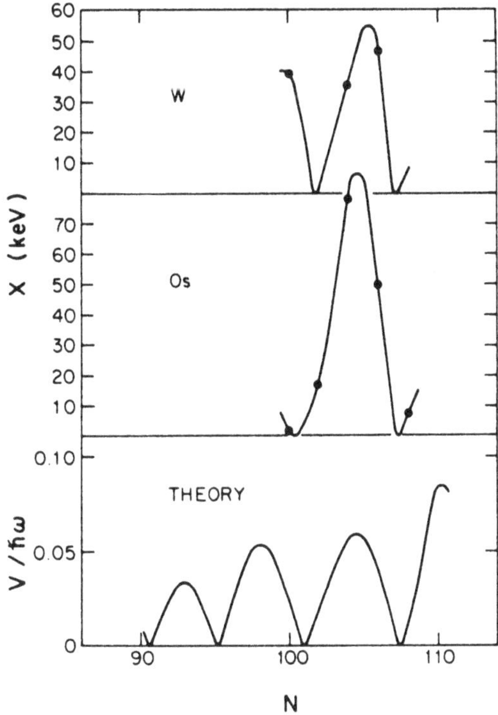

Fig. 14.3 The interaction strength V (as determined by model calculations with a superband with $K = 1$) for series of W and Os isotopes,
showing the theoretically predicted oscillatory behavior, which is shown at
the bottom. (Taken from Bonatsos (1985)).

Solving the secular equation

$$\begin{vmatrix} E(J) - W & V \\ V & E_s(J) - W \end{vmatrix} = 0, \tag{14.19}$$

we obtain two solutions for each J:

$$W_{\mp} = \frac{E(J) + E_s(J)}{2} \mp \sqrt{(\frac{E(J) - E_s(J)}{2})^2 + V^2}. \qquad (14.20)$$

The lower solution W_- corresponds to the Yrast band, the upper solution W_+ corresponds to the Yrare band. It must be emphasized that the variational conditions

$$\frac{\partial W_{\mp}(J)}{\partial \theta(J)}\Big|_J = 0 \qquad (14.21)$$

are equivalent to the variational condition (14.3), as long as the interband interaction remains independent of $\theta(J)$, which is true in the cases considered here. Thus by applying the condition (14.21) we minimize both the lower and the upper solutions simultaneously.

The quality of the results obtained by the model can be seen in Fig. 14.2, where the backbending plot for the Yrast band of ^{162}Er is shown. The solid line connects the model predictions, while the experimental points are indicated by x's.

The interband interaction strength is known to show an oscillatory behavior (Bengtsson 1980) as function of the neutron number N. This oscillatory behavior is well reproduced by the model, as it is shown in Fig. 14.3 for the W and Os isotopes. At the bottom of the figure, Bengtsson's theoretical prediction is shown for comparison.

References

Bengtsson, R, 1980. *J. Phys. (Paris) Colloq.*, **41**, C10.

Bhattacharya, S, and Sen, S, 1984. *Phys. Rev.*, **C30**, 1014.

Bonatsos, D, 1985. *Phys. Rev.*, **C31**, 2256.

Bonatsos, D, 1987. *Phys. Lett.*, **187B**, 1.

Bonatsos, D, and Klein, A, 1984a. *Phys. Rev.*, **C29**, 1879.

Bonatsos, D, and Klein, A, 1984b. *At. Data Nucl. Data Tables*, **30**, 27.

Das, T K, Dreizler, R M, and Klein, A, 1970. *Phys. Rev.*, **C2**, 632.

Klein, A, 1980a. *Phys. Lett.*, **93B**, 1.

Klein, A, 1980b. *Nucl. Phys.*, **A347**, 3c.

Mariscotti, M A J, Scharff-Goldhaber, G, and Buck, B, 1969. *Phys. Rev.*, **178**, 1864.

Riezebos, H J, Jansen, J F W, de Voigt, M J A, and Kaczarowski, R, 1985. *Mikołajki 1985*, 184.

Scharff-Goldhaber, G, 1984. *Gull Lake 1984*, 364.

Scharff-Goldhaber, G, and Dresden, M, 1980. *Trans. N. Y. Acad. Sci.*, Ser II **40**, 166.

Scharff-Goldhaber, G, and Goldhaber, A S, 1970. *Phys. Rev. Lett.*, **24**, 1349.

Scharff-Goldhaber, G, Dover, C B, and Goodman, A L, 1976. *Ann. Rev. Nucl. Sci.*, **26**, 239.

15

MICROSCOPIC FOUNDATIONS OF IBM

15.1 Introduction

The shell model is generally accepted as the fundamental model of nuclear physics. It is thus desirable to investigate possible connections between the shell model and the phenomenological IBM, as a means of microscopic justification for the latter. This is a formidable task, and much work has been done in this direction. Here we will not attempt a complete review of all existing methods, since such an effort could easily double the size of this book. Instead, we will give a brief qualitative discussion of the main ideas, suggesting to the interested reader to wander through the extensive list of references if he desires more excitement. The very recent review article by Iachello and Talmi (1987) can also be of great help.

The first major assumption of IBM is that nuclei are composed of a relatively inert doubly magic core, outside of which the valence nucleons are distributed in the relatively few single-j shells of the next major shell. With the exception of the core excitations described in Subsec. 5.10.3, the inert core is ignored in the IBM framework.

The second major assumption of IBM is that the low-energy properties of the nucleus, which are due to the valence nucleons, can be described in terms of bosons. At first it might seem quite surprising that a system of fermions can be described in terms of bosons, and one may wonder how the effects of the Pauli principle (valid for fermions) will be present in the boson description. However, descriptions of fermion systems in terms of bosons, the so-called **boson mappings** or **boson expansions** or **boson realizations**, were already known long before the introduction of IBM. A brief qualitative discussion of boson mappings will be given later.

The third major assumption of IBM is that only s and d bosons (or only $J=0$ and $J=2$ correlated fermion pairs) suffice for the description of low energy properties of medium and heavy nuclei. Since pairs of higher angular momentum can in general be present in the system of valence nucleons, a **truncation** to $J=0$ and $J=2$ pairs only has to be performed at some point, either before or after the boson mapping. Both approaches have been used, as we will see below.

234

15.2 Hermitian and non-hermitian boson mappings

In a boson mapping one corresponds to each fermion pair operator, as well as to each fermion multipole operator, a boson operator. In addition, one makes a correspondence between fermion states and boson states. Since Pauli principle restrictions exist in the fermion space, one has to take special care to preserve them in the boson space. There are three ways this can be achieved:

i) In mappings of the **Holstein–Primakoff** type (Holstein and Primakoff 1940), Pauli principle restrictions in the boson space are introduced by the existence in the boson operators of square roots, called the Pauli reduction factors. The fact that the quantities under the square roots have to remain positive induces restrictions analogous to these due to the Pauli principle. Unfortunately, these square roots do not contain only numbers; they contain operators as well. Thus the square roots have to be represented by their infinite Taylor expansions when matrix elements are to be calculated. Thus Holstein–Primakoff mappings have the drawback of representing the fermion operators by infinite boson images. In practice one has to truncate these infinite images in order to make calculations tractable. Then the Pauli principle restrictions are partially lost, and, as a consequence, one gets in the boson space more states than in the fermion space. The extra boson states, which do not have a fermionic analog, are called **spurious states** (Kim and Vincent 1985, Feng *et al.* 1985, Wu *et al.* 1985, Gambhir *et al.* 1985, Sheikh and Gambhir 1986, Geyer *et al.* 1986, Rowe and Carvalho 1986).

ii) An alternative approach is provided by **Dyson** mappings (Dyson 1956). In Dyson mappings the boson images of the fermion operators are finite. However, in contrast to the Holstein–Primakoff boson images, they are not hermitian. The non-hermiticity of Dyson boson mappings has long prevented their use in applications, although recently this situation is changing (Zirnbauer and Brink 1982, Zirnbauer 1984, Sugita *et al.* 1984, Takada 1985, 1986, Takada *et al.* 1985, Takada and Tazaki 1986, Takada and Yamada 1987, Tsukuma *et. al.* 1987).

iii) The third approach, the **Schwinger** mappings (Schwinger 1965), combines the advantages of the two previous methods. The boson images of the fermion pair and multipole operators are both finite and hermitian. However, in order to achieve this, one has to introduce an additional boson, not present in the previous two approaches. Since this extra boson does not correspond to a new physical degree of freedom, a constraint must accompany the mapping. This constraint is the conservation of the total number of bosons.

The use of Schwinger mappings in IBM-related problems has only recently started (Kaup 1987, Kaup and Ring 1987).

15.3 Traditional boson expansion theories

These older boson expansions were introduced with the aim of describing the anharmonicities mentioned in Chapter 0.

The first boson expansion method in nuclear physics was introduced by **Belyaev and Zelevinsky** (Belyaev and Zelevinsky 1962). They express the fermion pair operators as expansions in terms of suitable boson operators and determine the coefficients of the expansions by requiring that the boson images satisfy the commutation relations among their fermion archetypes. It turns out that their expansions contain a small parameter $\Omega^{-1/2}$, where 2Ω is roughly the average number of levels available to the active fermions, called the **degeneracy** of the shell. Thus for large enough degeneracies the expansions converge rapidly, so that in applications one can get good results by retaining only the first few terms. However, the original Belyaev and Zelevinsky (BZ) method suffers from a serious defect. Namely, the Lie algebra of the fermion operators does not contain all the constraints among these operators caused by the Pauli principle. Thus, an expansion satisfying only the commutation relations will, in general, fail to satisfy all the restrictions of the Pauli principle, giving rise to spurious boson states even before any truncation is attempted. A BZ type expansion which avoids the troubles associated with the Pauli principle has been introduced by Marshalek (Marshalek 1971, 1974a, 1974b, 1980b, 1981). We will refer to this expansion as the **Belyaev–Zelevinsky–Marshalek** boson expansion. The main advantage of boson expansions of this kind is their rapid convergence. They are also called **perturbative** boson expansions.

A different kind of boson expansion was introduced by Marumori *et al.*, known as the **Marumori** type boson expansion (Marumori, Yamamura, and Tokunaga 1964, Marumori, Yamamura, Tokunaga, and Takada 1964). They map a complete orthonormal basis of 2n fermions onto a subset of n states of an ideal boson space. Subsequently, they introduce a transformation which maps any operator in the fermion space onto a corresponding operator in the boson space. The boson expansions generated in this way satisfy all Pauli principle requirements. The coefficients in the expansions of operators are matrix elements in the fermion space. However, no well-defined small parameter has been identified in this method. As a result, Marumori boson expansions tend to converge very slowly, which is a great disadvantage if one wishes to use these expansions in applications. A modified Marumori type boson expansion with improved convergence properties was introduced by Li, Dreizler, and Klein (1971).

15.4 Number conserving boson mappings

In the previous section two different boson expansion techniques were discussed: the BZM expansion, which emphasizes the preservation of commutation relations, and the Marumori expansion, which emphasizes the mapping of states. In all traditional boson expansion theories, however, the mapping is achieved by first performing a BCS transformation from particles to quasiparticles, and then making the boson expansion in the quasiparticle basis. The quasiparticle pair operators are then obtained as normal-ordered polynomial expansions in terms of the boson creation and annihilation operators. However, the BCS quasiparticle transformation, involved in the beginning, has an important consequence: the resulting boson Hamiltonian does not conserve the boson number. Thus these traditional boson expansion theories are not suitable for investigating the microscopic foundations of IBM, since conservation of the boson number is a basic requirement of the model. Number conserving boson mappings have to be used instead.

15.4.1 The Otsuka–Arima–Iachello method

The first mapping procedure used in studying the shell model origins of IBM was the **Otsuka–Arima–Iachello** (OAI) method (Otsuka, Arima, Iachello, and Talmi 1978, Otsuka, Arima, and Iachello 1978). In this method one first carries out the required truncation in the fermion space, keeping only J=0 and J=2 fermion pairs. Then, one maps the truncated fermion space onto a boson space spanned by s and d bosons. The mapping of the states is based on the concept of seniority. By **seniority** we mean the number of fermion pairs (outside closed shells) not coupled to J=0. In the OAI procedure fermion states with definite seniority v are mapped onto boson states with a definite number of d bosons n_2, with $v = n_2$. Furthermore, one insists that matrix elements of the boson operators in the boson space be exactly equal to the matrix elements of the corresponding fermion operators in the original fermion space. Since no quasiparticle transformation is involved in the method, the resulting boson Hamiltonian does conserve the boson number, as required by IBM. As we have already mentioned, the OAI mapping is based on the preservation of matrix elements, thus it is a mapping of the Marumori type. It has been found that OAI mappings violate certain commutation relations (Matsuyanagi 1981).

15.4.2 The Bonatsos–Klein–Li method

A different approach has been introduced by **Bonatsos, Klein and Li** (BKL method) (Bonatsos, Klein, and Li 1984). The same seniority basis utilized by the OAI method is used in this approach, too. However, many differences exist. First, in the BKL method the full fermion space is mapped onto a boson space, i.e. the necessary truncation is not carried out in the fermion space but has to be done in the boson space. Second, the method is based on the preservation of the commutation relations of the fermion Lie algebra by the boson images of the fermion pair and multipole operators, not on the preservation of matrix elements. Thus the BKL method is a BZM type expansion, not a Marumori expansion. The boson images of the fermion operators are found as expansions in irreducible tensors multiplied by Pauli reduction factors. As expected, the BKL expansions of the various operators contain well-defined small parameters, allowing for rapid convergence of the infinite series.

As already mentioned, no truncation is performed in the fermion space in the BKL method. As a result, the boson mapping of a single-j shell algebra (which has the symmetry SO[2(2j+1)]) can contain higher bosons, in addition to the s and d bosons. It turns out that although the lowest order mapping contains only s and d bosons, g bosons enter at the next level of approximation. Thus the BKL method is suitable for investigating the microscopic origin of bosons other than s and d, while these additional bosons never occur in the OAI method, since an sd subspace is preselected. If many non-degenerate-j levels (with the symmetry SO[2 $\sum_i (2j_i + 1)$]) are considered in the BKL method (Bonatsos and Klein 1987), more than one s and one d boson occurs, along with higher bosons. Furthermore, bosons of negative parity, like the negative parity f boson used for the description of collective octupole states of negative parity, occur in a natural way, opening the road for the microscopic investigation of collective bosons with negative parity (Menezes, Bonatsos, and Klein, to be published in *Nucl. Phys. A*). Of course, one has to determine the collective bosons out of the many bosons arising in this approach. This has to be done dynamically. The need of determining dynamically the collective bosons has been repeatedly emphasized (Faessler and Morrison 1984).

15.4.3 The method of Klein and Vallières

One way of determining the collective bosons is provided by the method of **Klein and Vallières**, based on a trace variational principle for the boson Hamiltonian (Klein and Vallières 1981). The essence of this method is the following: there exists a collective sub-space of states (spanned by the collective s and d bosons) with an average energy which is lower than that of the remaining states of the given nucleus. Since the average energy of a set of states is proportional to the trace of the Hamiltonian over the subspace considered, a necessary condition for determining this subspace is the condition that the the variation of the trace of the Hamiltonian vanishes, i.e. $\delta(TrH) = 0$. This condition means that the average energy of the states in the collective subspace is stationary.

15.5 Deformed nuclei

15.5.1 Calculations in a deformed basis

Both the OAI method and the BKL method discussed above utilize the seniority concept. Thus they are expected to work well for spherical nuclei (nuclei near closed shells). The ground state of such nuclei is expected to contain only s-bosons. In addition, one expects that even at excited states at low energy the number of s bosons will be much greater than the numbers of other kinds of bosons. In deformed nuclei the situation is very different, since bosons other than s bosons have to be present in appreciable numbers in the ground state. In the fermion space one introduces a deformed pair operator Λ^+, which is expressed as the sum of spherical pair operators (Iachello and Talmi 1987)

$$\Lambda^+ = x_0 S^+ + x_2 D_0^+ + x_4 G_0^+ + \cdots, \qquad (15.1)$$

where S, D, G, mean pairs with J=0, J=2, J=4 respectively. By analogy, in the boson space one introduces the boson operator

$$\lambda^+ = x_0 s^+ + x_2 d_0^+ + x_4 g_0^+ + \cdots. \qquad (15.2)$$

The states of a deformed nucleus with N_π proton bosons and N_ν neutron bosons are then expressed as

$$(\lambda_\pi^+)^{N_\pi}(\lambda_\nu^+)^{N_\nu}|0> . \qquad (15.3)$$

This formulation of the problem has been used by many authors (see Iachello and Talmi (1987) for a list) in order to check the validity of the truncation into s and d bosons only, as well as the structure of the boson Hamiltonian and operators. Although much work has been done in this direction, we feel that further developments are necessary as far as the theoretical justification of IBM in the deformed limit is concerned. It is clear that in order to provide a full microscopic description of all collective states from the spherical to the deformed limit one has certainly to start from the spherical shell model.

15.5.2 Calculations starting from a spherical basis

As it has been mentioned above, in order to provide a rigorous justification of IBM in the rotational limit, one has to start from a spherical basis. Then one cannot start by considering a single-j shell (where the fermion pair and multipole operators form the Lie algebra $SO[2(2j+1)]$) or many non-degenerate-j shells (where the fermion pair and multipole operators form the Lie algebra $SO[2\sum_i(2j_i+1)]$), since these Lie algebras do not have an SU(3) subalgebra.

Instead of j–j coupling, it might be fruitful to use LST coupling when investigating the shell model foundations of IBM in deformed nuclei. Much work in this direction has been done by Bonatsos and Klein (Bonatsos and Klein 1984, 1985, 1986a, 1986b, Bonatsos, Klein, and Zhang 1987). The simplest problem one can consider is a proton or neutron p-shell, in which the fermion pair and multipole operators form the Lie algebra Sp(6), which does have an SU(3) subalgebra. The next major shell is the sd shell, where the relevant algebraic structure is

$$Sp(12) \supset U(6) \supset SU(3). \qquad (15.4)$$

Beyond the sd shell the argument breaks down, since the harmonic oscillator SU(3) symmetry is broken by the strong spin–orbit interaction.

As a result of the strong spin–orbit interaction the level with the highest j in each of the higher major shells is lowered in energy so that it goes into the major shell below its own shell. This can be clearly seen in Fig. 15.1, where an approximate level pattern for protons is shown. For example, we remark that in the sdg major shell the $1g_{9/2}$ level is lowered so that it goes into the pf shell. On the other hand, the sdg major shell acquires the $1h_{11/2}$ level of the pfh major shell, which is also lowered by the spin–orbit interaction. Thus each major shell beyond the sd shell is composed by some levels of the same parity (called the normal parity levels) and an intruder level of opposite parity (called the unique parity level).

In the case of major shells beyond the sd shell, an approximate SU(3) symmetry can be restored in the following way. Remark that the normal parity levels remaining in the sdg shell have the same j values as the levels of the full pf shell. One can then relabel the normal parity levels of the sdg major shell by the labels of the levels of the full pf shell. This relabelling has been checked experimentally and it has been found physically meaningful. The resulting approximate SU(3) symmetry is called **pseudo-SU(3)** (or **quasi-SU(3)**) symmetry (Ratna Raju, Draayer, and Hecht 1973, Draayer, Weeks, and Hecht 1982).

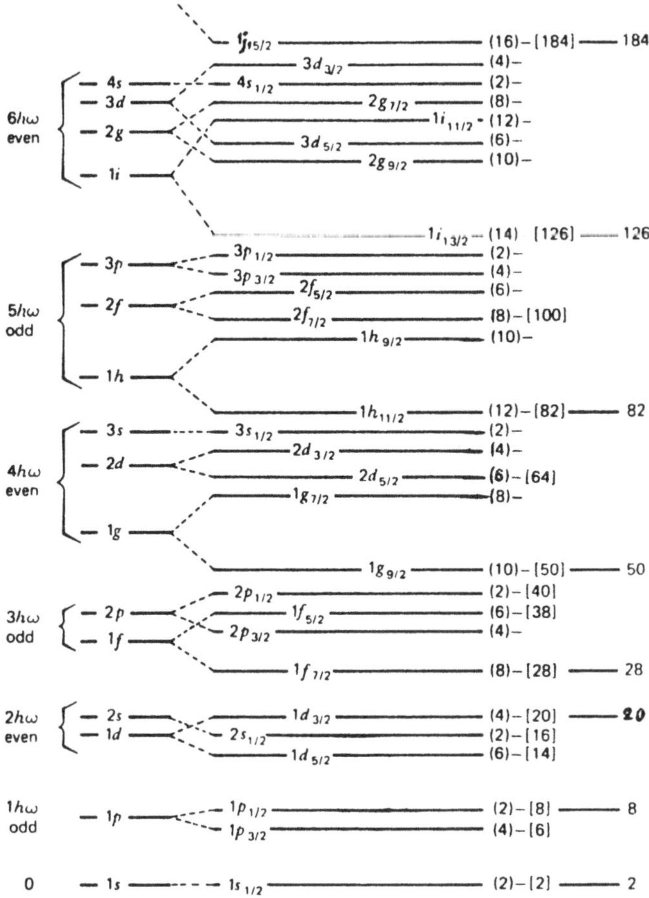

Fig. 15.1 Approximate level pattern for protons. The degeneracy of each level is shown next to it in parenteses, while the total occupation numbers are shown in square brackets. For neutrons the level pattern is the same as that for protons up to $N = 50$. Beyond this point it is found that the $2d_{5/2}$ level is below the $1g_{7/2}$ level, the $3s_{1/2}$ below $2d_{3/2}$, and the $2f_{7/2}$ close to the $1h_{9/2}$ level. (Taken from de Shalit and Feshbach (1974)).

Using the concept of pseudo-SU(3) one can extend the method

of Bonatsos and Klein beyond the sd shell. In the place of the sdg shell, for example, one obtains a pseudo pf shell with symmetry

$$Sp(20) \supset U(10) \supset SU(3), \tag{15.5}$$

plus the abnormal parity level $1h_{11/2}$. To lowest order approximation one can assume that the fermions occupying the abnormal parity level are all coupled into pairs of zero angular momentum, so that they do not influence the nuclear properties. Exact boson mappings in closed form have been provided for all the above mentioned symplectic algebras, as well as for the algebras

$$Sp(30) \supset U(15) \supset SU(3) \tag{15.6}$$

(which describes a pseudo-sdg shell, occurring in the place of the original pfh shell) and

$$Sp(42) \supset U(21) \supset SU(3) \tag{15.7}$$

(which describes a pseudo-pfh shell, occurring in the place of the original sdgi shell). However, the problem is far from finished, since one has to use the dynamics in order to select the collective bosons out of the many bosons occuring in the exact boson mappings. Furthermore, one has to identify suitable small parameters which will allow a truncation into s and d bosons only. In addition, in order to make contact with IBM, one has to keep only one- and two- body interactions in the Hamiltonian. Clearly much work is still required before a rigorous derivation of the IBM Hamiltonian in the SU(3) limit can be provided. A very successful algebraic approach to the description of deformed nuclei using the concepts mentioned in this subsection in the fermion basis has been given by Draayer and Weeks (Draayer and Weeks 1983, 1984, Castaños, Draayer, and Leschber 1987).

15.6 Ginocchio's model

An interesting model has been introduced by Ginocchio (Ginocchio 1979, 1980, Arima, Yoshida, and Ginocchio 1981). This model uses only s and d bosons (identical to the IBM-1 bosons) and provides exact boson mappings for the SO(8) Lie algebra, which is the algebra formed by fermions in a single $j = 3/2$ shell (or in many degenerate $j = 3/2$ shells). The great usefulness of this model lies in the fact that it can be used as a test ground for various number-conserving boson mapping techniques, i.e. one can compare the results gotten

with the approximate mapped operators to the corresponding exact
results of the SO(8) model.

References

As we have already mentioned, much work has been done on the
subject briefly discussed in this chapter. The interested reader will
find the recent review by Iachello and Talmi (1987) very informa-
tive. Similar accounts are given by Iachello (1986) and Talmi (1986).
Briefer and older reviews can be found in Elliott (1985), Dieperink
and Wenes (1985), Arima and Iachello (1984), and Barrett (1981).
A recent overview of boson mapping procedures has been given by
Ring (1986),while extended reviews of this subject are given by Garbaczewski (1978) and Dobaczewski (1981a, 1981b, 1982). Much recent work on the subject can be found in Dubrovnik 1986.

Akkermans, J N L, Loriaux, E, Allaart, K, and Bonsignori, G,
    1983. *Z. Phys.*, **A313**, 213.
Allaart, K, 1980. *Erice 1980*, 201.
Allaart, K, and Bonsignori, G, 1983. *Phys. Lett.*, **124B**, 1.
Allaart, K, Bonsignori, G, Savoia, M, and Paar, V, 1986a.
    Sorrento 1986, 237.
Allaart, K, Bonsignori, G, Savoia, M, and Paar, V, 1986b.
    Nucl. Phys., **A458**, 412.
Allaart, K, Bonsignori, G, Daman, M, Savoia, M, and Paar,
    V, 1986. *Dubrovnik 1986*, **2**, 837.
Amos, K, Morrison, I, and Smith, R, 1984. *J. Phys.*, **G10**, 331.
Arima, A, 1980a. *Erice 1980*, 87.
Arima, A, 1980b. *Nucl. Phys.*, **A347**, 339c.
Arima, A, 1981. *Granada 1981*, 46.
Arima, A, 1982. *Erice 1982*, 51.
Arima, A, 1986. *Beijing 1986*, 3.
Arima, A, and Iachello, F, 1984. In *Advances in Nuclear Physics*,
    (ed. J W Negele and E Vogt) Vol 13, p 139. Plenum, New York.
Arima, A, Yoshida, N, and Ginocchio, J N, 1981. *Phys. Lett.*,
    101B, 209.
Bao, C G, 1984. *Nucl. Phys.*, **A425**, 12.
Bao, C G, Wang, R L, and Pan, J Z, 1987. *Phys. Rev.*, **C35**,
    324.
Barrett, B R, 1981. *Rev. Mex. Fis.*, **27**, 533.
Barrett, B R, 1984. *Gull Lake 1984*, 22.
Barrett, B R, 1986. *Dubrovnik 1986*, **2**, 1010.
Barrett, B R, Pittel, S, and Duval, P D, 1982. *Erice 1982*, 535.

Barrett, B R, Pittel, S, and Duval, P D, 1983. *Nucl. Phys.*,
   A396, 267c.
Barrett, B R, Pittel, S, Duval, P D, and Druce,C H, 1984.
   Debrecen 1984, 649.
Belyaev, S T, and Zelevinsky, V G, 1962. *Nucl. Phys.*, **39**, 582.
Berard, M, and de Takacsy, N, 1979. *Phys. Rev.*, **C20**, 2439.
Bes, D R, Broglia, R A, Maglione, E, and Vitturi, A, 1982.
   Phys. Rev. Lett., **48**, 1001.
Bohr, A, and Mottelson, B R, 1980. *Phys. Scr.*, **22**, 468.
Bonatsos, D, 1985. *La Rábida 1985*, 615.
Bonatsos, D, and Klein, A, 1984. *Drexel 1984*, 635.
Bonatsos, D, and Klein, A, 1985. *Phys. Rev.*, **C31**, 992.
Bonatsos, D, and Klein, A, 1986a. *Dubrovnik 1986*, 1001.
Bonatsos, D, and Klein, A, 1986b. *Ann. Phys.*, **169**, 61.
Bonatsos, D, and Klein, A, 1987. *Nucl. Phys.*, **A469**, 253.
Bonatsos, D, Klein, A, and Li, C T, 1984. *Nucl. Phys.* , **A425**,
   521.
Bonatsos, D, Klein, A, and Zhang, Q Y, 1986a. *Phys. Lett.*,
   175B, 249.
Bonatsos, D, Klein, A, and Zhang, Q Y, 1986b. *Phys. Rev.*,
   C34, 686.
Bonsignori, G, Allaart, K, and van Egmond, A, 1982. *Erice
   1982*, 431.
Bonsignori, G, Savoia, M, Allaart, K, van Egmond, A, and Te
   Velde, G, 1985. *Nucl. Phys.*, **A432**, 389.
Broglia, R A, Maglione, E, and Vitturi, A, 1982. *Nucl. Phys.*,
   A376, 45.
Broglia, R A, Maglione, E, Sofia, H M, and Vitturi, A, 1982.
   Nucl. Phys., **A375**, 217.
Cambiaggio, M C, Dukelsky, J, and Zemba, G R, 1985a. *J.
   Phys.*, **G11**, L163.
Cambiaggio, M C, Dukelsky, J, and Zemba, G R, 1985b. *Phys.
   Lett.*, **162B**, 203.
Cambiaggio, M C, Dukelsky, J, and Zemba, G R, 1986.
   Dubrovnik 1986, **2**, 1042.
Castaños, O, and Moshinsky, M, 1985. *Oaxtepec 1985*, 43.
Castaños, O, Draayer, J P, and Leschber, Y, 1987. *Oaxtepec
   1987*, 11.
Castaños, O, Frank, A, Chacón, E, Hess, P O, and Moshinsky,
   M, 1982. *Phys. Rev.*, **C25**, 1611.
Catara, F, Insoglia, A, Maglione, E, and Vitturi, A, 1983.
   Phys. Lett., **123B**, 375.
Catara, F, Insoglia, A, Maglione, E, and Vitturi, A, 1984a.

Phys. Rev., **C29**, 1916.

Catara, F, Insoglia, A, Maglione, E, and Vitturi, A, 1984b. *Phys. Lett.*, **149B**, 41.

Catara, F, Insoglia, A, Sambataro, M, and Vitturi, A, 1985. *Legnaro 1985*, 153.

Catara, F, Sambataro, M, Insoglia, A, and Vitturi, A, 1986. *Phys. Lett.*, **180B**, 1.

Chakraborty, M, Kota, V K B, and Parikh, J C, 1981. *Phys. Lett.*, **100B**, 201.

Chasman, R R, 1982. *Phys. Rev.*, **C26**, 225.

Chen, H T, and Arima, A, 1983. *Phys. Rev. Lett.*, **51**, 477.

Chen, H T, Kiang, L L, Yang, C C, Chen, L M, Chen, T L, and Jiang, C W, 1986. *J. Phys.*, **G12**, L217.

Cizewski, J A, Flynn, E R, Brown, R E, Hanson, D L, Orbesen, S D, and Sunier, J W, 1981. *Phys. Rev.*, **C23**, 1453.

Cohen, T D, 1984. *Nucl. Phys.*, **A431**, 45.

Cohen, T D, 1985. *Phys. Lett.*, **158B**, 1.

Cohen, T D, and Klein, A, 1982. *Nucl. Phys.*, **A390**, 1.

de Shalit, A, and Feshbach, H, 1974. *Theoretical Nuclear Physics*, Vol. 1 : *Nuclear Structure*. Wiley, New York.

de Takacsy, N, 1980. *Nucl. Phys.*, **A339**, 54.

de Winter, L C, Walet, N R, Brussaard, P J, Allaart, K, and Dieperink, A E L, 1986. *Phys. Lett.*, **179B**, 322.

Dieperink, A E L, 1985. *La Rábida 1985*, 205.

Dieperink, A E L, and Wenes, G, 1985. *Annu. Rev. Nucl. Part. Sci.*, **35**, 77.

Dobaczewski, J, 1981a. *Nucl. Phys.*, **A369**, 213.

Dobaczewski, J, 1981b. *Nucl. Phys.*, **A369**, 237.

Dobaczewski, J, 1982. *Nucl. Phys.*, **A380**, 1.

Dönau, F, 1986. *Sorrento 1986*, 163.

Dönau, F, and Frauendorf, S, 1981. *Trieste 1981*, 57.

Dönau, F, and Janssen, D, 1973. *Nucl. Phys.*, **A209**, 109.

Draayer, J P, 1985. *Oaxtepec 1985*, 97.

Draayer, J P, and Weeks, K J, 1983. *Phys. Rev. Lett.*, **51**, 1422.

Draayer, J P, and Weeks, K J, 1984. *Ann. Phys.*, **156**, 41.

Draayer, J P, Weeks, K J, and Hecht, K T, 1982. *Nucl. Phys.*, **A381**, 1.

Druce, C H, and Moszkowski, S A, 1984. *Drexel 1984*, 672.

Druce, C H, and Moszkowski, S A, 1986. *Phys. Rev.*, **C33**, 330.

Druce, C H, Pittel, S, Barrett, B R, and Duval, P D, 1985. *Phys. Lett.*, **157B**, 115.

Druce, C H, Pittel, S, Barrett, B R, and Duval, P D, 1987. *Ann. Phys.*, **176**, 114.

Dukelsky, J, and Pittel, S, 1986a. *Sorrento 1986*, 285.

Dukelsky, J, and Pittel, S, 1986b. *Phys. Lett.*, **177B**, 125.

Dukelsky, J, Dussel, G G, and Sofia, H M, 1981. *Phys. Lett.*, **100B**, 367.

Dukelsky, J, Dussel, G G, and Sofia, H M, 1982. *Nucl. Phys.*, **A373**, 267.

Dukelsky, J, Dussel, G G, Perazzo, R P J, and Sofia, H M, 1983. *Phys. Lett.*, **130B**, 123.

Dukelsky, J, Pittel, S, Sofia, H M, and Lima, C, 1986. *Nucl. Phys.*, **A456**, 75.

Dukelsky, J, Dussel, G G, Perazzo, R P J, Reich, S L, and Sofia, H M, 1984. *Nucl. Phys.*, **A425**, 93.

Duval, P D, 1983. *Drexel 1983*, 199.

Duval, P D, and Barrett, B R, 1981a. *Phys. Rev. Lett.*, **46**, 1504.

Duval, P D, and Barrett, B R, 1981b. *Phys. Rev.*, **C24**, 1272.

Dyson, F J, 1956. *Phys. Rev.*, **102**, 1217.

Elliott, J P, 1985. *Rep. Prog. Phys.*, **48**, 171.

Evans, J A, Elliott, J P, and Szpikowski, S, 1985. *Nucl. Phys.*, **A435**, 317.

Faessler, A, and Morrison, I, 1984. *Nucl. Phys.*, **A423**, 320.

Feng, D H, Wang, K X, Wang, A L, and Wu, C L, 1985. *Phys. Lett.*, **155B**, 203.

Frank, A, and van Isacker, P, 1982. *Phys. Rev.*, **C26**, 1661.

Frank, A, Hess, P O, Castaños, O, and Pittel, S, 1987. *Phys. Rev.*, **C35**, 1896.

Gai, M, Arima, A, and Strottman, D, 1981. *Phys. Lett.*, **106B**, 6.

Galeão, A P, and Chen, H T, 1981. *Phys. Lett.*, **102B**, 221.

Gambhir, Y K, 1986. *Dubrovnik 1986*, **2**, 851.

Gambhir, Y K, and Sheikh, J A, 1986. *Sorrento 1986*, 265.

Gambhir, Y K, Ring, P, and Schuck, P, 1982a. *Phys. Rev.*, **C25**, 2858.

Gambhir, Y K, Ring, P, and Schuck, P, 1982b. *Nucl. Phys.*, **A384**, 37.

Gambhir, Y K, Sheikh, J A, Ring, P, and Schuck, P, 1985. *Phys. Rev.*, **C31**, 1519.

Garbaczewski, P, 1978. *Phys. Rep.*, **36**, 65.

Geyer, H B, 1986a. *Sorrento 1986*, 293.

Geyer, H B, 1986b. *Phys. Lett.*, **182B**, 111.

Geyer, H B, 1986c. *Phys. Rev.*, **C34**, 2373.

Geyer, H B, and Hahne, F J W, 1980a. *Phys. Lett.*, **97B**, 173.

Geyer, H B, and Hahne, F J W, 1980b. *Phys. Lett.*, **90B**, 6.

Geyer, H B, and Hahne, F J W, 1981. *Nucl. Phys.*, **A363**, 45.

Geyer, H B, and Lee, S Y, 1982. *Phys. Rev.*, **C26**, 642.

Geyer, H B, Hahne, F J W, and Scholtz, F G, 1987. *Phys. Rev. Lett.*, **58**, 459.

Geyer, H B, Engelbrecht, C H, and Hahne, F J W, 1986. *Phys. Rev.*, **C33**, 1041.

Ginocchio, J N, 1978a. *Erice 1978*, 111.

Ginocchio, J N, 1978b. *Phys. Lett.*, **79B**, 173.

Ginocchio, J N, 1979. *Phys. Lett.*, **85B**, 9.

Ginocchio, J N, 1980. *Ann. Phys.*, **126**, 234.

Ginocchio, J N, 1983. *Drexel 1983*, 81.

Ginocchio, J N, and Talmi, I, 1980. *Nucl. Phys.*, **A337**, 431.

Gomez, J M G, 1981. *Granada 1981*, 163.

Guidry, M W, Wu, C L, and Feng, D H, 1984. *Gull Lake 1984*, 181.

Hahne, F J W, 1981. *Phys. Rev.*, **C23**, 2305.

Halse, P, 1985. *Phys. Lett.*, **156B**, 1.

Halse, P, 1986. *Nucl. Phys.*, **A451**, 91.

Halse, P, 1987. *Phys. Rev.*, **C36**, 867.

Hasegawa, M, 1981. *Nucl. Phys.*, **A368**, 1.

Hasegawa, M, 1984. *Nucl. Phys.*, **A414**, 173.

Hasegawa, M, 1985a. *Nucl. Phys.*, **A435**, 152.

Hasegawa, M, 1985b. *Nucl. Phys.*, **A440**, 1.

Hasegawa, M, and Tazaki, S, 1987. *Phys. Rev.*, **C35**, 1508.

Hasselgren, L, and Cline, D, 1980. *Erice 1980*, 59.

Hecht, K T, 1985. *Oaxtepec 1985*, 165.

Hecht, K T, Hofmann, H M, and Zahn, W, 1981. *Phys. Lett.*, **103B**, 92.

Hess, P O, Frank, A, Castaños, O, and Pittel, S, 1987. *Oaxtepec 1987*, 143.

Heyde, K, and Sau, J, 1986. *Phys. Rev.*, **C33**, 1050.

Holstein, T, and Primakoff, H, 1940. *Phys. Rev.*, **58**, 1098.

Holzwarth, G, 1981. *Trieste 1981*, 1.

Iachello, F, 1986. *Sorrento 1986*, 213.

Iachello, F, and Talmi, I, 1987. *Rev. Mod. Phys.*, **59**, 339.

Iwasaki, S, and Hara, K, 1984. *Phys. Lett.*, **144B**, 9.

Iwasaki, S, Ring, P, and Schuck, P, 1980. *Nucl. Phys.*, **A339**, 365.

Janssen, D, Dönau, F, Frauendorf, S, and Jolos, R V, 1971. *Nucl. Phys.*, **A172**, 145.

Johnson, A B, and Vincent, C M, 1982. *Phys. Rev.*, **C25**, 1595.

Johnson, A B, and Vincent, C M, 1985. *Phys. Rev.*, **C31**, 1540.

Kaup, U, 1982. *Erice 1982*, 561.

Kaup, U, 1987. *Phys. Lett.*, **185B**, 249.

Kaup, U, and Ring, P, 1987. *Nucl. Phys.*, **A462**, 455.

Kaup, U, Vorwerk, R, Hippe, D, Schuh, H W, von Brentano, P, and Scholten, O, 1981. *Phys. Lett.*, **106B**, 439.

Kim, G K, and Vincent, C M, 1985. *Phys. Rev.*, **C32**, 1776.

Kim, G K, and Vincent, C M, 1987. *Phys. Rev.*, **C35**, 1517.

Kishimoto, T, and Tamura, T, 1983. *Phys. Rev.*, **C27**, 341.

Klein, A, 1980. *Nucl. Phys.*, **A347**, 3c.

Klein, A, 1986. *Sorrento 1986*, 127.

Klein, A, and Tanabe, K, 1984. *Phys. Lett.* , **135B**, 255.

Klein, A, and Vallières, M, 1981. *Phys. Lett.*, **98B**, 5.

Klein, A, Cohen, T D, and Li, C T, 1982. *Ann. Phys.*, **141**, 382.

Klein, A, Li, C T, Cohen, T D, and Vallières, M, 1982. *Erice 1982*, 183.

Kleppinger, E W, and Yates, S W, 1983. *Phys. Rev.*, **C27**, 2608.

Kuyucak, S, and Morrison, I, 1987. *Phys. Rev.*, **C36**, 774.

Leander, G A, Semmes, P B, and Dönau, F, 1984. *Gull Lake 1984*, 167.

Lee, S Y, Tsai, S F, Chen, H T, and Lin, L, 1983. *Phys. Rev.*, **C28**, 955.

Leviatan, A, 1984. *Phys. Lett.*, **143B**, 269.

Leviatan, A, 1985. *Z. Phys.*, **A321**, 467.

Leviatan, A, and Kirson, M W, 1984. *Nucl. Phys.*, **A419**, 358.

Li, C T, 1983. *Phys. Lett.*, **120B**, 251.

Li, C T, 1984a. *Nucl. Phys.*, **A417**, 37.

Li, C T, 1984b. *Phys. Rev.*, **C29**, 2309.

Li, C T, and Klein, A, 1979. *Phys. Rev.*, **C19**, 2023.

Li, C T, Pedrocchi, V G, and Tamura, T, 1985. *Phys. Rev.*, **C32**, 1745.

Li, C T, Pedrocchi, V G, and Tamura, T, 1986. *Phys. Rev.*, **C33**, 1762.

Li, S Y, Dreizler, R M, and Klein, A, 1971. *Phys. Rev.*, **C4**, 1571.

Li, C T, Chattopadhyay, P K, Klein, A, and Vassanji, M J, 1979. *Phys. Rev.*, **C19**, 2002.

Lie, S G, and Holzwarth, G, 1975. *Phys. Rev.*, **C12**, 1035.

Lipkin, H J, 1980. *Nucl. Phys.*, **A350**, 16.

Maglione, E, 1986. *Dubrovnik 1986*, **2**, 845.

Maglione, E, and Vitturi, A, 1982. *Erice 1982*, 87.

Maglione, E, and Vitturi, A, 1984. *Nucl. Phys.*, **A430**, 158.

Maglione, E, Catara, F, Insoglia, A, and Vitturi, A, 1983. *Nucl. Phys.*, **A397**, 102.

Maglione, E, Vitturi, A, Catara, F, and Insoglia, A, 1983.

Nucl. Phys., **A411**, 181.
Maglione, E, Vitturi, A, Catara, F, and Insoglia, A, 1984.
    Phys. Lett., **137B**, 1.
Maglione, E, Vitturi, A, Dasso, C H, and Broglia, R A, 1983.
    Nucl. Phys., **A404**, 333.
Marshalek, E R, 1971. *Nucl. Phys.*, **A161**, 401.
Marshalek, E R, 1974a. *Nucl. Phys.*, **A224**, 221.
Marshalek, E R, 1974b. *Nucl. Phys.*, **A224**, 245.
Marshalek, E R, 1980a. *Phys. Lett.*, **95B**, 337.
Marshalek, E R, 1980b. *Nucl. Phys.*, **A347**, 253c.
Marshalek, E R, 1981. *Nucl. Phys.*, **A357**, 398.
Marshalek, E R, 1982. *Ann. Phys.*, **143**, 191.
Marshalek, E R, 1987. *Phys. Rev.*, **C35**, 1900.
Marumori, T, Takada, K, and Sakata, F, 1984. *Trieste 1984*, **2**,
    897.
Marumori, T, Yamamura, M, and Tokunaga, A, 1964. *Progr.
    Theor. Phys.*, **31**, 1009.
Marumori, T, Yamamura, M, Tokunaga, A, and Takada, T,
    1964. *Progr. Theor. Phys.*, **32**, 726.
Marumori, T, Sakata, F, Maskawa, T, Une, T, and Hashimoto,
    Y, 1982. *Poiana Brasov 1982*, 1.
Matsuyanagi, K, 1981. *Trieste 1981*, 29.
McGrory, J B, 1978a. *Erice 1978*, 121.
McGrory, J B, 1978b. *Phys. Rev. Lett.*, **41**, 533.
Morrison, I, 1986. *Dubrovnik 1986*, **2**, 819.
Moshinsky, M, 1981. *Granada 1981*, 97.
Moshinsky, M, 1984a. *Drexel 1984*, 528.
Moshinsky, M, 1984b. *Nucl. Phys.*, **A421**, 81c.
Moshinsky, M, 1986. *Dubrovnik 1986*, **2**, 981.
Moszkowski, S A, 1984. *Gull Lake 1984*, 61.
Moszkowski, S A, 1985. *Phys. Rev.*, **C32**, 1063.
Okubo, S, 1974. *Phys. Rev.*, **C10**, 2048.
Otsuka, T, 1978. *Erice 1978*, 93.
Otsuka, T, 1980. *Erice 1980*, 73.
Otsuka, T, 1981a. *Phys. Rev. Lett.*, **46**, 710.
Otsuka, T, 1981b. *Nucl. Phys.*, **A368**, 244.
Otsuka, T, 1984a. *Phys. Lett.*, **138B**, 1.
Otsuka, T, 1984b. *Gull Lake 1984*, 3.
Otsuka, T, 1986a. *Sorrento 1986*, 255.
Otsuka, T, 1986b. *Dubrovnik 1986*, **2**, 828.
Otsuka, T, 1986c. *Phys. Lett.*, **182B**, 257.
Otsuka, T, and Arima, A, 1978. *Phys. Lett.*, **77B**, 1.
Otsuka, T, and Yoshinaga, N, 1986. *Phys. Lett.*, **168B**, 1.

Otsuka, T, Arima, A, and Iachello, F, 1978. *Nucl. Phys.*, **A309**, 1.

Otsuka, T, Arima, A, and Yoshinaga, N, 1982. *Phys. Rev. Lett.*, **48**, 387.

Otsuka, T, Arima, A, Iachello, F, and Talmi, I, 1978. *Phys. Lett.*, **76B**, 139.

Otsuka, T, Yoshida, N, van Isacker, P, Arima, A, and Scholten, O, 1987. *Phys. Rev.*, **C35**, 328.

Paar, V, and Brant, S, 1981. *Phys. Lett.*, **105B**, 81.

Pannert, W, and Ring, P, 1987. *Nucl. Phys.*, **A465**, 379.

Pannert, W, Ring, P, and Gambhir, Y K, 1985. *Nucl. Phys*, **A443**, 189.

Park, P, 1987. *Phys. Rev.*, **C35**, 807.

Pedrocchi, V G, and Tamura, T, 1983. *Phys. Rev.*, **C28**, 410.

Pedrocchi, V G, and Tamura, T, 1984. *Phys. Rev.*, **C29**, 1461.

Perazzo, R P J, and Sofia, H M, 1986. *Phys. Lett.*, **166B**, 249.

Pittel, S, 1983. *Drexel 1983*, 51.

Pittel, S, 1984. *Gull Lake 1984*, 37.

Pittel, S, and Dukelsky, J, 1983. *Phys. Lett.*, **128B**, 9.

Pittel, S, and Dukelsky, J, 1986a. *Oaxtepec 1986*, 315.

Pittel, S, and Dukelsky, J, 1986b. *Dubrovnik 1986*, **2**, 1016.

Pittel, S, and Frank, A, 1987. *Nucl. Phys.*, **A454**, 226.

Pittel, S, Duval, P D, and Barrett, B R, 1982a. *Oaxtepec 1982*, 274.

Pittel, S, Duval, P D, and Barrett, B R, 1982b. *Ann. Phys.*, **144**, 168.

Pittel, S, Duval, P D, and Barrett, B R, 1982c. *Phys. Rev.*, **C25**, 2834.

Pittel, S, Scholten, O, and Otsuka, T, 1985. *Phys. Lett.*, **157B**, 239.

Pittel, S, Dukelsky, J, Perazzo, R P J, and Sofia, H M, 1984. *Phys. Lett.*, **144B**, 145.

Pittel, S, Federman, P, Castaños, O, and Frank, A, 1981. *Oaxtepec 1981*, 382.

Ratna Raju, R D, Draayer, J P, and Hecht, K T, 1973. *Nucl. Phys.*, **A202**, 433.

Ring P, 1984. *Nucl. Phys.*, **A421**, 205c.

Ring, P, 1986. *Dubrovnik 1986*, **2**, 798.

Ring, P, and Pannert, W, 1984. *Drexel 1984*, 426.

Ring, P, and Schuck, P, 1977. *Phys. Rev.*, **C16**, 801.

Ring, P, Gambhir, Y K, Iwasaki, S, and Schuck, P, 1981. *Granada 1981*, 174.

Ring, P, Gambhir, Y K, Iwasaki, S, and Schuck, P, 1982.

Erice 1982, 465.

Rowe, D J, and Carvalho, J, 1986. *Phys. Lett.*, **175B**, 243.

Sage, K A, Goode, P R, and Barrett, B R, 1982. *Phys. Rev.*, **C26**, 668.

Sambataro, M, 1986a. *Sorrento 1986*, 303.

Sambataro, M, 1986b. *Dubrovnik 1986*, **2**, 857.

Sambataro, M, 1986c. *Phys. Rev. Lett.*, **57**, 1554.

Sambataro, M, 1987. *Phys. Rev.*, **C35**, 1530.

Sambataro, M, and Insoglia, A, 1986. *Phys. Lett.*, **166B**, 259.

Sambataro, M, Schaaser, H, and Brink, D M, 1986. *Phys. Lett.*, **167B**, 145.

Sarma, C R, and Sita, J, 1987. *Nucl. Phys.*, **A469**, 273.

Scholten, O, 1982. *Phys. Lett.*, **119B**, 5.

Scholten, O, 1983a. *Drexel 1983*, 223.

Scholten, O, 1983b. *Phys. Lett.*, **127B**, 144.

Scholten, O, 1983c. *Phys. Rev.*, **C28**, 1783.

Scholten, O, 1984a. *Drexel 1984*, 133.

Scholten, O, 1984b. *Gull Lake 1984*, 260.

Scholten, O, 1985. In *Progress in Particle and Nuclear Physics*, (ed. A. Faessler) Vol 14, p 189. Pergamon, Oxford.

Scholten, O, and Dieperink, A E L, 1980. *Erice 1980*, 343.

Scholten, O, and Kruse, H, 1983. *Phys. Lett.*, **125B**, 113.

Scholten, O, and Pittel, S, 1983. *Phys. Lett.*, **120B**, 9.

Scholten, O, Brant, S, and Paar, V, 1986. *Phys. Lett.*, **171B**, 335.

Schuck, P, Iwasaki, S, and Ring, P, 1981. *Trieste 1981*, 101.

Schwinger, J, 1965. *Quantum Theory of Angular Momentum.* Academic Press, New York.

Semmes, P B, Leander, G, Lewellen, D, and Dönau, F, 1986. *Phys. Rev.*, **C33**, 1476.

Sheikh, J A, 1987. *Phys. Rev.*, **C36**, 848.

Sheikh, J A, and Gambhir, Y K, 1986. *Phys. Rev.*, **C34**, 2344.

Silvestre-Brac, B, and Piepenbring, R, 1977. *Nucl. Phys.*, **A288**, 408.

Sørensen, B, 1967. *Nucl. Phys.*, **A97**, 1.

Sørensen, B, 1968. *Nucl. Phys.*, **A119**, 65.

Sørensen, B, 1970a. *Nucl. Phys.*, **A142**, 392.

Sørensen, B, 1970b. *Nucl. Phys.*, **A142**, 411.

Sørensen, B, 1973. *Nucl. Phys.*, **A217**, 505.

Sugawara-Tanabe, K, and Arima, A, 1982. *Phys. Lett.*, **110B**, 87.

Sugita, M, Sugawara-Tanabe, K, and Arima, A, 1984. *Phys. Lett.*, **148B**, 8.

Szpikowski, S, 1985. *Mikołajki 1985*, 193.
Szpikowski, S, 1986. *Sorrento 1986*, 275.
Takada, K, 1984. *Nucl. Phys.*, **A431**, 16.
Takada, K, 1985. *Nucl. Phys.*, **A439**, 489.
Takada, K, 1986. *Phys. Rev.*, **C34**, 750.
Takada, K, and Tazaki, S, 1986. *Nucl. Phys.*, **A448**, 56.
Takada, K, and Yamada, K, 1987. *Nucl. Phys.*, **A462**, 561.
Takada, K, Tamura, T, and Tazaki, S, 1985. *Phys. Rev.*, **C31**, 1948.
Talmi, I, 1978. *Erice 1978*, 79.
Talmi, I, 1980a. *Erice 1980*, 329.
Talmi, I, 1980b. *Drexel 1980*, 435.
Talmi, I, 1982a. *Oaxtepec 1982*, 301.
Talmi, I, 1982b. *Phys. Rev.*, **C25**, 3189.
Talmi, I, 1982c. *Erice 1982*, 27.
Talmi, I, 1984. *Trieste 1984*, **2**, 860.
Talmi, I, 1986. *Sorrento 1986*, 233.
Tamura, T, 1983a. *Phys. Rev.*, **C28**, 2154.
Tamura, T, 1983b. *Phys. Rev.*, **C28**, 2480.
Tamura, T, 1986a. *Sorrento 1986*, 311.
Tamura, T, 1986b. *Dubrovnik 1986*, **2**, 1063.
Tamura, T, Li, C T, and Pedrocchi, V G, 1985. *Phys. Rev.*, **C32**, 2129.
Tamura, T, Weeks, K J, and Kishimoto, T, 1979. *Phys. Rev.*, **C20**, 307.
Tamura, T, Weeks, K J, and Kishimoto, T, 1980. *Nucl. Phys.*, **A347**, 359c.
Tamura, T, Weeks, K J, and Pedrocchi, V G, 1982. *Phys. Rev.*, **C23**, 1297.
Tsukuma, H, Thorn, H, and Takada, K, 1987. *Nucl. Phys.*, **A466**, 70.
Vallières, M, Feng, D H, and Gilmore, R, 1983. *Drexel 1983*, 67.
van Egmond, A, and Allaart, K, 1982. *Erice 1982*, 405.
van Egmond, A, and Allaart, K, 1983a. *Phys. Lett.*, **131B**, 275.
van Egmond, A, and Allaart, K, 1983b. *Nucl. Phys.*, **A394**, 173.
van Egmond, A, and Allaart, K, 1984. *Nucl. Phys.*, **A425**, 275.
van Egmond, A, and Allaart, K, 1985. *Phys. Lett.*, **A164**, 1.
van Egmond, A, Allaart, K, and Bonsignori, G, 1985. *Nucl. Phys.*, **A436**, 458.
van Isacker, P, Pittel, S, Frank, A, and Duval, P D, 1986. *Nucl. Phys.*, **A451**, 202.
Vitturi, A, and Maglione, E, 1984. *Gull Lake 1984*, 194.
Weeks, K J, and Tamura, T, 1980a. *Phys. Rev.*, **C21**, 2632.

Weeks, K J, and Tamura, T, 1980b. *Phys. Rev.*, **C22**, 888.

Weeks, K J, and Tamura, T, 1980c. *Phys. Rev.*, **C22**, 1323.

Weeks, K J, and Tamura, T, 1980d. *Phys. Rev. Lett.*, **44**, 533.

Weeks, K J, Tamura, T, Udagawa, T, and Hahne, F J W, 1981. *Phys. Rev.*, **C24**, 703.

Wu, C L, and Feng, D H, 1980. *Phys. Lett.*, **96B**, 243.

Wu, C L, and Feng, D H, 1984. *Nucl. Phys.*, **A421**, 249c.

Wu, C L, Tang, W X, and Feng, D H, 1985. *Phys. Lett.*, **155B**, 208.

Wu, C L, Feng, D H, Chen, J Q, and Guidry, M W, 1982. *Phys. Rev.*, **C26**, 308.

Xu, G O, 1984. *Nucl. Phys.*, **A421**, 275c.

Xu, G O, and Wang, S J, 1982. *Nucl. Phys.* , **A380**, 529.

Yang, L M, 1980. *Drexel 1980*, 523.

Yang, L M, 1982. *Erice 1982*, 147.

Yang, L M, Lu, D H, and Song, T, 1986. *Dubrovnik 1986*, **2**, 867.

Yang, L M, Lu, D H, and Zhou, Z N, 1984. *Nucl. Phys.*, **A421**, 229c.

Yang, L M, Song, T, and Wang, X H, 1986. *Phys. Lett.*, **175B**, 6.

Yang, T S, and Yang, L M, 1980. *Erice 1980*, 229.

Yang, Z S, Liu, Y, and Qi, H, 1984. *Nucl. Phys.*, **A421**, 297c.

Yoshinaga, N, Arima, A, and Otsuka, T, 1984. *Phys. Lett.*, **143B**, 5.

Zelevinsky, V G, 1986. *Dubrovnik 1986*, **2**, 1125.

Zirnbauer, M R, 1984. *Nucl. Phys.*, **A419**, 241.

Zirnbauer, M R, and Brink, D M, 1982. *Nucl. Phys.*, **A384**, 1.

APPENDIX 1

LIE GROUPS AND LIE ALGEBRAS

A1.1 Definition

An abstract group G is a set of elements a, b, c, ..., for which a law of "multiplication" (which in general has nothing to do with ordinary multiplication) is defined, satisfying the following postulates:

i) Closure. If a and b are elements of the set, then so is their product

$$c = ab. \tag{A1.1}$$

ii) Associativity. The order in which multiplications of elements are performed is of no importance:

$$(ab)c = a(bc). \tag{A1.2}$$

iii) Identity. The set contains an element e (called the identity) such that

$$ea = ae = a. \tag{A1.3}$$

iv) Inverse. For every element a of the group, there is an (inverse) element a^{-1} such that:

$$aa^{-1} = a^{-1}a = e. \tag{A1.4}$$

A1.2 Matrix realizations of groups

Groups are abstract mathematical objects. However, one can obtain various "realizations" of a group. A useful realization of groups is the one in terms of matrices. To each element of the group there corresponds a square non-singular matrix. In addition, the group "multiplication" is corresponded to ordinary matrix multiplication. It is easy to verify that all the postulates given above are satisfied:

i) The product of two $n \times n$ square matrices is an $n \times n$ matrix.

ii) Matrix multiplication is associative.

iii) The identity is the well-known $n \times n$ matrix with diagonal elements equal to one and non-diagonal elements equal to zero.

$$1 = \begin{pmatrix} 1 & 0 & \cdots & 0 \\ 0 & 1 & \cdots & 0 \\ \vdots & \vdots & \ddots & \vdots \\ 0 & 0 & \cdots & 1 \end{pmatrix}. \qquad (A1.5)$$

iv) Since the matrices A are non-singular (i.e. $det|A| \neq 0$), their inverses exist.

A1.3 Continuous Lie groups

If a group has a finite set of elements it is called a **finite** group. If the set of elements of a group is infinite denumerable (such as the set of all integers), the group is called an **infinite discrete** group. If the elements of a group form a continuum (such as the set of all real numbers), the group is called a **continuous** group.

Discrete groups are widely used in solid state physics. In nuclear physics extensive use of continuous groups is made. Here we will deal with continuous groups only. A continuous group is characterized by the number r of the independent parameters it contains, which is called the **order** of the group. Then its elements are symbolized as $r_1, r_2, \ldots, r_n$.

A1.4 The Orthogonal groups O(n) and the Special Orthogonal groups SO(n)

Consider the set of orthogonal transformations in two dimensions (i.e. rotations around the z-axis). The corresponding group is SO(2). In matrix representation an element of this group can be written as

$$A = \begin{pmatrix} a_{11} & a_{12} \\ a_{21} & a_{22} \end{pmatrix}. \qquad (A1.6)$$

This looks like a 4-parameter expression, but it is not ! If we consider the action of A on a vector, we obtain

$$\begin{pmatrix} x' \\ y' \end{pmatrix} = \begin{pmatrix} a_{11} & a_{12} \\ a_{21} & a_{22} \end{pmatrix} \begin{pmatrix} x \\ y \end{pmatrix}. \qquad (A1.7)$$

However, the condition of invariance of the length of the vector under rotation (the orthogonality condition)

$$(x')^2 + (y')^2 = x^2 + y^2, \qquad (A1.8)$$

gives three constraints

$$a_{11}^2 + a_{21}^2 = 1, \qquad (A1.9)$$

$$a_{22}^2 + a_{12}^2 = 1, \qquad (A1.10)$$

$$a_{11}a_{12} + a_{21}a_{22} = 0. \qquad (A1.11)$$

Thus only one independent parameter is left, which is the angle of rotation around the z-axis. This is seen if one writes

$$\begin{pmatrix} x' \\ y' \end{pmatrix} = \begin{pmatrix} cos\phi & -sin\phi \\ sin\phi & cos\phi \end{pmatrix} \begin{pmatrix} x \\ y \end{pmatrix}. \qquad (A1.12)$$

Now the constraints are satisfied automatically, and the existence of only one free parameter is clear.

In a similar way one can study the group SO(3) of orthogonal transformations in 3 dimensions. The elements of the group can be represented as 3×3 matrices, having 9 parameters. The orthogonality condition

$$x^2 + y^2 + z^2 = (x')^2 + (y')^2 + (z')^2 \qquad (A1.13)$$

imposes six constraints, leaving us with 3 free parameters only, the three Euler angles.

In general, the group of orthogonal transformations in n dimensions is called SO(n). Its characteristic property is that it leaves the quantity $x_1^2 + x_2^2 + \ldots + x_n^2$ invariant. It turns out that the group SO(n) of orthogonal transformations in n dimensions has $n(n-1)/2$ free parameters.

Notice that in the special orthogonal groups studied above the matrices representing them have determinant equal to +1. If this restriction is lifted, one obtains the orthogonal groups O(n).

A1.5 The Unitary groups U(n) and the Special Unitary groups SU(n)

Consider the problem of rotations in a two-dimensional complex space (i.e. a generalization of the problem of rotations in a two-dimensional real space, which was considered in the previous section). A vector in two-dim complex space will be characterized by two complex components, z_1 and z_2, in the same way that a vector in a two-dim real space is characterized by two real components. Consider the set of transformations in complex space which leaves the quantity $|z_1|^2 + |z_2|^2$ invariant. The corresponding group is U(2). It can be realized in terms of 2×2 complex matrices. Since these matrices contain 4 complex numbers, this appears to be a group of 8 parameters (one for the real part and one for the imaginary part of each of the four complex numbers). One writes

$$\begin{pmatrix} z_1' \\ z_2' \end{pmatrix} = \begin{pmatrix} a_{11} & a_{12} \\ a_{21} & a_{22} \end{pmatrix} \begin{pmatrix} z_1 \\ z_2 \end{pmatrix}. \qquad (A1.14)$$

However, it is easy to check that the above mentioned condition of invariance of $|z_1|^2 + |z_2|^2$ imposes four constraints

$$a_{11}^* a_{11} + a_{21}^* a_{21} = 1, \qquad (A1.15)$$

$$a_{11}^* a_{12} + a_{21}^* a_{22} = 0, \qquad (A1.16)$$

$$a_{12}^* a_{11} + a_{22}^* a_{21} = 0, \qquad (A1.17)$$

$$a_{12}^* a_{12} + a_{22}^* a_{22} = 1. \qquad (A1.18)$$

Thus U(2) has only $8 - 4 = 4$ free parameters. It can be seen that the unitary group U(n), which acts in an n-dimensional complex space leaving the quantity $|z_1|^2 + |z_2|^2 + \ldots + |z_n|^2$ invariant, has n^2 free parameters.

If we impose the additional restriction that the matrices representing the group have determinant equal to $+1$, we obtain the Special Unitary group SU(n). In this case an additional constraint exists. For SU(2), for example, the additional constraint is

$$a_{11} a_{22} - a_{12} a_{21} = 1. \qquad (A1.19)$$

Thus SU(2) is a 3-free-parameter group, while in general SU(n) has $n^2 - 1$ free parameters.

The matrices of SU(2) can be written in the form

$$\begin{pmatrix} a & b \\ -b^* & a^* \end{pmatrix} \qquad (A1.20)$$

where

$$|a|^2 + |b|^2 = 1. \qquad (A1.21)$$

A1.6 The Symplectic groups Sp(2n)

Consider two real vectors,

$$(x_1, x_2, \ldots, x_n, \tilde{x}_1, \tilde{x}_2, \ldots, \tilde{x}_n), \qquad (A1.22)$$

$$(y_1, y_2, \ldots, y_n, \tilde{y}_1, \tilde{y}_2, \ldots, \tilde{y}_n). \qquad (A1.23)$$

Clearly, these vectors belong to a 2n-dimensional space. The transformations which leave invariant the skew-symmetric bilinear form $\sum_{i=1}^{n}(x_i\tilde{y}_i - \tilde{x}_i y_i)$ form the symplectic group Sp(2n).

As an example, consider the group Sp(2). Since n=1, in this case we deal with a real two-dim space. The group will be represented by 2×2 matrices. The two vectors, $(x_1, \tilde{x}_1)$ and $(y_1, \tilde{y}_1)$, transform as follows

$$\begin{pmatrix} x_1' \\ \tilde{x}_1' \end{pmatrix} = \begin{pmatrix} a_{11} & a_{12} \\ a_{21} & a_{22} \end{pmatrix} \begin{pmatrix} x_1 \\ \tilde{x}_1 \end{pmatrix}, \qquad (A1.24)$$

$$\begin{pmatrix} y_1' \\ \tilde{y}_1' \end{pmatrix} = \begin{pmatrix} a_{11} & a_{12} \\ a_{21} & a_{22} \end{pmatrix} \begin{pmatrix} y_1 \\ \tilde{y}_1 \end{pmatrix}. \qquad (A1.25)$$

Imposing the condition

$$x_1\tilde{y}_1 - \tilde{x}_1 y_1 = x_1'\tilde{y}_1' - \tilde{x}_1' y_1' \qquad (A1.26)$$

one can easily check that the following constraints occur

$$a_{11}a_{22} - a_{21}a_{12} = 1, \qquad (A1.27)$$

$$a_{12}a_{21} - a_{22}a_{11} = -1, \qquad (A1.28)$$

$$a_{11}a_{21} - a_{21}a_{11} = 0, \qquad (A1.29)$$

$$a_{12}a_{22} - a_{22}a_{12} = 0. \qquad (A1.30)$$

Since the last two are identities and the first two are the same, we conclude that only one independent constraint occurs. (Notice that this constraint says that the determinant of the matrix is equal to $+1$). Thus Sp(2) is a group with $4 - 1 = 3$ independent parameters. In general it can be seen that Sp(2n) is a group with $2n(2n + 1)/2$ independent parameters.

A1.7 Cartan classification of Lie groups

The groups studied above are special cases of Lie groups. All possible Lie groups have been classified by Cartan. They are divided in four classes:

i) Class A_n: The special unitary groups SU(n), having order $n^2 - 1$.

ii) Class B_n: The special orthogonal groups SO(n) of odd n, having order $n(n-1)/2$.

iii) Class C_n: The symplectic groups Sp(2n) which have order $2n(2n+1)/2$.

iv) Class D_n: The special orthogonal groups SO(n) of even n, having order $n(n-1)/2$.

In addition, there are five exceptional groups (G_2, F_4, E_6, E_7, E_8). We will not deal with them here.

A1.8 Isomorphism and homomorphism

The groups listed above are not unrelated to each other. In some cases the elements of a group can be put in an one-to-one correspondence with the elements of another group. The two groups are then called **isomorphic**. An example of isomorphic groups is provided by SO(6) and SU(4). If a different correspondence exists, which is not one-to-one, the two groups are called **homomorphic**. An example of homomorphic groups is given by SU(2) and SO(3). We will study this example in more detail below.

As we have already seen, the elements of SU(2) can be put in the form of a 2×2 matrix of the form

$$\begin{pmatrix} a_{11} & a_{12} \\ -a_{12}^* & a_{11}^* \end{pmatrix} \tag{A1.31}$$

with

$$a_{11}a_{11}^* + a_{12}a_{12}^* = 1. \tag{A1.32}$$

When an element of SU(2) acts on a 2-dim complex vector, (u_1, u_2), we have

$$u_1' = a_{11}u_1 + a_{12}u_2, \tag{A1.33}$$

$$u_2' = -a_{12}^*u_1 + a_{11}^*u_2. \tag{A1.34}$$

Now form the quantities

$$x_1 = u_1^2, \tag{A1.35}$$

$$x_2 = u_1u_2, \tag{A1.36}$$

$$x_3 = u_2^2. \tag{A1.37}$$

The corresponding quantities after the transformation are

$$x_1' = (u_1')^2 = a_{11}^2 x_1 + 2a_{11}a_{12}x_2 + a_{12}^2 x_3, \tag{A1.38}$$

$$x_2' = u_1'u_2' = -a_{11}a_{12}^* x_1 + (a_{11}a_{11}^* - a_{12}a_{12}^*)x_2 + a_{11}^* a_{12}x_3, \tag{A1.39}$$

$$x_3' = (u_2')^2 = a_{12}^* x_1 - 2a_{11}^* a_{12}^* x_2 + (a_{11}^*)^2 x_3. \tag{A1.40}$$

If we introduce the quantities

$$y_1 = \frac{x_1 - x_3}{2}, \tag{A1.41}$$

$$y_2 = \frac{x_1 + x_3}{2i}, \tag{A1.42}$$

$$y_3 = x_2, \tag{A1.43}$$

the corresponding transformed quantities read

$$y_1' = \frac{1}{2}(a_{11}^2 - a_{12}^{*2} - a_{12}^2 + a_{11}^{*2})y_1 + \frac{i}{2}(a_{11}^2 - a_{12}^{*2} + a_{12}^2 - a_{11}^{*2})y_2$$

$$+(a_{11}a_{12} + a_{11}^* a_{12}^*)y_3, \tag{A1.44}$$

$$y_2' = -\frac{i}{2}(a_{11}^2 + a_{12}^{*2} - a_{12}^2 - a_{11}^{*2})y_1 + \frac{1}{2}(a_{11}^2 + a_{12}^{*2} + a_{12}^2 + a_{11}^{*2})y_2$$

$$-i(a_{11}a_{12} - a_{11}^* a_{12}^*)y_3, \tag{A1.45}$$

$$y_3' = -(a_{11}^* a_{12} + a_{11}a_{12}^*)y_1 + i(a_{11}^* a_{12} - a_{11}a_{12}^*)y_2$$

$$+(a_{11}a_{11}^* - a_{12}a_{12}^*)y_3. \tag{A1.46}$$

One can "easily" check that the orthogonality condition

$$y_1^2 + y_2^2 + y_3^2 = (y_1')^2 + (y_2')^2 + (y_3')^2 \tag{A1.47}$$

is satisfied. Thus the transformation from y_1, y_2, y_3 to y_1', y_2', y_3' is an orthogonal transformation in three dimensions, i.e. an element of SO(3).

As an example of the correspondence between the elements of SU(2) and the elements of SO(3), consider rotations by an angle a around the z-axis. In SU(2) these are described by the matrix

$$\begin{pmatrix} e^{ia/2} & 0 \\ 0 & e^{-ia/2} \end{pmatrix}. \tag{A1.48}$$

From the relations given above one finds that the same rotation in SO(3) is expressed by the matrix

$$\begin{pmatrix} cosa & -sina & 0 \\ sina & cosa & 0 \\ 0 & 0 & 1 \end{pmatrix}.$$
$$(A1.49)$$

The correspondence between the elements of SU(2) and those of SO(3) is not an one-to-one correspondence, though. Consider for example rotations around the z-axis by 0 and by 2π. Both of them correspond to

$$\begin{pmatrix} 1 & 0 & 0 \\ 0 & 1 & 0 \\ 0 & 0 & 1 \end{pmatrix}$$
$$(A1.50)$$

in SO(3), while in SU(2) the 0 rotation corresponds to

$$\begin{pmatrix} 1 & 0 \\ 0 & 1 \end{pmatrix},$$
$$(A1.51)$$

while the 2π rotation corresponds to

$$\begin{pmatrix} -1 & 0 \\ 0 & -1 \end{pmatrix}.$$
$$(A1.52)$$

Thus the relation between SU(2) and SO(3) is homomorphic and not isomorphic.

A list of a few homomorphisms is given below

$$SU(2) \approx SO(3) \approx Sp(2),$$
$$(A1.53)$$

$$SO(5) \approx Sp(4),$$
$$(A1.54)$$

$$SO(4) \approx SU(2) \otimes SU(2) \approx SO(3) \otimes SO(3) \approx Sp(2) \otimes Sp(2),$$
$$(A1.55)$$

$$SU(4) \approx SO(6).$$
$$(A1.56)$$

A1.9 Lie algebras

Consider a Lie group of order r. This means by definition (Sec. A1.3) that the elements of the group depend on r independent parameters $a_1, a_2, \ldots, a_r$. Thus for the elements of the group we can use the symbol $c(a_1, a_2, \ldots, a_r)$. One can prove (Joshi 1977) that the elements of any Lie group of order r can be put into the form

$$c(a_1, a_2, \ldots, a_r) = exp(\sum_{j=1}^{r} i a_j x_j), \qquad (A1.57)$$

where the r operators x_1, x_2, $\ldots$, x_r are called the **generators** of the Lie group. They satisfy the following postulates

i) Closure. The commutator of two operators is an operator in the set

$$[x_k, x_l] \equiv x_k x_l - x_l x_k = c_{kl}^m x_m, \qquad (A1.58)$$

with

$$c_{kl}^m = -c_{lk}^m. \qquad (A1.59)$$

The coefficients c_{kl}^m are called Lie structure constants.

ii) Jacobi identity.

$$[x_k, [x_l, x_m]] + [x_l, [x_m, x_k]] + [x_m, [x_k, x_l]] = 0. \qquad (A1.60)$$

These are the postulates satisfied by a **Lie algebra**. The operators x_1, x_2, $\ldots$, x_r are called the **generators** of the r-dimensional Lie algebra corresponding to the above mentioned Lie group of order r. The "multiplication" properties of the elements of the Lie group are fully determined by the values of the Lie structure constants of the corresponding Lie algebra.

The study of Lie algebras is of fundamental importance in the study of Lie groups. Actually in most problems in physics one is more interested in the algebraic than in the group structure. In what follows, no distinction will be made between the Lie group and its corresponding Lie algebra.

Example: Consider the 3 operators L_x, L_y, L_z, obeying the commutation relations

$$[L_x, L_y] = iL_z, \qquad (A1.61)$$

$$[L_y, L_z] = iL_x, \qquad (A1.62)$$

$$[L_z, L_x] = iL_y. \qquad (A1.63)$$

These are the familiar commutation relations of angular momentum operators. These 3 operators form the Lie algebra SO(3).

A1.10 Realizations of Lie algebras

A1.10.1 Various realizations of SO(3)

An abstract Lie algebra can be realized in many ways. As an example, consider the Lie algebra SO(3). One possible realization of this algebra is in terms of differential operators

$$L_x = \frac{1}{i}(y\frac{\partial}{\partial z} - z\frac{\partial}{\partial y}), \qquad (A1.64)$$

$$L_y = \frac{1}{i}(z\frac{\partial}{\partial x} - x\frac{\partial}{\partial z}), \qquad (A1.65)$$

$$L_z = \frac{1}{i}(x\frac{\partial}{\partial y} - y\frac{\partial}{\partial x}), \qquad (A1.66)$$

acting on wave functions $\psi(x, y, z)$. One can check that the commutation relations of the previous section are indeed satisfied.

Another realization of the same algebra can be obtained in terms of creation and annihilation operators in a two-dimensional space. Consider, for example, proton and neutron creation (annihilation) operators a_p^+, a_n^+ (a_p, a_n). Construct the operators

$$T_+ = a_p^+ a_n, \qquad (A1.67)$$

$$T_- = a_n^+ a_p, \qquad (A1.68)$$

$$T_0 = \frac{1}{2}(a_p^+ a_p - a_n^+ a_n), \qquad (A1.69)$$

and then form the operators

$$T_x = \frac{1}{2}(T_+ + T_-), \qquad (A1.70)$$

$$T_y = \frac{1}{2i}(T_+ - T_-), \qquad (A1.71)$$

$$T_z = T_0. \qquad (A1.72)$$

Using the anticommutation relations of the proton and neutron operators one can check that these operators satisfy the commutation relations of the algebra SO(3). Thus the operators T_x, T_y, T_z (which are actually the isospin operators) constitute a different realization of the algebra SO(3). They act on a Fock space spanned by the vectors $(a_p^+)^k (a_n^+)^i |0>$. The operators T_+, T_-, and T_0 form the isospin algebra SU(2), which is homomorphic to SO(3).

A1.10.2 The Wigner supermultiplet algebra SU(4)

We just saw that the isospin operators form the algebra SU(2). In a similar way one can define the algebra SU(2) of spin, by introducing in the place of a_p and a_n operators $a_\uparrow$ and $a_\downarrow$, which create particles with spin up and spin down, respectively. One can then form the operators $S_\uparrow$, $S_\downarrow$, and S_0, which have forms analogous to T_+, T_-, and T_0. Together the 3 spin operators and the 3 isospin operators form the algebra $SU(2) \otimes SU(2)$. If one adds to the 3 spin operators and the three isospin operators the nine bilinear products $S_i T_j$, where $i = \uparrow, \downarrow, 0$ and $j = +, -, 0$, an SU(4) algebra is formed, called the **Wigner supermultiplet algebra** (Irvine 1972).

A1.11 Subalgebras

A subset of the operators of a Lie algebra L_1 which closes under commutation (i.e. satisfies the postulates mentioned above) is said to form a **subalgebra** L_2 of the algebra L_1.

As an example consider the operator L_z. Since $[L_z, L_z] = 0$, this operator forms a subalgebra of SO(3), which is actually the SO(2) subalgebra.

A1.12 Casimir operators

For each Lie algebra there exist one or more operators C which commute with all the generators X_i of the Lie algebra, i.e.

$$[C, X_i] = 0 \qquad (A1.73)$$

for any $i = 1, 2, \ldots, r$. The operators C are called **Casimir operators** (or **Casimir invariants**). They can be linear, quadratic, cubic, ...combinations of the generators. A Casimir operator containing products of p generators is called a Casimir operator of **order** p. Beware that the order of a Casimir operator has nothing to do with the order of the algebra. The number of linearly independent Casimir operators in a Lie algebra is called the **rank** of the algebra.

Example: Consider the algebra SO(3). It can be seen that this algebra has only one Casimir operator, which happens to be quadratic in the generators

$$C = L_1^2 + L_2^2 + L_3^2. \qquad (A1.74)$$

This operator is thus a second order Casimir operator of SO(3). Remember that SO(3) is an algebra of order 3. Since it has only one Casimir operator, it is an algebra of rank 1.

It is clear that if C is a Casimir operator of a given algebra, so is C multiplied by any constant. This implies that particular care has to be taken in the definition of Casimir operators, because various authors use definitions differing by multiplicative constants. It is also clear that if C is a Casimir operator of an algebra, so is any power of C. These additional operators are not counting as separate Casimir operators. In the case of an algebra with two or more linearly independent Casimir operators it is clear that any product of powers of these operators will also be a Casimir operator of the algebra. These operators are not counting as separate Casimir operators, either.

References

This appendix has been based on Iachello (1979) and on the unpublished lecture notes by F. Iachello *Lie groups, Lie algebras and some applications* (1982, University of Trento, Italy). Extended accounts of group theory can be found in Lipkin (1966), Joshi (1977), Gilmore (1974), Wybourne (1974), Hamermesh (1962), Wigner (1959). The inexperienced reader might find the first two of these books easier to deal with.

Gilmore, R, 1974. *Lie Groups, Lie Algebras, and some of their Applications.* Wiley, New York.
Hamermesh, M, 1962. *Group Theory.* Addison-Wesley, Reading.
Iachello, F, 1979. *Gull Lake 1979*, 140.
Irvine, J M, 1972. *Nuclear Structure Theory.* Pergamon, New York.
Joshi, A W, 1977. *Elements of Group Theory for Physicists*, (2nd ed.) Wiley Eastern Limited, New Delhi.
Lipkin, H J, 1966. *Lie Groups for Pedestrians.* North-Holland, Amsterdam.
Wigner, E P, 1959. *Group Theory and its Applications to the Quantum Mechanics of Atomic Spectra.* Academic Press, New York.
Wybourne, B G, 1974. *Classical Groups for Physicists.* Wiley, New York.

APPENDIX 2

REPRESENTATIONS OF LIE GROUPS

A2.1 Representations of groups in terms of irreducible tensors

For applications in physics we need to represent an abstract group G in some linear vector space. This representation will provide the quantum numbers necessary to classify the states. For the ordinary Lie groups mentioned above, the representations usually used are these in terms of irreducible tensors of rank r.

Consider a vector (x_1, x_2, x_3). Under a linear transformation this transforms as

$$x_i' = \sum_k a_{ik} x_k. \qquad (A2.1)$$

A vector is a tensor of rank 1. It has 3^1 components.

Consider now two vectors, (x_1, x_2, x_3) and (y_1, y_2, y_3). Under the same linear transformation the product $x_i y_j$ (which is called the Kronecker product) transforms as

$$x_i' y_j' = \sum_{k,l} a_{ik} a_{jl} x_k y_l. \qquad (A2.2)$$

A tensor of rank 2 is a quantity obeying the same transformation properties as the Kronecker product:

$$F_{ij}' = \sum_{k,l} a_{ik} a_{jl} F_{kl}. \qquad (A2.3)$$

The tensor F_{ij} has $3^2 = 9$ components. Tensors of higher rank can be constructed following the same procedure. Notice that the **rank** of a tensor has nothing to do with the **rank** of a Lie algebra (defined in sec. A1.12).

It is convenient to introduce the concept of **irreducible tensors**, i.e. to classify the tensors according to their symmetry properties. (For a more precise definition of irreducible tensors, see any of the group theory books given in the references). In the case of the previous example, the tensor F_{ij} has both symmetric (S_{ik}) and antisymmetric (A_{ik}) parts. Separated they read

$$S_{ik} = S_{ki} = \frac{F_{ik} + F_{ki}}{2}, \qquad (A2.4)$$

$$A_{ik} = -A_{ki} = \frac{F_{ik} - F_{ki}}{2}. \qquad (A2.5)$$

The symmetric part has 6 components (11,22,33,12,13,23), while the antisymmetric part has 3 components (12,13,23). These two parts are irreducible tensors with respect to the linear transformations considered

$$S_{ik} = \frac{x_i y_k + x_k y_i}{2}, \qquad (A2.6)$$

$$A_{ik} = \frac{x_i y_k - x_k y_i}{2}. \qquad (A2.7)$$

The various irreducible representations (irreps) of a group are similarly classified according to their symmetry character. Below we give some results.

A2.2 The Unitary groups U(n) and the Special Unitary groups SU(n)

In order to specify the possible irreps of rank r of the group U(n), one uses the following procedure:

i) Partition r into n integers

$$r = f_1 + f_2 + \ldots + f_n, \qquad (A2.8)$$

with

$$f_1 \geq f_2 \geq \ldots \geq f_n \geq 0. \qquad (A2.9)$$

ii) To each partition corresponds a diagram, called the **Young tableau**, with f_1 boxes in the first row, f_2 boxes in the second row, ..., f_n boxes in the nth row. The symbol $[f_1, f_2, \ldots, f_n]$ is used for this Young tableau. Two examples are shown in Fig. A2.1.

iii) The irrep is symmetric under interchanges occurring in a row of the corresponding Young tableau, while they are antisymmetric under interchanges occuring in a column of the Young tableau.

Example: Consider $r = 2$, any n ($n \geq 2$). The possible partitions are $2 = 2 + 0$ and $2 = 1 + 1$. Thus the corresponding Young tableaux are [2] (for brevity we use the symbol [2] instead of [2,0]) and [11]. [2] is the symmetric irrep, while [11] is the antisymmetric one.

Fig. A2.1 Young tableaux for the [3,2,2] and [3,3,1] irreps. (Taken from Wybourne (1974)).

In the case of the Special Unitary groups SU(n), it turns out that the irreps $[f_1, f_2, \ldots, f_n]$ and $[f_1 + s, f_2 + s, \ldots, f_n + s]$, where s is any integer, are equivalent. Thus for the groups SU(n) Young diagrams of n−1 rows suffice to specify the irreps, since $[f_1, f_2, \ldots, f_n]$ and $[f_1 - f_n, f_2 - f_n, \ldots, f_{n-1} - f_n, 0]$ are equivalent. As a consequence, the irreps of SU(2) are fully determined by one label, while the irreps of SU(3) are fully determined by two labels. For example, in SU(3) the irreps [542] and [32] are equivalent.

A2.3 The Special Orthogonal groups SO(n)

In order to specify the irreps of rank r of the group SO(n), one partitions r into ν integers,

$$r = \mu_1 + \mu_2 + \ldots + \mu_\nu, \qquad (A2.10)$$

with

$$\mu_1 \geq \mu_2 \geq \ldots \geq \mu_\nu, \qquad (A2.11)$$

where $\nu = n/2$ for $n = $ even, and $\nu = (n - 1)/2$ for $n = $ odd. For example, in the case of SO(3) one has $n = 3$, $\nu = 1$, thus only one label is needed, while in the case of SO(4) one has $n = 4$, $\nu = 2$, and thus two labels are necessary.

A2.4 The Symplectic groups Sp(n)

In order to specify the irreps of rank r of the group Sp(n), one partitions r into ν integers,

$$r = \mu_1 + \mu_2 + \ldots + \mu_\nu, \qquad (A2.12)$$

with

$$\mu_1 \geq \mu_2 \geq \ldots \geq \mu_\nu, \qquad\qquad (A2.13)$$

where $\nu = n/2$.

A2.5 The Spinor groups Spin(n)

The representations discussed so far, involved partitions of r into a sum of integers. These are called **tensor representations**. In the case of the special orthogonal groups SO(2n) and SO(2n+1), representations involving partitions of r into a sum of n half-integers are also possible. These are called **spinor representations**, and are labelled by $[l_1, l_2, \ldots, l_n]$, where l_1, l_2, ..., l_n are half-integers. In the particular case of SO(n), the irreps with $l_n \neq 0$ separate into two conjugate irreps $[l_1, l_2, \ldots, l_n]$ and $[l_1, l_2, \ldots, l_{-n}]$.

When one includes the spinor irreps, the special orthogonal groups are called **spinor groups**, for distinction. The spinor groups of lowest dimensionality are isomorphic to other Lie groups, as listed below

$$Spin(3) \approx SU(2), \qquad\qquad (A2.14)$$

$$Spin(4) \approx SU(2) \otimes SU(2), \qquad\qquad (A2.15)$$

$$Spin(5) \approx Sp(4), \qquad\qquad (A2.16)$$

$$Spin(6) \approx SU(4). \qquad\qquad (A2.17)$$

Thus usually they are not used explicitly. Above Spin(6), however, the spinor groups are not any more isomorphic with other Lie groups.

A2.6 Outer products of irreducible representations

In applications, it is often necessary to calculate the outer product of two irreps. This can be done by combining together the corresponding Young tableaux, in the way explained below through an example (Wybourne 1974).

Suppose we want to form the outer product $[2, 1] \otimes [2, 1]$. We first draw the Young diagram corresponding to the first factor. This will form the upper left corner of the final result. Then we draw the Young diagram for the second factor, assigning a label α to each box of the first row, and a label β to each box of the second row. In the general case where more than two rows are present, we go on assigning a label γ to each box in the third row, etc. Then we enlarge the Young diagram of the first factor by adding to it the labelled boxes of the Young diagram of the second factor, forming all possible results subject to the following limitations:

i) No identical labels appear in the same column of the resulting diagram.

ii) If we count the α's, β's, γ's, ..., from right to left starting from the top right corner, then at all times during the count , the number of α's must be not less than the number of β's, which must be not less than the number of γ's, and so on.

iii) The graph obtained after the addition of each label must be regular, i.e. it must satisfy the condition $[f_1 \geq f_2 \geq \ldots \geq f_n]$.

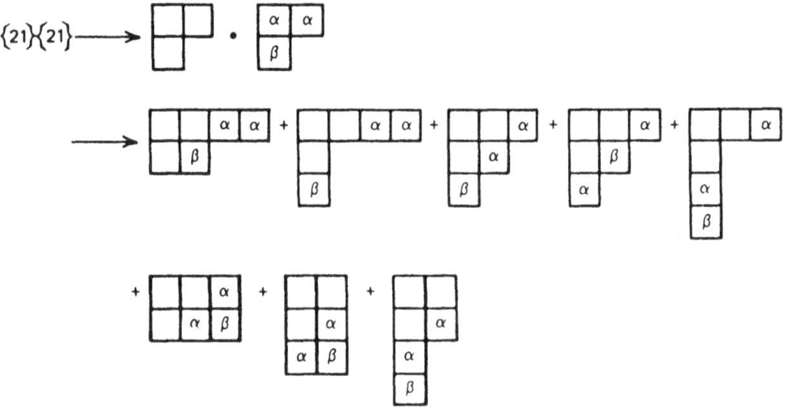

Fig. A2.2 Young tableau representation of the outer product $[2,1] \otimes [2,1]$. (Taken from Wybourne (1974)).

Using these rules one finds that

$$[2,1] \otimes [2,1] = [4,2] \oplus [4,1,1] \oplus 2[3,2,1]$$

$$\oplus [3,1,1,1] \oplus [3,3] \oplus [2,2,2] \oplus [2,2,1,1]. \qquad (A2.18)$$

The factor of two in front of [3,2,1] indicates that two [3,2,1] irreps are obtained, in which the relative positions of α's and β's differ, as it can be seen in Fig. A2.2.

References

This appendix has been based on Iachello (1979) and on the unpublished notes by F. Iachello *Lie groups, Lie algebras and some applications* (1982, University of Trento, Italy). Extended accounts of these topics can be found in the group theory books listed below.

Gilmore, R, 1974. *Lie Groups, Lie Algebras, and some of their Applications.* Wiley, New York.

Hamermesh, M, 1962. *Group Theory.* Addison-Wesley, Reading.

Iachello, F, 1979. *Gull Lake 1979*, 140.

Joshi, A W, 1977. *Elements of Group Theory for Physicists*, (2nd ed.) Wiley Eastern Limited, New Delhi.

Lipkin, H J, 1966. *Lie Groups for Pedestrians.* North-Holland, Amsterdam.

Wigner, E P, 1959. *Group Theory and its Applications to the Quantum Mechanics of Atomic Spectra.* Academic Press, New York.

Wybourne, B G, 1974. *Classical Groups for Physicists.* Wiley, New York.

DECOMPOSITION OF THE IRREPS OF A GROUP INTO IRREPS OF A SUBGROUP

In applications it is usually needed to know the irreps of a subgroup G' occuring in each irrep of the original group $G \supset G'$, as a means of classifying the states of the physical system under study. An elementary example is provided by the group chain $SO(3) \supset SO(2)$. The irreps of SO(3) are characterized by the integer L, the orbital angular momentum quantum number, while the irreps of SO(2) are labelled by the integer M_L, the quantum number for the z-component of the orbital angular momentum. It is well known that for given L the allowed values of M_L are

$$M_L = -L, -L+1, \ldots, L-1, L. \qquad (A3.1)$$

The states are then characterized by $|LM>$.

The method usually used for determining the irreps of the subgroup G' occuring in each irrep of the original group G (in other words: for decomposing the irreps of G into irreps of its subgroup G') is the **building up process**, which we will illustrate here through an example.

Example: Consider the group chain $SU(3) \supset SO(3)$. We want to determine the irreps of SO(3) (labelled by the integer L) contained in each irrep of SU(3) (represented by the Young tableau $[f_1, f_2]$).

We start with the elementary irrep of SU(3), which is [1]. This contains the $L = 1$ irrep of SO(3), since $2L + 1 = 3$.

Next we consider the product of two [1] SU(3) irreps

$$[1] \otimes [1] = [2, 0] \oplus [1, 1]. \qquad (A3.2)$$

This corresponds to the product of two $L = 1$ SO(3) irreps

$$(L = 1) \otimes (L = 1) \Rightarrow L = 0, 1, 2. \qquad (A3.3)$$

Remember at this point that Young tableaux are symmetric under interchanges occurring in a given row, while they are antisymmetric under interchanges occuring in a given column. Since it is well known that $L = 0, 2$ are symmetric while $L = 1$ is antisymmetric, we conclude that the [2,0] irrep of SU(3) contains the SO(3) irreps $L = 0, 2$, while the [1,1] irrep contains $L = 1$.

Next we consider the SU(3) product

$$[1,1] \otimes [1] = [2,1] \oplus [1,1,1], \qquad (A3.4)$$

which corresponds to the SO(3) product

$$(L = 1) \otimes (L = 1) \Rightarrow L = 0, 1, 2. \qquad (A3.5)$$

Remember at this point that in SU(n) Young tableaux with $n - 1$ rows suffice. Here, in particular, $[1,1,1] \equiv [0]$. Since the SU(3) irrep [0] contains only $L = 0$, we conclude that [2,1] contains $L = 1, 2$.

Next we consider the SU(3) product

$$[2] \otimes [1] = [3] \oplus [2,1], \qquad (A3.6)$$

which corresponds to the SO(3) product

$$(L = 0, 2) \otimes (L = 1) \Rightarrow L = 1, 1, 2, 3. \qquad (A3.7)$$

Since we already know that [2,1] contains $L = 1, 2$, we conclude that [3] contains $L = 1, 3$.

Next we consider the SU(3) product

$$[1,1] \otimes [2] = [3,1] \oplus [2,1,1], \qquad (A3.8)$$

which corresponds to the SO(3) product

$$(L = 1) \otimes (L = 0, 2) \Rightarrow L = 1, 1, 2, 3. \qquad (A3.9)$$

But [2,1,1] is equivalent to [1], which is already known to contain $L = 1$. Thus [3,1] contains $L = 1, 2, 3$.

Next we consider the SU(3) product

$$[3] \otimes [1] = [4] \oplus [3,1], \qquad (A3.10)$$

which corresponds to the SO(3) product

$$(L = 1, 3) \otimes (L = 1) \Rightarrow L = 0, 1, 2, 2, 3, 4. \qquad (A3.11)$$

Since it is already known that [3,1] contains $L = 1, 2, 3$, we conclude that [4] contains $L = 0, 2, 4$.

Next we consider the SU(3) product

$$[2,1] \otimes [1] = [3,1] \oplus [2,2] \oplus [2,1,1], \qquad (A3.12)$$

which corresponds to the SO(3) product

$$(L = 1,2) \otimes (L = 1) \Rightarrow L = 0,1,2,1,2,3. \qquad (A3.13)$$

Since we already found that [3,1] contains $L = 1,2,3$ and $[2,1,1] \equiv [1]$ contains $L = 1$, we conclude that [2,2] contains $L = 0,2$. Remember that [2] contains $L = 0,2$, too. Thus we conclude that $[2,2] \equiv [2]$.

The results found so far are given in the following list, where r is the rank of the irrep

r	[f]	L
0	[0]	0
1	[1]	1
2	[2]	0,2
	$[1,1] \equiv [1]$	1
3	[3]	1,3
	[2,1]	1,2
	$[1,1,1] \equiv [0]$	0
4	[4]	0,2,4
	[3,1]	1,2,3
	$[2,2] \equiv [2]$	0,2
	$[2,1,1] \equiv [1]$	1

Continuing the same procedure, for $r = 5$ we find

[f]	L
[5]	1,3,5
[4,1]	1,2,3,4
$[3,2] \equiv [3,1]$	1,2,3
$[3,1,1] \equiv [2]$	0,2
$[2,2,1] \equiv [1,1]$	1

For $r = 6$ we find

[f]	L
[6]	0,2,4,6
[5,1]	1,2,3,4,5
[4,2]	$0,2^2,3,4$
$[4,1,1] \equiv [3]$	1,3
$[3,3] \equiv [3]$	1,3
$[3,2,1] \equiv [2,1]$	1,2
$[2,2,2] \equiv [0]$	0

As it can be seen in the above list, it turns out that the [4,2] irrep contains two $L = 2$ irreps. Thus, an additional quantum number is needed in order to distinguish these two irreps. The identification and use of "missing" quantum numbers, like the one required here, is one of the most difficult problems in group representation theory. This specific problem is dealt with in subsec. 1.6.2.

References

This chapter has been based on Iachello (1979) and on the unpublished notes by F. Iachello *Lie groups, Lie algebras and some applications* (1982, University of Trento, Italy). More details on the subject can be found in the group theory books listed below.

Gilmore, R, 1974. *Lie Groups, Lie Algebras, and some of their Applications.* Wiley, New York.

Hamermesh, M, 1962. *Group Theory.* Addison-Wesley, Reading.

Iachello, F, 1979. *Gull Lake 1979* 140.

Joshi, A W, 1977. *Elements of Group Theory for Physicists*, (2nd ed.) Wiley Eastern Limited, New Delhi.

Lipkin, H J, 1966. *Lie Groups for Pedestrians.* North-Holland, Amsterdam.

Wigner, E P, 1959. *Group Theory and its Applications to the Quantum Mechanics of Atomic Spectra.* Academic Press, New York.

Wybourne, B G, 1974. *Classical Groups for Physicists.* Wiley, New York.

APPENDIX 4

EIGENVALUES OF THE CASIMIR OPERATORS
OF LIE ALGEBRAS

In many problems one needs to know the eigenvalues of Casimir operators in various representations. This can be done by standard group theoretical techniques. Here we will only quote the formulas generating the eigenvalues of the Casimir operators of various interesting groups and we will clarify their use through a few examples.

A4.1 The Unitary groups U(n) and the Special Unitary groups SU(n)

The irreps of U(n) and SU(n) can be denoted by the Young tableau $[f_1, f_2, \ldots, f_n]$. In the case of SU(n) only, the following equivalences hold

$$[f_1, f_2, \ldots, f_n] = [f_1 - f_n, f_2 - f_n, \ldots, f_{n-1} - f_n, 0], \qquad (A4.1)$$

$$[f_1, f_2, \ldots, f_n] = [f_1 - f_n, f_1 - f_{n-1}, \ldots, f_1 - f_2, 0]. \qquad (A4.2)$$

Then construct the quantities

$$S_k = \sum_{i=1}^{n} (\lambda_i^k - \rho_i^k), \qquad (A4.3)$$

$$\rho_i = n - i, \qquad (A4.4)$$

$$\lambda_i = m_i + n - i, \qquad (A4.5)$$

where

$$m_i = f_i \qquad (A4.6)$$

for U(n), while for SU(n)

$$m_i = f_i - \frac{f}{n} \qquad (A4.7)$$

with

$$f = f_1 + f_2 + \ldots + f_n. \qquad (A4.8)$$

Now construct the function

$$\phi(z) = \sum_{k=2}^{\infty} a_k z^k, \tag{$A4.9$}$$

where

$$a_k = \sum_{l=1}^{k-1} \frac{(k-1)!}{l!(k-l)!} S_l. \tag{$A4.10$}$$

Define the quantities B_p by

$$e^{-\phi(z)} = 1 - \sum_{p=0}^{\infty} B_p z^{p+1}, \tag{$A4.11$}$$

with

$$B_0 = 0. \tag{$A4.12$}$$

The expectation value of the Casimir operator of order p is given by

$$C_p = B_p - nB_{p-1}. \tag{$A4.13$}$$

Using this procedure, we obtain for U(n)

$$C_1 = S_1, \tag{$A4.14$}$$

$$C_2 = S_2 - (n-1)S_1, \tag{$A4.15$}$$

$$C_3 = S_3 - (n - \frac{3}{2})S_2 - \frac{1}{2}S_1^2 - (n-1)S_1, \tag{$A4.16$}$$

$$C_4 = S_4 - (n-2)S_3 - S_2 S_1 - \frac{1}{2}(3n-4)S_2 - \frac{1}{2}(n+2)S_1^2 - (n-1)S_1, \tag{$A4.17$}$$

$$\ldots$$

If we put $S_1 = 0$ in these formulas, we obtain the eigenvalues of the Casimir operators of SU(n)

$$C_1 = 0, \tag{$A4.18$}$$

$$C_2 = S_2, \tag{$A4.19$}$$

$$C_3 = S_3 - (n - \frac{3}{2})S_2, \tag{$A4.20$}$$

$$C_4 = S_4 - (n-2)S_3 - \frac{1}{2}(3n-4)S_2. \qquad (A4.21)$$

$$C_5 = S_5 - (n-\frac{5}{2})S_4 - \frac{1}{2}S_2^2 - \frac{2}{3}(3n-5)S_3 - \frac{1}{2}(4n-5)S_2, \quad (A4.22)$$

$$\cdots$$

In addition to these Casimir operators, symmetrized Casimir operators exist. The lowest order symmetrized Casimir operators of SU(n) are

$$I_2 = C_2, \qquad (A4.23)$$

$$I_3 = C_3 - \frac{n}{2}C_2, \qquad (A4.24)$$

$$I_4 = C_4 - nC_3 + \frac{n^2}{6}C_2, \qquad (A4.25)$$

$$I_5 = C_5 - \frac{3n}{2}C_4 + \frac{1}{2}C_2^2 + \frac{7n^2}{12}C_3 - \frac{n^3}{24}C_2. \qquad (A4.26)$$

Example 1: We calculate the eigenvalues of the first order Casimir operator of U(3) ($n = 3$). We have

$$C_1 = S_1 = \lambda_1 - \rho_1 + \lambda_2 - \rho_2 + \lambda_3 - \rho_3, \qquad (A4.27)$$

$$\rho_1 = 2, \qquad (A4.28)$$

$$\rho_2 = 1, \qquad (A4.29)$$

$$\rho_3 = 0, \qquad (A4.30)$$

$$\lambda_1 = m_1 + 2, \qquad (A4.31)$$

$$\lambda_2 = m_2 + 1, \qquad (A4.32)$$

$$\lambda_3 = m_3, \qquad (A4.33)$$

$$m_1 = f_1, \qquad (A4.34)$$

$$m_2 = f_2, \qquad (A4.35)$$

$$m_3 = f_3, \qquad (A4.36)$$

$$C_1 = f_1 + f_2 + f_3 = f. \qquad (A4.37)$$

Thus the eigenvalue of the first order Casimir invariant of U(3) is the rank of the group !

Example 2: We calculate the eigenvalues of the second order Casimir invariant of SU(3) ($n = 3$). Remember that in this case $f_3 = 0$, because of the equivalence noted above. We have

$$C_2 = S_2 = \lambda_1^2 - \rho_1^2 + \lambda_2^2 - \rho_2^2 + \lambda_3^2 - \rho_3^2, \tag{A4.38}$$

$$\rho_1 = 2, \tag{A4.39}$$

$$\rho_2 = 1, \tag{A4.40}$$

$$\rho_3 = 0, \tag{A4.41}$$

$$\lambda_1 = m_1 + 2, \tag{A4.42}$$

$$\lambda_2 = m_2 + 1, \tag{A4.43}$$

$$\lambda_3 = m_3, \tag{A4.44}$$

$$m_1 = f_1 - \frac{f_1 + f_2}{3}, \tag{A4.45}$$

$$m_2 = f_2 - \frac{f_1 + f_2}{3}, \tag{A4.46}$$

$$m_3 = -\frac{f_1 + f_2}{3}, \tag{A4.47}$$

$$C_2 = \frac{2}{3}(f_1^2 + f_2^2 - f_1 f_2 + 3 f_1). \tag{A4.48}$$

In the case of SU(n) in general an alternative notation (the **Elliott notation**) (Elliott 1958a, 1958b) is obtained if one denotes the irreps not by the lengths of the rows f_1, f_2, ..., f_{n-1}, $f_n = 0$, but by the differences of the lengths of the rows, denoted by

$$\lambda = f_1 - f_2, \tag{A4.49}$$

$$\mu = f_2 - f_3, \tag{A4.50}$$

$$\nu = f_3 - f_4, \tag{A4.51}$$

$$\ldots$$

These are called the **Elliott quantum numbers**. In the present case of SU(3) we have

$$\lambda = f_1 - f_2, \tag{A4.52}$$

$$\mu = f_2. \tag{A4.53}$$

In this notation we get

$$C_2 = \frac{2}{3}[\lambda^2 + \mu^2 + \lambda\mu + 3(\lambda + \mu)]. \tag{A4.54}$$

Example 3: In a similar way one can find that the third order symmetrized Casimir invariant of SU(3) has eigenvalues

$$I_3 = \frac{1}{9}(\lambda - \mu)(2\lambda + \mu + 3)(\lambda + 2\mu + 3). \tag{A4.55}$$

A4.2 The Special Orthogonal groups SO(2n+1)

Construct the quantities

$$S_k = \sum_{i=-n}^{+n} (\lambda_i^k - \rho_i^k), \tag{A4.56}$$

$$\rho_i = \lambda_i - f_i, \tag{A4.57}$$

$$\lambda_i = f_i + n + i - \theta_{0i}, \tag{A4.58}$$

$$\lambda_{-i} = -\lambda_i + 2n - 1, \tag{A4.59}$$

$$\lambda_0 = n, \tag{A4.60}$$

$$f_{-i} = -f_i, \tag{A4.61}$$

$$f_0 = 0, \tag{A4.62}$$

$$S_0 = S_1 = 0, \tag{A4.63}$$

where θ_{ji} is 1 for $j < i$ and 0 for $j \geq i$.
Construct the function

$$\phi(z) = \sum_{k=3}^{\infty} a_k z^k, \tag{A4.64}$$

where

$$a_k = \sum_{l=2}^{k-1} \frac{(k-1)!}{l!(k-l)!} S_k. \tag{A4.65}$$

Define the quantities B_p by

$$e^{-\phi(z)} = 1 - \sum_{p=2}^{\infty} B_p z^{p+1}, \qquad (A4.66)$$

$$B_0 = B_1 = 0. \qquad (A4.67)$$

Then, the eigenvalue of C_p in the representation $[f_n, f_{n-1}, \ldots, f_1]$ (notice the inverse order !) is

$$C_p = (2n+1)\delta_{p0} + B_p - (n+\frac{1}{2})B_{p-1} - \sum_{q=1}^{p-1}[B_q - (n+\frac{1}{2})B_{q-1}]n^{p-q}. \qquad (A4.68)$$

This procedure gives

$$C_0 = 2n + 1, \qquad (A4.69)$$

$$C_1 = 0, \qquad (A4.70)$$

$$C_2 = S_2, \qquad (A4.71)$$

$$\cdots$$

Example 1: We calculate the eigenvalues of the second order Casimir invariant of the group SO(3), the well known group of rotations in a three-dimensional space. In this case $n = 1$. We then have

$$C_2 = S_2 = \sum_{i=-1}^{+1}[\lambda_i^2 - (\lambda_i - f_i)^2] = \sum_{i=-1}^{+1}(2\lambda_i f_i - f_i^2), \qquad (A4.72)$$

$$\lambda_1 = f_1 + 1, \qquad (A4.73)$$

$$\lambda_{-1} = -f_1, \qquad (A4.74)$$

$$\lambda_0 = 1, \qquad (A4.75)$$

$$f_{-1} = -f_1, \qquad (A4.76)$$

$$f_0 = 0, \qquad (A4.77)$$

$$C_2 = 2(f_1 + 1)f_1. \qquad (A4.78)$$

For the irrep [L] this becomes

$$C_2 = 2L(L + 1), \qquad (A4.79)$$

the eigenvalue of the square of the angular momentum operator, up to a multiplicative constant.

Example 2: We calculate the eigenvalues of the second order Casimir invariant of SO(5). In this case $n = 2$. We then have

$$C_2 = S_2 = \sum_{i=-2}^{+2} (2\lambda_i f_i - f_i^2), \qquad (A4.80)$$

$$\lambda_2 = f_2 + 3, \qquad (A4.81)$$

$$\lambda_1 = f_1 + 2, \qquad (A4.82)$$

$$\lambda_0 = 2, \qquad (A4.83)$$

$$\lambda_{-1} = -f_2, \qquad (A4.84)$$

$$\lambda_{-2} = -f_1 + 1, \qquad (A4.85)$$

$$f_{-2} = -f_2, \qquad (A4.86)$$

$$f_{-1} = -f_1, \qquad (A4.87)$$

$$f_0 = 0, \qquad (A4.88)$$

$$C_2 = 2f_2(f_2 + 3) + 2f_1(f_1 + 1). \qquad (A4.89)$$

For the irrep $[v, 0]$ we get

$$C_2 = 2v(v + 3). \qquad (A4.90)$$

A4.3 The Symplectic groups Sp(2n)

Construct the quantities

$$S_k = \sum_{i=-n, i\neq 0}^{+n} (\lambda_i^k - \rho_i^k), \qquad (A4.91)$$

$$\rho_i = \lambda_i - f_i, \qquad (A4.92)$$

$$\lambda_i = f_i + n + i, \qquad (A4.93)$$

$$\lambda_{-i} = -\lambda_i + 2n, \qquad (A4.94)$$

$$f_{-i} = -f_i, \qquad (A4.95)$$

$$S_0 = S_1 = 0, \qquad (A4.96)$$

and the function $\phi(z)$ as before. Then

$$C_p = 2n\delta_{p0} + (B_p - nB_{p-1}) - \sum_{q=1}^{p-1}(B_q - nB_{q-1})(n + \frac{1}{2})^{p-q}. \quad (A4.97)$$

This procedure gives

$$C_0 = 2n, \quad (A4.98)$$

$$C_1 = 0, \quad (A4.99)$$

$$C_2 = S_2, \quad (A4.100)$$

$$\cdots\cdots$$

A4.4 The Special Orthogonal groups SO(2n)

Construct the quantities

$$S_k = \sum_{i=-n, i\neq 0}^{+n} (\lambda_i^k - \rho_i^k), \quad (A4.101)$$

$$\rho_i = \lambda_i - f_i, \quad (A4.102)$$

$$\lambda_i = f_i + n + i - (1 + \epsilon_i), \quad (A4.103)$$

$$\lambda_{-i} = -\lambda_i + 2n - 2, \quad (A4.104)$$

$$f_{-i} = -f_i, \quad (A4.105)$$

$$S_0 = S_1 = 0, \quad (A4.106)$$

where ϵ_i is equal to $+1$ for $i > 0$, -1 for $i < 0$, and 0 for $i = 0$. After constructing the function $\phi(z)$ as before, one finds

$$C_p = 2n\delta_{p0} + (B_p - nB_{p-1}) - \sum_{q=1}^{p-1}(B_q - nB_{q-1})(n - \frac{1}{2})^{p-q}. \quad (A4.107)$$

This procedure gives

$$C_0 = 2n, \quad (A4.108)$$

$$C_1 = 0, \quad (A4.109)$$

$$C_2 = S_2, \quad (A4.110)$$

$$C_3 = (S_3 + \frac{3}{2}S_2) - (2n - \frac{1}{2})S_2, \qquad (A4.111)$$

$$\cdots$$

Example 1: We calculate the eigenvalues of the second order Casimir operator of SO(2). In this case $n = 1$. One has

$$S_2 = \sum_{i=-1, i\neq 0}^{+1} (2\lambda_i f_i - f_i^2), \qquad (A4.112)$$

$$\lambda_1 = f_1, \qquad (A4.113)$$

$$\lambda_{-1} = -f_1, \qquad (A4.114)$$

$$f_{-1} = -f_1, \qquad (A4.115)$$

$$C_2 = 2f_1^2. \qquad (A4.116)$$

In the irrep $[M_L]$ one then has

$$C_2 = 2M_L^2. \qquad (A4.117)$$

Example 2: We calculate the eigenvalues of the second order Casimir invariant of SO(4). In this case $n = 2$. One has

$$S_2 = \sum_{i=-2, i\neq 0}^{+2} (2\lambda_i f_i - f_i^2), \qquad (A4.118)$$

$$\lambda_2 = f_2 + 2, \qquad (A4.119)$$

$$\lambda_{-2} = -f_2, \qquad (A4.120)$$

$$\lambda_1 = f_1 + 1, \qquad (A4.121)$$

$$\lambda_{-1} = -f_1 + 1, \qquad (A4.122)$$

$$f_{-1} = -f_1, \qquad (A4.123)$$

$$f_{-2} = -f_2, \qquad (A4.124)$$

$$C_2 = 2f_2(f_2 + 2) + 2f_1^2. \qquad (A4.125)$$

In the irrep $[\omega, 0]$ one then has

$$C_2 = 2\omega(\omega + 2). \qquad (A4.126)$$

References

This appendix has been based on Iachello 1979, Perelomov and Popov (1966), Popov and Perelomov (1967), and on the unpublished lecture notes by F. Iachello *Lie groups, Lie algebras and some applications* (1982, University of Trento, Italy). More details can be found in the group theory books listed below.

Elliott, J P, 1958a. *Proc. R. Soc. London* Ser A, **245**, 128.
Elliott, J P, 1958b. *Proc. R. Soc. London* Ser A, **245**, 562.
Gilmore, R, 1974. *Lie Groups, Lie Algebras, and some of their Applications.* Wiley, New York.
Hamermesh, M, 1962. *Group Theory.* Addison-Wesley, Reading.
Iachello, F, 1979. *Gull Lake 1979*, 140.
Joshi, A W, 1977. *Elements of Group Theory for Physicists*, (2nd ed.) Wiley Eastern Limited, New Delhi.
Lipkin, H J, 1966. *Lie Groups for Pedestrians.* North-Holland, Amsterdam.
Perelomov, A M, and Popov, V S, 1966. *Sov. J. Nucl. Phys.*, **3**, 676.
Popov, V S, and Perelomov, A M, 1967. *Sov. J. Nucl. Phys.*, **5**, 489.
Wigner, E P, 1959. *Group Theory and its Applications to the Quantum Mechanics of Atomic Spectra.* Academic Press, New York.
Wybourne, B G, 1974. *Classical Groups for Physicists.* Wiley, New York.

LIST OF CONFERENCES

Beijing 1980. (1981). *Topics in Nuclear Physics II* (ed. T T S Kuo and S S M Wong). Springer-Verlag, Berlin.

Beijing 1986. (1987). *Physics at Tandem* (ed. C Jiang, S Li, Z Sun, and H Zhang). World Scientific, Singapore.

Chester 1984. (1985). *Clustering Aspects of Nuclear Structure* (ed. J S Lilley and M A Nagarajan). Reidel, Dordrecht.

Debrecen 1984. (1984). *In-beam Nuclear Spectroscopy* (ed. Zs Dombrádi and T Fényes). Akadémiai Kiadó, Budapest.

Drexel 1980. (1982). *Contemporary Research Topics in Nuclear Physics* (ed. D H Feng, M Vallières, M W Guidry, and L L Riedinger). Plenum, New York.

Drexel 1983. (1984). *Bosons in Nuclei* (ed. D H Feng, S Pittel, and M Vallières). World Scientific, Singapore.

Drexel 1984. (1985). *Nuclear Shell Models* (ed. M Vallières and B H Wildenthal). World Scientific, Singapore.

Dronten 1980. (1981). *Nuclear Structure* (ed. K Abrahams, K Allaart, and A E L Dieperink). Plenum, New York.

Dubrovnik 1986. (1986). *Nuclear Structure, Reactions, and Symmetries* (ed. R A Meyer and V Paar). World Scientific, Singapore.

Erice 1978. (1979). *Interacting Bosons in Nuclear Physics* (ed. F Iachello). Plenum, New York.

Erice 1980. (1981). *Interacting Bose–Fermi Systems in Nuclei* (ed. F Iachello). Plenum, New York.

Erice 1982. (1983). *Progr. Part. Nucl. Phys.* (ed. D Wilkinson) Vol 9. Pergamon, Oxford.

Florence 1983. (1983). *Proceedings of the International Conference on Nuclear Physics Florence 1983* (ed. P Blasi and R A Ricci). Tipografia Compositori, Bologna.

Granada 1981. (1982). *Interacting Bosons in Nuclei* (ed. J S Dehesa, J M G Gomez, and J Ros). Springer-Verlag, Berlin.

Gull Lake 1979. (1980). *Nuclear Spectroscopy* (ed. G F Bertsch and D Kurath). Springer-Verlag, Berlin.

Gull Lake 1984. (1984). *Interacting Boson-Boson and Boson-Fermion Systems* (ed. O Scholten). World Scientific, Singapore.

Harrogate 1986. (1987). *Proceedings of the International Nuclear Physics Conference Harrogate 1986* (ed. J L Durell, J M Irvine, and G C Morrison) Vol 2. Institute of Physics, Bristol.

Jyväskylä 1984. (1984). *5th Nordic Meeting on Nuclear Physics* (ed. J Hattula).

La Rábida 1985. (1985). *Theory of Nuclear Structure and Reactions* (ed. M Lozano and G Madurga). World Scientific, Singapore.

Legnaro 1985. (1986). *Conference on Nuclear Structure with Heavy Ions* (ed. R A Ricci and C Villi). Societá Italiana di Fisica, Bologna.

Mikołajki 1985. (1985). *Trends in Nuclear Physics* (ed. Z Wilhelmi, G Szeflińska, and M Kicińska-Habior). Warsaw University, Warsaw.

Oaxtepec 1979. (1979). *Proceedings of the 2nd Oaxtepec Symposium on Nuclear Physics* (ed. A Dacal, P Federman, P A Mello and M E Ortiz). Instituto de Fisica, Mexico.

Oaxtepec 1981. (1981). *Proceedings of the 4th Oaxtepec Symposium on Nuclear Physics* (ed. G Garcia-Calderón). Instituto de Fisica, Mexico.

Oaxtepec 1982. (1982). *Proceedings of the 5th Oaxtepec Symposium on Nuclear Physics* (ed. M E Ortiz and J Lomnitz). Instituto de Fisica, Mexico.

Oaxtepec 1983. (1983). *Proceedings of the 6th Oaxtepec Symposium on Nuclear Physics* (ed. M E Ortiz and J Lomnitz). Instituto de Fisica, Mexico.

Oaxtepec 1984. (1984). *Proceedings of the 7th Oaxtepec Symposium on Nuclear Physics* (ed. M E Ortiz and J Lomnitz). Instituto de Fisica, Mexico.

Oaxtepec 1985. (1985). *Proceedings of the 8th Oaxtepec Symposium on Nuclear Physics* (ed. E Chávez, C Guerra, E Hernández and M E Ortiz). Instituto de Fisica, Mexico.

Oaxtepec 1986. (1986). *Proceedings of the 9th Oaxtepec Symposium on Nuclear Physics* (ed. E Hernández, A Menchaca-Rocha and M E Ortiz). Instituto de Fisica, Mexico.

Oaxtepec 1987. (1987). *Proceedings of the 10th Oaxtepec Symposium on Nuclear Physics* (ed. A Menchaca-Rocha). Instituto de Fisica, Mexico.

Osaka 1984. (1984). *Nuclear Spectroscopy and Nuclear Interactions* (ed. H Ejiri and T Fukuda). World Scientific, Singapore.

Poiana Brasov 1982. (1983). *Nuclear Collective Dynamics* (ed. D Bucurescu, V Ceausescu, and N V Zamfir). World Scientific, Singapore.

Rhodes 1979. (1980). *Structure of Medium-Heavy Nuclei 1979* (ed. G S Anagnostatos, C A Kalfas, S Kossionides, T Paradellis, L D Skouras, and G Vourvopoulos). Institute of Physics, Bristol.

Sorrento 1986. (1986). *Microscopic Approaches to Nuclear Structure Calculations* (ed. A Covello). Società Italiana di Fisica, Bologna.

Strasbourg 1980. (1980). *Nuclear Behaviour at High Angular Momentum.* Centre de Recherches Nucléaires, Strasbourg.

Trieste 1981. (1982). *Nuclear Physics* (ed. C H Dasso, R A Broglia, and A Winther). North Holland, Amsterdam.

Trieste 1984. (1985). *Winter College on Fundamental Nuclear Physics* (ed. K Dietrich, M Di Toro, and H J Mang). World Scientific, Singapore.

Varenna 1985. (1985). *4th International Conference on Nuclear Reaction Mechanisms* (ed. E Gadioli). Universitá degli Studi di Milano, Milano.

SUBJECT INDEX

The manufacturer's authorised representative in the EU for product safety is Oxford University Press España S.A. of el Parque Empresarial San Fernando de Henares, Avenida de Castilla, 2 – 28830 Madrid (www.oup.es/en or product. safety@oup.com). OUP España S.A. also acts as importer into Spain of products made by the manufacturer.

www.ingramcontent.com/pod-product-compliance
Lightning Source LLC
Chambersburg PA
CBHW061440050726
47637CB00001B/6